# AVIATION PSYCHOLOGY IN PRACTICE

# Aviation Psychology in Practice

edited by

**Neil Johnston**
**Nick McDonald**
**Ray Fuller**

Routledge
Taylor & Francis Group

LONDON AND NEW YORK

First published 1994 by Ashgate Publishing

2 Park Square, Milton Park, Abingdon, Oxfordshire OX14 4RN
52 Vanderbilt Avenue, New York, NY 10017

*Routledge is an imprint of the Taylor & Francis Group, an informa business*

First issued in paperback 2019

**British Library Cataloguing in Publication Data**
Aviation Psychology in Practice
  I. Johnston, Neil
  363.124019

**Library of Congress Cataloguing-in-Publication Data**
  Aviation Psychology in Practice/edited by Neil Johnston,
  Nick McDonald, Ray Fuller.
  p. cm
  Includes index.
  ISBN 0-291-39808-1
  1. Aeronautics - Psychology.   2. Psychology, Industrial
  3. Aeronautics, Commercial- Safety measures.   4. Aeronautics-
  Human factors.  I. Johnson, Neil, 1947- . II. McDonald,
  Nick, 1949- .  III. Fuller, Ray 1943- .
  TL555.A954  1994
  629.132'52'019--dc20

ISBN 13: 978-1-84014-133-7 (hbk)

# Contents

# List of contributors

**Clint A. Bowers** is Associate Director in the Team Performance Laboratory at the University of Central Florida. He holds a PhD in Clinical/Counselling Psychology from the University of South Florida. His research interests include aircrew coordination training and the role of team communication in team performance.

**Sheryl L. Chappell** is the Principal Scientist of the Aviation Safety Reporting System at the National Aeronautics and Space Administration's Ames Research Center, Moffett Field, California. Previously, she was the principal investigator for the Traffic Alert and Collision Avoidance System and the Data Link Programs at NASA. She is a PhD candidate in Experimental Psychology at Ohio State University and an instrument-rated commercial pilot and flight instructor.

**Asaf Degani** received a Practical Engineer degree in Architecture and Design from Ort Givataim, Israel, and a BSc from Florida International University. He received an MSc in Environmental Health and Safety (with a human factors concentration) from the University of Miami in 1989. He served as a weapon systems officer in the Israeli Navy and holds a private pilot's licence. Since 1989 he has worked as a researcher at the Flight Human Factors Branch at NASA-Ames Research Center under a grant from San Jose State University Foundation, conducting research in the area of cockpit automation and procedures. He is currently a graduate student in the School of Industrial and System Engineering at Georgia Institute of Technology.

**Ray Fuller** is Head of the Department of Psychology and a Fellow of Trinity College, Dublin. He is a former President of the Psychological Society of Ireland and representative for Ireland on the EC Committee on Toxic and Psychological Factors in Road Traffic Accidents. He has directed several research projects funded by the US Army Research Institute for the Social and Behavioral Sciences (on truck driver fatigue and road accidents), the Tobacco Research Council (UK) and ITMAC (on behavioural motivation) and Eolas and Dublin Bus (on the ergonomic design of fleet control systems). His current research is focused on human behaviour and safety, with particular reference to transportation and to airport ramp safety training.

**R. Curtis Graeber**, since receiving his PhD in Experimental Psychology, has served as a research scientist at three US government laboratories. In 1981 he joined NASA's Ames Research Center as Principal Investigator for research on flight crew fatigue and jet lag and was appointed Chief, Flight Human Factors Research Branch in 1989. He is a Fellow of the Aerospace Medical Association and in 1986 served as the Human Factors Specialist for the Presidential Commission on the Space Shuttle Challenger Accident. In December 1990, he joined the Boeing Commercial Airplane Group as Manager of Flight Deck Research for Avionics/Flight Systems and in 1993 became Chief, Human Engineering. He is responsible for human factors considerations in aircraft design, operational procedures and training.

**Peter G. Harle** is a graduate mechanical engineer and pilot. He joined the Royal Canadian Air Force in 1959, serving for seven years in the US and Europe, and retiring in 1985 as the Director of Air Operations and Training. He joined the Canadian Aviation Safety Board in 1985, and is currently the Director of Accident Prevention with the Transportation Safety Board of Canada. His responsibilities include the identification of safety deficiencies in the marine, rail and aviation modes of the national transportation systems, including those deriving from breakdowns in expected human performance. He has participated in the ICAO Flight Safety and Human Factors Study Group since its inception.

**Neil Johnston** is a Boeing 737 captain with Aer Lingus, where he has responsibility for training development. He was the founding chairman of the Human Performance Committee of IFALPA (the International Federation of Airline Pilots' Associations) and was chairman of the Human Factors Working Group of IATA (the International Air Transport Association). He represents IATA on the Flight Safety and Human Factors Study Group of ICAO (the International Civil Aviation Organization). He is Associate Editor of the *International Journal of Aviation Psychology*.

**George L. Kaempf** entered the aviation system 20 years ago as an avionics technician for the air force. Since that time he has been an active participant in the aviation community in a variety of capacities: working in a fixed base operation, flying as a private pilot, and conducting and managing research as an experimental psychologist. He has led research teams investigating numerous issues relevant for commercial and military aviation and for fixed-wing as well as rotary-wing aircraft. His research interests have focused on pilot and flight crew training. He has investigated training effectiveness and the utility of various training devices and training programmes. He has extensive experience investigating the use of flight simulation for sustaining both flight and combat skills in military aircraft. Recently, his research has focused on decision processes and requirements for operators in time-pressured, high-risk settings. He has led decision research in complex domains including commercial air transport flight crews, helicopter flight crews, AEGIS Combat System operators, and military command and control.

**Gary Klein** is the Chairman and Chief Scientist at Klein Associates. He has performed research on critical decision making and has developed significant new models of proficient decision making under conditions such as time pressure and uncertainty. He has furthered the development and application of a decision-centred design approach, and studies applications of case-based reasoning for domains such as the cost/benefit evaluation of training devices. As research psychologist for the Air Force Human Resources Laboratory from 1974–1978, he initiated his investigations into the use of comparison cases as an integral part of the problem-solving process. Prior to 1974, Dr Klein was Assistant Professor of Psychology at Oakland University.

**David A. Marx**, after receiving his BS in Mechanical Engineering, joined Boeing Commercial Aircraft Group (BCAG) as a mechanical systems designer and analyst on the 737 and 757 programmes. Since 1989, he has been a special programmes engineer within Maintenance Engineering. Focusing on airline maintenance operations, he has provided consultancy to airlines on maintenance programme development, and issues relating to ageing aeroplanes and extended twin engine operations (ETOPS). He is currently responsible for maintenace safety process development and the integration of human factors considerations in the maintenance aspects of BCAG products.

**Nick McDonald** is Senior Lecturer in Psychology at Trinity College, Dublin. He is author of two books: *The Stresses of Work* and *Fatigue, Safety and the Truck Driver*. He is a founder member and chairman of the Aerospace Psychology Research Group, a broad-based research organization dedicated to fostering human factors research in aviation. This group works closely with the IATA IGHC Ramp Safety Group in the development of a human factors approach to airport safety. He is currently directing an international consortium dedicated to the design, production and evaluation of safety training materials for ground handling (Project SCARF).

**Helen Muir** is the Director of the Department of Applied Psychology in the College of Aeronautics at Cranfield University. She is a Chartered Psychologist, a member of the Occupational Division of the British Psychological Society and a Fellow of the Royal Aeronautical Society. She obtained a degree in Psychology from St Andrews University, a doctorate from the University of London and has a diploma in counselling. She has been involved in applied research in the field of transportation since 1973. Her research interests include pilot workload assessment, and cabin safety and aircraft emergencies. Recently she has extended her research and teaching activities into the field of post-traumatic stress and victim counselling.

**Mike O'Leary** is an airline pilot and a psychologist. He has a degree in Electronic Engineering and a PhD in Psychology, both from the University of London. He joined British Airways in 1972 flying a variety of aircraft before taking a command on the Airbus 320 on which he is still current. He is an honorary research fellow at Birkbeck College with interests in visuo-spatial information processing and stimulus-response compatibility. His two interests, flying and psychology, were happily married in early 1992 when he helped British Airways start up its Air Safety Human Factors Reporting Programme.

**Judith M. Orasanu** Joined the Aerospace Human Factors Research Division at NASA's Ames Research Center in 1991, where she now conducts research on crew problem solving and decision making. Previously she worked for the US Army Research Institute, managing their Basic Research Program on Education and Training. In 1989–1990 she was a Visiting Fellow at Princeton University and began her aeronautical decision-making research. Prior to joining the Army Research Institute, she managed research programmes on literacy and bilingualism at the National Institute of Education. Dr Orasanu received her PhD in Experimental Psychology from Adelphi University. She was a Postdoctoral Fellow and Research Associate at the Rockefeller University, where she studied culture, language and cognition.

**Nick Pidgeon** completed his PhD in Psychology in 1986. Subsequently he conducted interdisciplinary research into structural failures in the UK construction industry, and then joined the Department of Psychology, Birbeck College, University of London, as a lecturer in 1989. His current research interests include human judgement and decision making, risk perception and assessment, the organizational causes of disasters, and qualitative research methods in psychology.

**Carolyn Prince** is Research Psychologist at the Naval Training Systems Center in Orlando, Florida. She holds a PhD in Industrial/Organizational Psychology from the University of South Florida. Her research interests include aircrew coordination training and team situational awareness.

**Richard E. Redding** is a research psychologist and Project Director at Human Technology, Incorporated, where he has directed a major cognitive task analysis study of air traffic control for the US Federal Aviation Administration. He has also been a principal scientist on US Air Force and Navy training research and development programmes. He holds a Juris Doctor degree with honours from Washington and Lee University, where he is an adjunct professor. He also holds an MS degree in Psychology from Vanderbilt University, and was a doctoral candidate in psychology. He has written approximately 25 journal articles, book chapters, and monographs on cognitive task analysis, cognitive approaches to instruction, cognitive development, and law and psychology.

**Eduardo Salas** is Senior Research Psychologist in the Human Factors Division of the Naval Training Systems Center in Orlando, Florida, where he has been since 1984. Dr Salas also has courtesy appointments at the University of South Florida and the University of Central Florida. He earned his PhD from Old Dominion University in Industrial/Organizational Psychology in 1984. He is on the editorial boards of Human Factors and Personnel Psychology, and has co-authored over 50 journal articles and book chapters, and has co-edited four books. Among Dr Salas's research interests are team training and performance, training effectiveness, tactical decision making under stress, team decision making, human performance measurement and modelling, and learning strategies for teams.

**Thomas L. Seamster** is a senior research scientist and consultant in cognitive and human factors. He received his PhD from Indiana University in Instructional Systems Technology in 1986. He has conducted research on expertise in air traffic control, computer programming, flight crew resource management assessment, fighter pilot weapons deployment, and spacecraft control. He has also developed and tested user interfaces for both military and commercial systems; developed guidelines for user interface languages, usability and prototyping processes, and user interfaces for expert systems; and has developed expert systems and intelligent tutoring systems. His research interests include the analysis of expertise, modelling cognitive activities in complex environments, training and educational systems analysis and development, and user interface research and development.

**William R. Taggart** is a training specialist and consultant in developing Crew Resource Management (CRM) training programmes. Since 1984, he has worked with airlines around the world to develop basic and advanced CRM programmes for line pilots and check-airmen, and is a member of the NASA/University of Texas/FAA Crew Aerospace Research Team. Together with Dr Robert Helmreich, he is co-author of the NASA/University of Texas set of CRM behavioural markers which are now the industry standard for assessing CRM crew performance, and was part of the team that revised the FAA's Advisory Circular on CRM training. Taggart's university training is in computer science and systems analysis. His graduate level studies were completed at Carnegie Mellon's School of Industrial Engineering, concentrating on organizational behaviour.

**Ross Telfer** is Director of the Institute of Aviation, Foundation Professor, and Head of the Department of Aviation at the University of Newcastle, Australia. He has teaching and research experience over three decades in school and university. He has carried out pilot judgement training research for the Australian CAA. His jointly written book *Psychology and Flight Training* was translated into both German and French. *Aviation Instruction and Training*, edited by Professor Telfer, was published in 1993. He has lectured throughout the world, and his methods and materials have been used by Cathay Pacific Airways, Lufthansa, Sabena and Air Canada.

**Earl L. Wiener** is Professor of Management Science at the University of Miami. He received his BA in Psychology from Duke University, and his PhD in Psychology and Industrial Engineering from Ohio State University. He served as a pilot in the US Air Force and US Army, and is rated in fixed-wing and rotary-wing aircraft. Since 1979 he has been active in the aeronautics and cockpit automation research of NASA's Ames Research Center. Dr Wiener is a fellow of the Human Factors Society and the American Psychological Association, and has served as president of the Human Factors Society. He is a co-editor of two books: *Human Factors in Aviation*, published in 1988 by Academic Press, and *Cockpit Resource Management*, published in 1993 by Academic Press.

# Foreword

Few human industrial endeavours dedicate such vigorous energy, concentrated attention and allocation of resources to safety as the air transportation industry. Safety has become a trademark of aviation, almost a myth deeply rooted in the beliefs of most involved in this industry. That the realities of a typical day in the life of aviation suggest that beliefs might not always be carried into deed is beyond the point. What is important is a perception that has developed for decades: *in aviation, safety is first*. This is dogmatic, a creed entrenched in aviation professionals since the first moments of apprenticeship.

The aviation industry may occasionally have been reluctant, but has always been prepared to accept outside help when sensing dangers to its safety frontiers. The proverbial distrust of aviation technical people for outsiders has endured a reasonably balanced relationship with the need and acceptance of external help when the answers to safety questions overcome the know-how of those within the industry. The contributions of ergonomics, physiology, the engineering sciences and psychology, among others, to elevate aviation to its present status as the safest mode of mass transportation are beyond discussion. However, notwithstanding all the foregoing, it is puzzling to see how many renewed concerns about safety in aviation loom over the horizon as we approach the end of the first century of manned, heavier-than-air flight.

Looking back without abusing the benefit of hindsight sometimes proves to be helpful in understanding what *really* went wrong in the past. We know the outcome of events. If we are able to discriminate the real symptoms and circumstances from dogmatic decree, we can then learn valuable lessons for designing the ways of the future. Although safety in aviation amounts to an impressive succession of achievements, there were undoubtedly flaws and false steps along the way. To be critical, if one were to single out a major flaw in safety endeavours, such would undoubtedly be a piecemeal approach, underpinned by the pontification of safety.

Over time, progress in aviation safety has been hindered by piecemeal approaches. Pilots, controllers, designers, engineers, researchers, trainers and others in the safety community have advocated solutions to safety deficiencies which were undoubtedly biased by their professional backgrounds. Such approaches have neglected to look into the big picture of aviation safety, and have thus produced dedicated solutions to observed deficiencies and conveyed the notion that different activities within aviation take place in isolation. On the other hand, safety became a hobbyhorse, sometimes abused over the years. Aberrant endeavours have hidden under its skirts. Ignoring years of research on human error, exhortations to professional, error-free behaviours – as pathetic as futile – have been daily currency in safety practices. The fundamental issue, simple as it is, has been consistently dodged: while safety is yet another means to achieve aviation production goals, mythical beliefs perpetuated since World War II have fostered the perception of safety as an end in itself.

The contribution of psychology to aviation safety has, in many ways, been no exception to this. Indeed, it would be unfair towards the discipline and its practitioners to expect otherwise. For many years, psychologists who contributed to aviation safety followed particular schools of thought predominant at the time. Thus, from the early 1940s to the mid-1970s clinical, behavioural and cognitive psychologists made significant contributions to the progress of safety in aviation by fostering understanding of human capabilities and limitations. Such academic knowledge gradually made its way into the realms of practical life through personnel selection procedures, the design of equipment, training methods, flight and duty time limitations and a myriad of other applications. Ominously, all this wealth of knowledge was directed to the 'tip of the arrow', to the front-line operators: pilots, mechanics, dispatchers, controllers, etc. The context within which operational personnel discharged their responsibilities, the system which provided the framework and by force bounded them, remained largely untouched – and unimproved.

The late 1970s and the 1980s witnessed the blossoming of social psychology in aviation. An offspring of this school of thought, Crew Resource Management, became a world-class citizen, a training technology adopted throughout the world. While cognitive and behavioural psychologists continued to amaze the community by providing more and more data on how humans make decisions, on the limitations and capabilities of human cognition and its relevance to task analysis, on how to improve the process of learning and so forth, social psychologists dedicated their attention to small-group dynamics. The aviation community thus began to learn how small groups make decisions, about issues involved in the flow of information among members of small teams, about leadership and followership, and about group problem-solving strategies. CRM, as the application of this knowledge, has been credited with further improving aviation safety. A fundamental proposal of social psychology, however, went unheeded in most quarters. Social psychologists behind CRM went to considerable pains to explain that human behaviour and therefore human error seldom – if ever – take place in a social vacuum, and they went on to emphasize the importance of organizational commitment to CRM if such programmes were to be successful. For the first time, safety endeavours went one step beyond operational personnel, to address the importance of the contribution of high-level, strategic decision makers to aviation safety and effectiveness. The message of safety as one means to achieve organizational development rather than being an end in itself became loud and clear. Although nodding in agreement, the community paid lip service to these proposals, and responding to solidly-entrenched biases, turned its back and continued brandishing professionalism as the key to aviation safety. The consequences were unfair to the professionals involved and denied the realities of social psychology: in many instances, 'lack of CRM skills' or similar phrases have merely replaced the World War II-vintage term 'pilot error' in accident reports. Again, the system and its flaws escaped unscathed.

As the industry approached the end of the 1980s, it was obvious that neither 'traditional' psychology-based endeavours nor CRM were the final answer to the pervasiveness of human error in aviation. The vision of human factors as the last frontier of the aviation system began to crumble under the cold numbers in accident statistics. The arrival of technology, in the form of advanced cockpit displays and devices, brought little relief. Human error continued to circumvent such finely engineered technology with the same ease with which it had dodged previous attempts based on training and regulatory interventions. However, rather than questioning the wisdom of the approaches being pursued, the industry redoubled efforts along the same avenues, through what amounted to nothing more than an escalation of commitment. More technology was introduced. The cockpits and other workstations became more automated. Paradoxically, while engineers tried to design human error out of aviation by displacing humans out of the control loop, human factors enjoyed its heyday, advocating a human-centred automation. Training budgets skyrocketed. The industry stopped one foot short of enacting regulations forbidding humans to err . . .

Then, on Friday 10 March, 1989, Air Ontario flight 1363, a Fokker F-28 with 65 passengers and four crew members on board crashed while attempting take-off from a small provincial airport at Dryden, Ontario,

Canada. In all, 21 passengers and three crew members died. It was not, by any standards, a major disaster; however, it deeply shook the Canadian aviation community and Canadian public opinion. A Royal Commission of Inquiry was appointed, which after three years of hard work produced a benchmark report, truly a turning point in up-to-then established approaches to accident investigation, prevention and safety in aviation. Although early into the inquiry it was evident that operational human error had been a significant factor, the Royal Commissioner rejected what was as evident as inconsequential in terms of prevention value. Assisted by a multi-disciplinary team including pilots, engineers, doctors, psychologists, ergonomists, human factors professionals and accident investigators, he tore apart the Canadian aviation system, looking for the systemic flaws and failures in the safety network which either fostered, or failed to contain, the accident. The philosophy under which the inquiry was conducted is best exemplified by the following statement excerpted from the four-volume, 1500 page report:

> . . . The pilot-in-command made a flawed decision, but that decision was not made in isolation. It was made in the context of an integrated air transportation system that, if it had been functioning properly, should have prevented the decision to take off . . . there were significant failures, most of them beyond the captain's control, that had an operational impact on the events in Dryden . . . *the regulatory, organizational, physical and crew components must be examined to determine how each may have influenced the captain's decision* [italics added].

The report represents the first non-controversial systemic approach to the investigation of an aviation accident. It also represents the stepping stone to a much-needed development: the advent of organizational psychology into aviation. In a sense, the *Final Report of the Commission of Inquiry into the Air Ontario Crash at Dryden, Ontario* is an accomplished exercise of applied organizational and social psychology in practice. Since Dryden, the influence of organizational psychology in aviation safety thinking has become evident. Accident investigations increasingly look at the whole rather than at the parts. Issues like corporate culture and its influence on individual behaviour, the issue of safe and unsafe cultures, the impact of organizational design on operational performance and the contribution of strategic decision makers on all the foregoing, are but a few of the subjects that have started to be openly discussed in both accident reports and international fora. These subjects have also defined an agenda for research for the scientific community. There should be little doubt that by pursuing an avenue of action nurtured by organizational psychology and constantly fed back by behavioural, cognitive and social psychology, one of the maladies of the past – piecemeal approaches – will be overcome.

The organizational perspective also provides a framework for safety, allowing it to be viewed realistically within the goals of aviation organizations, thereby defeating the insidious pontification and out-cropping myths the industry endured for years.

Is organizational psychology the new 'last frontier' in aviation safety? I think not. Furthermore, I think that such a 'last frontier' simply does not exist. The passage of time will bring unforeseen challenges which will demand new solutions. I do believe that organizational psychology will be at the centre of safety endeavours for years to come, but we must continue to draw upon human factors and all its tributary disciplines in a truly systemic approach. On the other hand, it would be naive to presume that there is any way back from technology. There are too many interests at stake. The real issue, however, is that we should never sit back, assuming that, once and forever, we have found the solution for aviation safety. Such a magical solution was implicit in presenting human factors as the 'last frontier'; and it was dogmatically asserted in a thousand-and-one opportunities during the past 50 years. The message I am trying to convey is that the battle for safety is an endless quest against a relentless enemy having infinite resources, and who will never give up.

This book is a step in the right direction. I believe it is yet another contribution to the changing of the guard which the aviation industry requires as we are about to fold the twentieth century. The adoption of systemic, organizational approaches to aviation safety can be regarded more like a minefield than a garden path because it involves considerable change. Change, if it is to be implemented though the commitment of those involved, must rest on education. This book fosters such education. It blends a unique combination of talents and expertise. There are chapters which inform us about the latest developments in what can be regarded as 'traditional' topics in psychology: learning, mental models, task analysis and decision making. Such chapers blend into the day-to-day realities of aviation through contributors who discuss accident investigation, safety deficiency reporting systems, aircraft maintenance and operations. As a protective envelope, further chapters discuss organizations and cultures.

In short, this book satisfies two prerequisites for contemporary aviation safety. Firstly, the marriage of theory and practice, the teaming of research and academia with practitioners who can act as a bridge and apply scientific knowledge into operational settings. Secondly, it advances and defends through its various chapters a systemic view of aviation safety. *Aviation Psychology in Practice* is beyond doubt another example that there is more to be gained by facing challenges and endeavouring to provide solutions, even with caveats, doubts and vacillations, than by refraining from action by decrying the difficulties involved in the challenge.

*Captain Daniel Maurino*
*Project Manager, Flight Safety and*
*Human Factors Programme, ICAO*

# List of abbreviations

| | |
|---|---|
| A&P | Airframe and Powerplant |
| AAIB | Air Accidents Investigation Branch (UK) |
| AAR | Aviation Accident Report (NTSB) |
| AAS | Advanced Automation System |
| AC | Advisory Circular (FAA) |
| ACSNI | Advisory Committee on the Safety of Nuclear Installations |
| ADM | Aeronautical Decision Making |
| ADREP | Accident Data Reporting System (ICAO) |
| AMT | Aviation Maintenance Technician |
| APRG | Aviation Psychology Research Group |
| AQP | Advanced Qualification Program (US) |
| ASRS | Aviation Safety Reporting System (NASA) |
| ATA | Air Transport Association (US) |
| ATC | Air Traffic Control |
| ATIS | Automatic Terminal Information Service |
| AWACS | Airborne Warning and Control System |
| BCAG | Boeing Commercial Airplane Group |
| CAA | Civil Aviation Administration |
| CDU | Control Display Unit |
| CMAQ | Cockpit Management Attitudes Questionnaire (NASA/UT) |
| COMETT | Community Programme for Education and Training in Technology (EC) |
| CRM | Crew Resource Management |
| CSERIAC | Crew System Ergonomics Information Analysis Center |
| CTA | Cognitive Task Analysis |
| CVR | Cockpit Voice Recorder |
| DME | Distance Measuring Equipment |
| EBA | Elimination By Aspects |
| EC | European Community |

| | |
|---|---|
| ERAU | Embry-Riddle Aeronautical University |
| ETOPS | Extended Twin-Engine Operations |
| EWO | Electronic Warfare Officer |
| FAA | Federal Aviation Administration (US) |
| FMAQ | Flight Management Attitudes Questionnaire |
| HSE | Health and Safety Executive |
| IATA | International Air Transport Association |
| IGHC | IATA Ground Handling Council |
| ICAO | International Civil Aviation Organization |
| IFALPA | International Federation of Air Line Pilots' Associations |
| ILS | Instrument Landing System |
| INSAG | International Nuclear Safety Advisory Group |
| ISD | Instructional Systems Development |
| ITMAC | Irish Tobacco Manufacturers Advisory Committee |
| KHUFAC | KLM Human Factors Course |
| LLC | Line/LOFT Checklist (NASA/UT) |
| LOFT | Line Oriented Flight Training |
| LRU | Line-Replaceable Units |
| MAC | Military Airlift Command (US) |
| MCP | Mode Control Panel |
| NASA | National Aeronautics and Space Administration (US) |
| NDM | Naturalistic Decision Making |
| NOTAM | Notice to Airmen |
| NPRM | Notice of Proposed Rule Making (US) |
| NTSB | National Transportation Safety Board |
| OECD | Organization for Economic Cooperation and Development |
| PF | Pilot Flying |
| PLPQ | Pilot Learning Process Questionnaire |
| PNF | Pilot Not Flying |
| RPD | Recognition-Primed Decision Model |
| RTO | Rejected Take-Off |
| SAINT | Systems Analysis of Integrated Networks of Tasks |
| SAT | Systems Approach to Training |
| SCARF | Safety Courses for Airport Ramp Functions (IATA) |
| SFAR | Special Federal Air Regulation (US) |
| SHEL | Software, Hardware, Environment, Liveware (human factors model) |
| SOP | Standard Operating Procedure |
| TCAS | Traffic Collision Avoidance System |
| TOD | Top of Descent |
| UK | United Kingdom |
| USD | United States Dollar |
| UT | University of Texas at Austin |

| | |
|---|---|
| VHF | Very High Frequency |
| VNAV | Vertical Navigation |
| V/S | Vertical Speed |
| VOR | Very High Frequency Omni-directional Range |
| WEAAP | Western European Association of Aviation Psychologists |

# Introduction

# 1 Applied psychology and aviation: Issues of theory and practice

*Nick McDonald and Neil Johnston*

## Introduction

The focus of this book is on the role of psychology in the aviation system – the ways in which psychological ideas and theories support and inform how things are done, or might be done, in aviation. Aviation psychology is not unique, however, and many of the issues it confronts as a developing applied scientific discipline are parallelled in other fields. This introductory chapter locates the book's argument in a broader landscape of theory and practice in applied psychology. The articulation of the issues in the following chapters address themselves specifically to psychology in aviation, but they may also play a role as a paradigm for a wider context of applied psychology.

The immediate occasion for producing this book was the planning of the 21st conference of the Western European Association of Aviation Psychologists. This planning forced an appraisal of the state of aviation psychology, both in general but particularly within a European context. Western Europe is in a current conjuncture of developing the role of transnational institutions in regulating and administering various areas of life, including aviation. Aviation human factors has not been prominent on this agenda, and even at a national level in Europe there are few coordinated research programmes. This contrasts with the United States, where the Federal Aviation Authority (FAA) has promulgated a detailed national plan for aviation human factors. In a wider world context, aviation psychology is in a state of considerable growth and development.

These different considerations provided a stimulus to explore the issues which should perhaps be on a developmental agenda for aviation psychology, whether in Europe or elsewhere.

## The tension between research and practice

The ideas for the book were formed in the discussion of a fairly unusual group, the Aerospace Psychology Research Group, which comprises both university-based researchers and aviation professionals who are furthering their mutual interest in aviation psychology. Underlying many of our discussions is a tension between two points of view, objectives or professional agendas – those of the practitioner and those of the researcher. This tension is common to most, if not all, applied disciplines. It can be a source of productive creativity in the development of the discipline. Too often it is a source of frustration.

Frustration arises from the difficulty of building productive links between applied research and the application of research to practice. The researcher seeks a better understanding of the topic under investigation, while the practitioner seeks a credible basis upon which to plot a course of action. Although both goals might seem entirely complementary, even symbiotic, it does not take too much scratching beneath the surface for the researcher to discover the inability of almost any theory to account for the entire operational situation, as set within its natural context. It is a rare piece of research which can claim to further the aims of science and simultaneously appeal to the practitioner (though see Rosekind *et al.*, 1992). Practitioners almost invariably experience theory as an irrelevant straightjacket which prevents critical issues from being addressed productively. In practice the design of operational, procedural or training tasks is rarely initiated by reviewing the relevant research literature. Similarly, there is often a disjunction between the findings of research and the drafting of regulations (this has been discussed in a different transportation context by McDonald, 1984, 1985).

Our optimism, which stimulated this book, is founded on the notion that such frustrations might be the springboard for the creation of new and better theories and a more effective basis for practice.

Underlying many of the arguments put forward in this book are two interrelated issues which confront the evolving discipline of aviation psychology and which will shape its future development. The first concerns the somewhat selective range of topics with which aviation psychology has traditionally been concerned. The second issue concerns the 'ecological validity' of the theoretical and empirical work which comprises aviation psychology. By way of an introduction to these issues it may be fruitful to consider the role of aviation psychology as an applied science.

## An applied science

An applied discipline like aviation psychology is inevitably a hybrid. The very term 'applied psychology' can have a variety of meanings including the following:

- the application of general psychological knowledge, principles or laws to a specific context or area of practice (e.g. applying general principles of small group dynamics to crew interactions on the flight deck);
- the conduct of research in a particular context to develop general psychological theory (e.g. how does the behaviour of flight crews on transmeridian flights inform our understanding of the principles of circadian patterns of activation?);
- carrying out research in a specific context (what are the characteristics of *aeronautical* decision making?);
- carrying out research on the practice of psychology – or related professional disciplines – in a particular context (e.g. what is the most effective way to modify crew interpersonal attitudes and communication behaviour?).

In very crude terms it might seem that an 'applied' scientific discipline like aviation psychology exists at the interface between 'pure science' – which seeks to derive generalizations and laws which are independent of any particular context – and 'practice' – the application of psychological knowledge to particular sets of circumstances. In the development of psychology, the understanding of scientific enquiry implicit in this distinction has had some very unfortunate consequences. Generations of psychologists were inculcated with the doctrine of the founding fathers of experimental psychology that the way to establish a scientific discipline of psychology was to remove that enquiry from its social context into the neutral environment of the laboratory. This positivist doctrine was the cornerstone of the development of the great movements of behaviourism and information-processing cognitive psychology.

Subsequently, human factors engineering and ergonomics grew alongside the developing information processing theory, becoming more complex as both the human-machine interface became more complex, and as more complex computational metaphors became available to help develop the theory of the human operator. Although avowedly applied and contextually driven, it shared with mainstream scientific psychology the fundamental concern that this context should be so narrowly defined as to exclude any analysis of the overall interaction between the operator and his or (rarely) her task, let alone the social and organizational system in which that work is done. It frequently had a narrow focus on aspects of performance and a strict adherence to a rigorous experimental methodology. The result is a discipline which is piecemeal and atomized; where there is a constant unresolved tension between the specifics of scientific knowledge and the context of its application.

One result of this unresolved contadiction between the process of research and the application of its findings is a growing crisis between promise and achievement. In 1975, the 20th IATA Technical Conference on 'Safety in Flight Operations' – what was for the industry a watershed conference – set the scene for attacking what many saw as 'the last frontier' in aviation safety (Shaw, 1976). This phrase encapsulated the optimistic mood that the application of the increasingly systematized human factors knowledge from central areas of psychology would make aviation safe for all time. Just one year later the core theoretical edifice underpinning the most notable advances in aviation psychology, cognitive psychology, began to creak when Ulrich Neisser published his book *Cognition and Reality* (1976). The essence of his argument is captured in the following quote:

> the study of information processing has momentum and prestige, but it has not yet committed itself to any conception of human nature which could apply beyond the confines of the laboratory.

Neisser articulated an increasing uneasiness among many researchers that their theoretical concepts did not allow them to address practical and relevant 'real-world' issues in an effective manner. Fifteen years later this unease had transmitted itself to the highest level of policy formation responsible for the 'real world' of aviation – the United States' FAA. Thus, for example, in presenting the FAA's National Plan for Aviation Human Factors, the Chief Scientific and Human Factors Advisor, Clayton Foushee, admitted 'we are not in any better position today than we were 10 years ago to say what "human-centered" automation is or isn't' (Foushee, 1991).

It would be wrong to identify the source of this crisis as being simply the lack of good theory. In one sense there is no lack of theoretical models or ideas which are applicable to the problems of aviation. The development of Crew (or Cockpit) Resource Management (CRM) is an example of the introduction of a new theoretical perspective (the social psychology of small group interaction and decision making), which facilitated an entirely novel and applied training approach. The very terminology of CRM is presumably drawn from or inspired by Human Resource Management. This is a generic management strategy or approach which takes a total organization perspective in focusing on developing an organization's members as a productive resource (see, for example, Storey, 1992). In the future this may become an influential source of ideas on the development of CRM.

The process of doing first rate applied science is the potential victim of processes in the two main interest groups – the research community and the industry community. The commercial imperative in industry tends to make for a short-term perspective in in relation to investment and results. There is often a perception that many of the human factors problems in aviation and other applied contexts can be solved by the application of existing expertise. While this perception is often justifiable in particular cases, this is at best a small part of what is involved in an applied discipline. Too often it takes a decade or more to discover that some intervention based upon the conventional scientific wisdom just does not work in practice. Often what is lacking is a systematic needs analysis, or a thoroughgoing evaluation. While at the time of initiation of an intervention this might seem to involve an excessive investment of time or resources, in the long run that investment can seem trivial compared to the overall resources which are at stake.

On the other hand, in the research sector, applied research tends to be treated as the 'poor relation' of 'basic' or 'pure' research. Applied research is not, by and large, the avenue to a prestigious academic career. Career paths in applied and pure research are often quite separate, with only the exceptional researcher being able to span an agenda which is both pure and applied.

When both these factors are put together they can make for a lack of serious rigour and commitment to the achievement of strategic goals of research and development in applied psychology. Often the difficulties of doing worthwhile applied research are attributed to problems of methodology – the intractability of the complex and multifaceted nature of the 'real world'. The quest to reduce operational and policy issues of considerable pragmatic importance to a set of discrete independent and objective measures can at times seem as difficult as dissolving oil in water. However, we strongly believe that this seeming incompatibility between the very terms of research and practice should be seen as a stimulus for methodological innovation and the development of appropriate methodologies.

## Setting an agenda

There are many ways in which both 'pure' and 'applied' psychologists have responded to this crisis. Briefly, it is possible to summarize some of the directions and departures which have been initiated and must be intensified if aviation psychology is to continue to deliver a useful service to the aviation community.

The role of organizational factors in the genesis of accidents is becoming much better understood and more widely accepted (see, for example, Reason, 1990, Chapter 7). What is needed is a greater understanding of how organizations can play a proactive and preventative role in managing the human factors of safety. Safety is strongly influenced by the culture and climate of an organization (the goals, values, attitudes, beliefs and shared practices that pervade an organization). However, the means by which organizational culture can be influenced to achieve safety goals are still not well understood.

For too long theoretical models applied in practical situations have been derived from laboratory research though never validated in the context of their application. The driving force of much applied psychology has tended to be the search for some instantiation for a pre-existing theory. The problem is therefore made to fit the theory rather than the theory being chosen, or developed, to fit the problem. Theoretical models which underlie aviation psychology need to be grounded in a more comprehensive and detailed analysis of practice. This requires new methodologies and new research goals. One source of data, among many others, which can provide a more naturalistic understanding of the human factors underlying safety failures, comes from the material provided by incident reports and accident investigation.

A range of potentially incompatible theoretical formulations need to be permitted to coexist and interact more freely and fruitfully than they do at present. Organization theory, cognitive theory and behaviour theory are well represented in this volume. Their compatibilities are emphasized in the book, but at a fundamental level they also have incompatibilities. These incompatibilities are also a potentially fruitful stimulus for innovative research.

This agenda implies the loosening up of the relationship between pure and applied science. The requirement for psychological theory *per se* to be clearly related to the everyday world of normative behaviour and interaction will place a premium on data and evidence from practical situations. This may throw up some challenges to existing theoretical paradigms. For example, the development of an ecological perspective has given rise to theories of 'situated cognition' (Lave, 1988; Lave and Wenger, 1991; Suchman, 1987). What is meant by an 'ecological' approach or 'ecological' theory? For the purposes of the present argument, 'ecological' is used in the weak sense that the theory must be related to its normal situational context in a clear and explicit way, with some claims to the validity of that relationship. A more rigorous notion of ecological theory derives from the work of Gibson (1979), who developed the first fully blown ecological theory in psychology. Ecological in this sense refers to a systematic account of individuals in their environment, with that mutual relationship, 'the ecological niche', being the central focus of analysis (as opposed to cognitions or mental representations in cognitive theory). Ironically, although Gibson's theory of perception was developed in the context of military aviation psychology during the Second World War, his ecological approach has had no appreciable influence on later developments of this sub-discipline.

It is quite possible that the dynamic caused by the discrepancy between promise and performance alluded to above could provide the stimulus for a more thoroughgoing ecological theory. This could relate to the concerns of aviation psychology in the same way as the development of 'situated cognition' has addressed the issues of learning how to become a competent practitioner (Lave and Wenger, 1991). However this is more appropriately the subject for a different book.

## An outline of the book

### The aviation socio-technical system

Human factors impacts upon virtually every part of the aviation system. However, the major focus of aviation psychology to date has been on the flight phase of aviation or on the capacities or characteristics of the individual aviator. This may seem to be a natural priority for research efforts, allowing aviation psychologists to specialize in what is peculiar to their subject and leaving other issues to organizational, industrial, clinical or other professional psychologists. There are two reasons why this is not a desirable state of affairs. Firstly, different levels of the system (for example, organizational and individual behaviour) interact and depend on each other. Secondly, and perhaps more importantly, critical areas of aviation have been relatively neglected by aviation psychology.

The global organizational context for safety in aviation is developed in Pidgeon and O'Leary's chapter which describes aviation as a socio-technical system. Needless to say the socio-technical system which is described is not the familiar 'man–machine interface' of classic ergonomics, where the emphasis is on the precise interaction between an individual and a piece of equipment. Instead it encompasses the whole organization and social practices of work. Socio-technical systems theory has contributed substantially to our understanding of the preconditions of large-scale accidents in major industries. The role of organizational factors has been increasingly brought to the fore in the investigation of aviation accidents. This has developed with the growing disillusionment with the overused, and abused, explanatory category – pilot error – which should, in fact, be seen as the starting point, rather than the end point, of the investigative process. The search for explanations as to why people in organizations behave unsafely has led to the development of the notion of organizational safety culture. This refers to the norms, roles, values and common practices which govern the way in which the organization performs its work, and, most crucially, the organization's members' shared understanding of the meaning of these norms, rules and practices. In Pidgeon and O'Leary's chapter the characteristics of safety culture are discussed under the four principle headings of responsibility at strategic management level, distributed attitudes of care and concern throughout an organization, appropriate norms and rules for handling hazards, and ongoing reflection upon safety practice.

Developing the organizational perspective creates dilemmas for the practical aviation psychologist. First, as Pidgeon and O'Leary clearly state, the relationship between the organization's culture and the ways in which its members behave is not a simple direct causal influence. It is not possible to predict the behaviour of individuals from the shared understandings and collective meanings that make up an organization's culture. Secondly, diagnosing a deficient safety culture is a very different matter from finding a cure. The principles of engineering a change in safety culture are, so far, unknown territory. However the remaining chapters of the book do address the link between organizational culture in its broadest sense and what actually happens in practice.

Degani and Wiener approach the issue from the starting point of the organization's philosophy regarding how it wishes its aircraft to be operated. From this philosophy appropriate operating policies and procedures are derived. Degani and Wiener first review the significance of operational philosophies, policies and procedures before pointing out that we must have methods of addressing the gap between the reality of operational practices and management's intentions. The following observation encapsulates the key issue:

> It is not . . . much good relying on the protective effect of standard drills when the evidence shows that the standard drills are being disregarded on a not-insignificant number of occasions (Report of the Public Inquiry into the causes and circumstances of the accident near Staines on 18 June 1972).

This chapter discusses the variety of reasons why there are deviations from procedures (or variations of practice within procedures) and how these should be managed. The authors argue for a strong feedback loop to moderate the development and ongoing evaluation of procedures and their effective application in practice. This is an instantiation of the process of 'ongoing reflection' which Pidgeon and O'Leary identify as one of the core characteristics of a good safety culture.

The issue of standard operating procedures is also one of the central themes of McDonald and Fuller's chapter on the management of safety in aircraft ground handling, though the context is radically different. The airport ramp does not have the same culture of safety as flight operations. As both this chapter and the following one by Marx and Graeber on aircraft maintenance make clear, aviation is a set of interdependent socio-technical systems in which airborne safety relies on safe practices in maintainance and ground handling. The nexus of the problem in ground handling concerns technology and standardization. The core technology here, the aircraft, is not designed for optimal ground handling. This in turn throws up a series of fundamental ergonomic problems in the safe handling of equipment and aircraft. Standardization of operating procedures has been one of the prime methods of controlling this problem. However the ramp accident statistics appear to indicate that this approach is not working. Among the problems for management are: ensuring that the organization learns from the accidents which do happen; using information appropriately to enhance accident prevention; and developing the capacity to assess and implement a range of countermeasures within a framework which is longer term and preventive rather than short term and reactive.

Marx and Graeber start with a bottom-up approach – the nature of error in aircraft maintenance and the ways in which systems and procedures deal with error. Two themes run through their chapter: first, the problems and difficulties of classifying and investigating the human factors of error in maintenance in a way that is intellectually coherent and objective – and which simultaneously meets the practical needs of the industry. Secondly, they raise the issue of the goals of procedures and systems. Error reduction is but one of three strategies to minimize the risk of system failure. The other two are error capturing (or detection) and systemic error tolerance. They draw out two implications for the practice of aviation psychology. The first of these is the importance of a systems perspective as opposed to simply focusing on the individual operator and on what he or she may or may not do. Following from this is the need for psychologists to integrate themselves into an interdisciplinary team with experts from other disciplines including industry professionals who also play a role in system design.

The safety goals of the aviation system are brought most clearly into focus by Muir's chapter on passenger behaviour. The implicit goal of much of aviation psychology (to reduce accidents) has tended to eclipse a complementary goal of increasing passenger survival (both physical and psychological) in those accidents which do happen. This chapter describes a programme of research which has done much to elucidate the human factors of accident survival, particularly in relation to aircraft evacuation. Perhaps the central finding here is that in a simulated disaster scenario one can motivate people in a way which will produce behaviour which is a functional analogue of the behaviour of people in actual emergency evacuation situations. Muir used this to investigate the role of a variety of factors including the configuration of seats, smoke, various signals and the role of cabin crew. Another issue raised in this research programme is the preparedness of passengers for emergency procedures in terms of their knowledge, the information which is presented to them and how it is presented. This research programme has major implications for aircraft certification procedures.

## Learning from accidents

The following two chapters address, in a very practical way, the issue of learning from accidents, mishaps and incidents. Accidents, almost by definition, identify faults in the system. The identification and under-standing of those faults by accident investigators is one of the prime ways of improving the system. Harle's chapter is very important because it summarizes the work done by the Canadian Air Safety Board in taking a comprehensive and systematic approach to the investigation of the human factors of accident causation. Although human factors specialists play an increasingly important role in major accident investigations, the primary accident investigators are rarely psychologists. The accident investigator has to play a multidisciplinary role if the multiplicity of causal and contributory factors which have to be taken into account in an accident investigation are to be fully investigated. Harle's chapter provides an authoritative guide for the non-psychologist on the principles of conducting an investigation of human factors. The investigative orientation is clearly set within the socio-technical context of aviation organizations, and the 'why?' questions – which originate with what happened and what was done by whom – are only terminated by the pragmatic considerations of generating effective answers which can be used in prevention. Of particular value is the discussion of how to overcome the difficulty of conclusively inferring the role of human factors in accident causation from the *post hoc* perspective of the investigator. This difficulty arises from the inherent intractability of much human factors material to precise quantification, measurement or prediction.

The aviation industry has played a leading role in developing confidential feedback systems for learning from incidents where safety may have been compromised but no adverse consequences ensued. Confidential incident reporting schemes operate at many different levels. Their most obvious role is to identify directly specific hazards which can then be avoided or rectified. Their feedback function does much to inform, educate and create a general awareness of hazards. Confidential reporting systems are also part of a contract of professional trust (which takes different forms in different countries). Here the normal expectation that we be held culpably responsible for our actions, inactions or errors is suspended, and frank disclosure is accompanied by complete confidentiality, if not immunity from prosecution (Johnston, in press). On these grounds alone such systems play an indispensable role in aviation safety. Chappell's chapter discusses the role of these systems, building on their archives as a rich source of naturalistic research data. This is a practical, pragmatic guide for the researcher through the territory, too often unfamiliar to the human factors specialist, of the analysis of qualitative data of verbal reports, mapping out the biases which may influence the data generation process and pitfalls of analysis and interpretation. She illustrates the ways in which the data can be used in a specific investigation or as an invaluable first stage, 'scoping the problem' for a more extensive programme of research. Degani and Wiener's chapter also provides some excellent illustrations of the role of incident reports in the evaluation of procedures in practice.

## New theoretical models

The practice of research should lead to the development of theory. Fuller's chapter is one of several which are concerned with developing and extending the core theoretical base of aviation psychology. The behavioural approach has been revitalized by a newfound enthusiasm for systematic analysis of naturalistic contingencies. It has a focused concern on the consequences of behaviour as an important explanatory mechanism. This enables processes of motivation, change and incentive to be dealt with more effectively than in other approaches. Fuller argues that the behavioural approach should be seen as being quite complementary to other theoretical orientations, particularly the cognitive approach, where, to take one example, each theory treats different aspects of procedural rules (or condition action sequences).

Most of the advances in the core areas of aviation psychology have come within the cognitive theoretical paradigm. However it is only recently that this has resulted in a thoroughgoing and systematic analysis of the core tasks within the aviation system. Redding and Seamster's chapter describes the process of cognitive task analysis as applied to air traffic control (ATC). The central focus of cognitive task analysis is to develop a cognitive model of the task. This is achieved by a process rather like triangulation: the cross-comparison of data from parallel and complementary methodologies. The outcome is the identification of salient events which are processed by the controller's dynamic mental model of the situation and underlying knowledge structures. This analysis has been incorporated in the the new FAA ATC training curriculum. Perhaps a further stage of this important work would be a validation of this whole process through a full training evaluation study. The outcomes of cognitive task analysis can also be applied to evaluate new technological developments, and this technique has been applied, though not so comprehensively, to other key aviation tasks. Redding and Seamster's chapter also provides a thorough overview of cognitive task analysis methodology and teases out many of the conceptual and methodological problems involved in studying mental models and expertise in applied environments.

In recent years the 'mental model' has become one of the key theoretical notions of cognitive psychology. In common with most cognitive constructs, mental models are posited as attributes of the individual's cognitions – since mainstream cognitive psychology has no developed theory of intersubjectivity. The increasing acceptance that many of the human factors problems underlying major safety failures could be traced back to problems in crew coordination, communication and shared decision making has forced cognitive psychology to broaden its individual focus and develop the notion of shared understanding. The burden of Orasanu's chapter is to clarify the characteristics of crew communication as a component of effective crew decision making. One of the functions of communication is to build a 'shared problem model' and this is facilitated by communication which is explicit and contextually appropriate. However it is also important to develop the ability of crews to think about problems, not only in terms of their situational awareness but also in terms of awareness of appropriate decision and task management strategies. The implications for training that Orasanu draws are: that crew training has to be training *as a crew*; and that training for effective decision making has to be embedded in a meaningful context, so that the exercise of communication and decision skills becomes a normal part of the operational environment.

Kaempf and Klein's chapter gives a clear understanding of the nature of the shift in cognitive theorizing which underlies analyses such as that of Orasanu. They first provide a valuable and comprehensive review of research into pilot judgement. They then contrast two models of aeronautical decision making as frameworks on which to base ADM training. The 'classical' model of decision theory sees the decision maker as generating an exhaustive list of options, evaluating each and choosing the one with the most advantageous outcome. Naturalistic decision making, on the other hand, starts from the analysis of what people do in natural settings and suggests that in many operational contexts it is the diagnosis of the situation which is the critical task of the decision maker. This led to the development of the model of Recognition-Primed Decision making (RPD). Where decision makers are forced to evaluate alternative courses of action they frequently do this using a satisficing strategy, choosing the first adequate option, rather than an optimizing one. As the authors suggest, this type of analysis needs to be taken further to identify the underlying skills (feature matching, mental simulation, for example) which can define training objectives more clearly.

One of the fundamental issues that needs to be confronted if there is to be a comprehensive analysis of the task of the flight crew (or of any of the other tasks of the aviation system) is the problem of integrating theoretical models and research findings from diverse areas, to counteract the atomism and narrow focus of much psychological theorizing. Orasanu's chapter illustrates the beginnings of an integration of cognitive theory with the social communications theory, which has underpinned the conceptualization of CRM. Prince, Bowers and Salas take this a stage further by introducing stress into the equation of decision making by crews. They pose the issue as a series of challenges: to develop a theoretical model of stress effects on aeronautical decision making to guide training; to identify the types of stressors confronting flight crews and to develop methods for their mitigation; and to integrate training on decison making with managing and coping with stress, recognizing the specific roles of different crew members. The implications for training of their analysis are the development of prevention-focused behaviours, which seek to minimize stress by anticipating problems, and problem-focused behaviours to respond to unavoidable problems.

## The delivery of training

One of the major spurs to the development and refinement of increasingly precise models of crew functioning has been the introduction of Crew Resource Management (CRM) training. This has been a novel and innovative way of addressing the human factor issues in the cockpit and has great potential to be exported to other high risk situations of team decision making. Taggart discusses the rationale for the introduction of CRM training, some of the history of its evolution and its increasing acceptance by the world aviation community. He sets out a range of pragmatic issues to do with the organization and presentation of CRM courses, and with their acceptance by the target audience. Evaluation studies are beginning to demonstrate the success of CRM in creating awareness and changing attitudes. Perhaps this is the first stage of an agenda to evaluate such interventions in terms of their impact on the actual patterns of crew interaction.

The issue of training delivery which is raised by Taggart is taken forward in a systematic analysis by Telfer. Aviation instruction has tended to be well grounded in the operational context, with instructors drawn from the cadre of expert professional aviators, rather than career-track professional trainers. However, there are increasing trends towards more formal education and training based in the classroom and training institution. One of the consequences of the former characteristics of aviation instruction has been a relative lack of systematic investigation of the training process. Parenthetically however, one should not overlook the possible advantages of such a system for effective training transfer. Telfer provides an invaluable guide for the training practitioner to the three main stages of aviation instruction: (i) task analysis, planning and preparation (labelled presage); (ii) the process of instruction; and (iii) the product of the instruction process. This very practical and focused analysis is again an illustration of the advantages of drawing from diverse theoretical orientations (in this case cognitive and behavioural).

## Loose ends

The chapters in this book are diverse in content, organization and style. A number of them provide a unique contribution to the aviation psychology literature. This arises naturally out of the role which the book is attempting to play. Aviation psychology is at the interface of diverse areas of psychological theory and a variety of contexts for the application of that theory. The fundamental messages of the book are that aviation psychology must take the perspective of the whole aviation system and begin to understand more thoroughly the social situation in which work takes place.Within this broad context the trend towards the increasing grounding of theory within specific operational settings must be fostered and developed. This has several implications:

- there needs to be a greater interchange of ideas between researchers from different theoretical orientations;
- researchers need to involve themselves more deeply in practice;
- practitioners need to see more clearly how the issues they confront may be addressed productively by research and existing theory;
- aviation psychology and other branches of applied psychology

    need to become more permeable to each other's developments of
    theory and practice;
- aviation psychologists need to become more interdisciplinary as
  their role is extended deeper into a system which is designed
  primarily by engineers and systems analysts;
- more non-psychologists will have to develop proficiency in the
  human and organizational factors which govern the ways in
  which people act or behave; these will include accident investigators, trainers, managers or regulators.

If this book is in any way a catalyst in promoting the diverse multilateral
communications which all of this implies, then it will have succeeded.


# References

Foushee, C. (1991), *Overview of the National Plan for Aviation Human Factors. Proceedings of the Sixth International Symposium on Aviation Psychology*, 29 April–2 May, 1991, Columbus, OH, Ohio State University, Department of Aviation.

Gibson, J. J. (1979), *The Ecological Approach to Visual Perception*, Boston, Houghton Mifflin.

Johnston, A.N. (in press), 'Blame, punishment and risk management', in C. Hood, D. Jones, N. Pidgeon and B. Turner (eds), *Accident and Design*, London, University College Press.

Lave, J. (1988), *Cognition in Practice*, Cambridge, Cambridge University Press.

Lave, J. and Wenger, E. (1991), *Situated Learning*, Cambridge, Cambridge University Press.

McDonald, N. (1984), *Fatigue, Safety and the Truck Driver*, London, Taylor and Francis.

McDonald, N. (1985), 'Regulating hours of work in the road haulage industry: the case for social criteria', *International Labour Review*, **124** (5), 577–592.

Neisser, U. (1976), *Cognition and Reality*, San Francisco, Freeman.

Reason, J. (1990), *Human Error*, Cambridge, Cambridge University Press.

Report of the Public Inquiry into the causes and circumstances of the accident near Staines on 18 June 1972 (1973), CAAR 4/73, London, HMSO.

Rosekind, M.R., Graeber, R.C., Dinges, D.F., Connell, L.J., Rountree, M.S., Spinweber, C.L. and Gillen, K. (1992), *Crew factors in flight operations: IX. Effects of planned cockpit rest on crew performance and alertness in long haul operations*, NASA Technical Memorandum 103884, Mountain View, CA, NASA.

Shaw, R. R. (1976), ' "The Last Frontier": an overview of the 20th IATA Technical Conference on "Safety in Flight Operations" ', Lecture to the Dublin Branch of the Royal Aeronautical Society, 5 February 1976, Montreal, IATA.

Storey, J. (1992), *Developments in the Management of Human Resources*, Oxford, Blackwell.

Suchman, H. (1987), *Plans and Situated Actions*, Cambridge, Cambridge University Press.

# Part 1

# The Aviation Socio-technical System

# 2 Organizational safety culture: Implications for aviation practice

*Nick Pidgeon and Mike O'Leary*

## Introduction

What role might organizational and management factors play in enhancing or undermining safety in aviation systems? Despite a long and successful tradition of work into the important relationship between safety and individual aspects of behaviour and attitudes, under the general heading of human factors, wider organizational factors have only recently been clearly identified as contributing significantly to accident causation, and hence as a topic of concern for both aviation safety researchers and practitioners.

This does not, of course, necessarily mean that organizational causes of accidents are in themselves a fundamentally new phenomenon in aviation; these factors have almost certainly, to a greater or lesser extent, been present since the earliest days of civilian and military aviation. What, however, has undoubtedly changed in recent years has been our thinking about the human origins of large-scale accidents and incidents. This development derives in part from a number of prominent accidents and disasters that have occurred internationally over the past decade, although the majority of the research work prompted by these events has been concerned with safety in contexts other than aviation, such as the chemical process, nuclear and surface transportation industries.

Examples of significant accidents, across a wide variety of large-scale hazardous systems, include: in the United Kingdom, the King's Cross Underground fire, the Clapham Junction rail crash and the Piper Alpha oil rig disaster; in India the Bhopal Chemical disaster; in the relatively low-technology arena of marine transportation the Zeebrugge Ferry capsize as well as the *Exxon Valdez* and the *Braer* oil-tanker disasters; and in aerospace, perhaps the most vivid image of the 1980s will remain that of the destruction of the Space Shuttle *Challenger* in January 1986.

We would argue that some of the recent hard lessons learned concerning the role of organizational factors and safety cannot be ignored by aviation practitioners, a point now explicitly recognized by the International Civil Aviation Organization, amongst others, in a recent digest on *Human Factors in Management and Organization* (ICAO, in press). In addressing this important issue the chapter introduces, and is structured around, two very general theoretical notions. The first is that of socio-technical system failure, and the second the idea of organizational safety culture.

A part of our argument is that these theoretical ideas are not solely of academic interest, but are also of direct practical importance. Socio-technical systems theory has provided new and wide ranging insights into the preconditions of large-scale accidents, and in doing so has suggested expanded approaches to accident and incident analysis and the diagnosis of fundamental background causes of such events. More recently, work on the concept of safety culture points to a number of ways of understanding, and in turn possibly influencing, some of the high level social factors that serve to undermine safety in aviation, as well as in a wide range of other contexts where the management of risk and hazard is the responsibility of large organizations.

The chapter begins with an illustration of why the current broadening of perspectives on the human element and safety is essential, and must include organizational as well as individual factors. As frameworks for understanding these issues, two theoretical models of large-scale accidents, namely Turner's (1978) disaster incubation model and Perrow's (1984) complexity-coupling account of failures in socio-technical systems, are outlined.

We then go on to discuss the more recent concept of organizational safety culture. This term first arose as the result of a European analysis of the specific human and organizational factors underlying the 1986 Chernobyl disaster in the former Soviet Union (OECD, 1987). Moreover, safety culture can be related to a number of more general social science treatments of culture, as well as to parallel literature in the safety field.

In the final section we consider some of the implications of the concept of safety culture for aviation practice, relating this to the question of institutional or organizational design for safety. However, in doing so we seek to adopt a critical perspective with respect to the recent discussions of poor and good safety cultures. It will be no simple matter either to translate the many theoretical treatments of the concept into practical action, or to resolve a number of the generic dilemmas which arise in any attempt at institutional design.

## From human factors to socio-technical systems

Since the Second World War, aviation in the developed world has been marked out from many other high-technology activities by its early and increasingly successful commitment to the application of human factors psychology to questions both of safety, and to other more general ergonomics problems (e.g. Hawkins, 1987). The reasons for this are not difficult to discern. Members of the public perceive many high-consequence/low-probability technological hazards, including those associated with flying, in complex and often subtle ways (see Pidgeon *et al.*, 1992a for a review). This is one of the reasons why, whenever an accident does occur, there will invariably be strong social and political pressures for thorough investigation and remedial action as well as a collective desire to apportion blame after the event. In addition to this, flight deck crew, air traffic control staff and maintenance personnel play out critical roles and responsibilities, as the front-line actors within a set of highly structured and visible human–machine processes. This, coupled with the many opportunities for learning that are presented when things do go wrong in either an actual accident or a significant incident, has inevitably drawn (and continues to draw) the investigative focus towards the ways in which individual human errors contribute to such events.

A number of writers within the aviation research community have recently argued that there is now an urgent need to complement analyses of individual human error by moving towards an understanding of the role played by broader system factors in accidents. Murphy (1992) notes that while civilian passenger risk, measured as deaths per passenger mile flown, has decreased steadily in the past decades, the year-on-year numbers of accidents involving commercial aircraft have remained remarkably stable. And the ICAO (in press) state that:

> The late 70's, the 80's and 90's will undoubtedly be remembered as the golden era of aviation Human Factors. Cockpit (and then Crew) Resource Management (CRM), Line-Oriented Flight Training (LOFT), Human Factors training programmes, attitude-development programmes and similar efforts have multiplied, and a sustained campaign to increase the awareness of human error in aviation safety has been initiated. But much to the consternation of safety practitioners and the entire aviation community, human error continues to be at the forefront of accident statistics (p. 1).

The authors of the digest then go on to describe, with the aid of case-study illustrations, how 'human error' is often precipitated by more systemic, background management and organizational factors. Similarly, Adams and Payne (1992) review the contribution of pilot-errors to air ambulance accidents, making a distinction between pilot-generated (e.g. individual abilities, attitudes and judgements) and system-generated (e.g. training, procedures, supervision and air crew selection, and general management) causes. They point out that one implication of this for risk management is that 'we can achieve only limited success in reducing pilot-error accident rates if the pilot is the only part of the operational problem being fixed' (Adams and Payne, 1992, p. 40).

Enders (1992) makes the additional point that most aviation accidents have several causes (a feature in common with accidents in many other hazardous technologies, to which we return later) and that to seek a single 'probable' cause of an event, such as pilot or maintenance error, therefore misses opportunities for learning. He goes on to suggest that causes involving 'management or supervisory inattention at all levels' are the most prevalent category, and perhaps contribute as much to accidents as the total numbers of pilot and maintenance errors put together.

In a similar vein, Johnston (1991) argues that the preoccupation in current aircraft accident investigations with the immediately visible causes of accidents, typically technical malfunction and front-line operator errors, diverts attention away from consideration of whether underlying organizational causes may be present. Johnston is particularly concerned that aircraft accident analysts adopt a wider 'investigative reality'. For example, where individuals are found to have failed to follow Standard Operating Procedures, any conclusion 'the accident occurred because X failed to follow procedure Y' should not be the end of the matter, but should always be accompanied by the question 'and why was this so?'

A recent case study example, which illustrates a number of these concerns, is the sudden in-flight structural break-up and crash, with the loss of all 14 lives aboard, of a twin-engined Continental Express Embraer 120 on 11 September 1991 near Eagle Lake, Texas. The catastrophic structural failure of the aircraft occurred without warning during a descent in good weather through approximately 12 000 feet *en route* to landing at Houston Intercontinental Airport. Analysis of the cockpit voice and flight data recorders, together with the pattern of wreckage, revealed that neither pilot actions or weather contributed to the accident. Rather, the sudden loss of control and subsequent structural break-up was triggered by the separation of the leading edge assembly from the left side of the horizontal stabilizer on top of the aircraft's T-type tail.

The Embraer 120 leading edge assembly is normally fixed to the horizontal stabilizer by two rows of screws, at the top and bottom.

However, the accident damage was consistent with the top row of 49 screws being missing during the whole of the flight, and a consequent sudden separation of the partially attached assembly under the peak (but nevertheless within normal limits) dynamic loads present during the descent. Since the top of the horizontal stabilizer is not visible from the ground, the fact that screws were missing would not have been apparent to the crew during their pre-flight checks.

The US National Transportation Safety Board report (NTSB, 1992) into the accident documents how the failure was not purely a 'mechanical' circumstance, but the result of deficiencies rooted in the maintenance, management and regulatory systems surrounding the operation of the aircraft. The immediate reasons for the missing screws were found to reside in the events of the evening prior to the crash, when the Embraer 120 had undergone scheduled maintenance operations to replace the deicing assemblies, known as deice 'boots', installed on both the left and right leading edges of the horizontal stabilizer. The operations to change the deice boots required separation of the leading edge assemblies from the aircraft by removal of both the top and bottom rows of screws, respectively.

During the course of two shifts over the evening and night, maintenance personnel successfully replaced and resecured the right-hand leading edge assembly and boot. However, work on the left-hand assembly was started but not completed. On the first 'evening' shift the top rows of screws for both left and right assemblies were removed in preparation by an inspector, while two mechanics began work to remove the old boot on the right-hand leading edge assembly. However, when the second 'midnight' shift took over the aircraft they successfully completed the right-hand deice boot change, but did no further work on the left-hand assembly.

A number of factors contributed to this state of affairs, and the NTSB accident report notes that 'the reasons for the errors and the overall failure of the maintenance programme are complex and are not simply related to the failures of any single individual' (NTSB, 1992, p. 38). Significant factors included poor supervision and inspection of the work coupled with a complex pattern of errors of communication between the mechanics, supervisors and inspectors of the respective outgoing and incoming shift teams. In particular, there were ambiguities over who held practical responsibility for the work, together with failures to follow set document-ation and verbal procedures for reporting any work done. These in combination conspired to conceal the incomplete work on the left-hand assembly.

The NTSB report also points to the operation of more general systemic factors, arguing that there was also evidence that these violations of approved maintenance procedures were not isolated instances but consistent with a 'general disregard for following established procedures on the part of the maintenance department personnel' (NTSB, 1992, p. 43). As a consequence, the Safety Board concluded that 'the lax attitude in the hanger suggests that management *did not establish an effective safety orientation* for its employees' (p. 44, emphasis added), and that this was in all probability a contributory cause of the accident. A related factor concerned the failure of management to properly classify the deice boot change as a safety critical procedure, despite it being integral to, and involving removal of, the leading edge assembly. The latter clearly was a safety critical item, and the correction of this oversight would have resulted in more stringent (Required Inspection Item) checking to be carried out on all such work. Over and above all of this, the final line of monitoring, by the US Federal Aviation Administration (FAA) in its role as supervisor of correct hanger practice, is criticized in the accident report as having been insufficient to enable detection of what were in all probability ongoing violations of approved maintenance procedures at the company.

The Embraer case is not an isolated incidence, since the pattern of errors, ambiguities and mistakes share many features in common with those found by research into large-scale accidents more generally. A first finding of this work, and one that has been long recognized in aviation practice, is that failures in large-scale technological systems cannot be described in technical terms alone. Behavioural causes are often predominant in disasters. This is perhaps not surprising given that individuals, their organizations and groups, and ultimately their cultures are all implicated in the design, construction, operation and monitoring of a technology.

Following on from this, many of the behavioural causes of disasters can be traced to the social and organizational arrangements of the *sociotechnical systems* associated with large-scale hazards. The concept of a socio-technical system stresses the close inter-dependence between a technology and the human resources (individual, group, and organizational) necessary for its use. The social and technical components interact with, and over time change each other in complex and often unforeseen ways. Viewing human-made hazards in these terms is valuable in that we are forced to look beyond overly narrow technical explanations for accidents, to consider the technical failures in relation to a range of human causes.

However, it is also now clear that the human causes of disasters in complex socio-technical systems cannot all be adequately classified under the rather catch-all label of 'human error'. The behavioural contributions to disasters are often more subtle and diverse in nature (Pidgeon, 1988;

Pidgeon and Turner, 1986). They range from simple individual human errors such as slips, lapses, mistakes and faulty decision making (Reason, 1990); to those associated with small-group social structure and dynamics, many involving failures of communication or collective decision making (perhaps stemming from inadequate evaluation of the available information, as documented by Janis, 1982, in his classic account of the Groupthink phenomenon); through to those involving more systemic organizational or management issues (e.g. Bignell *et al.*, 1977; Horlick-Jones, 1990; Turner, 1978).

For any particular disaster there will typically be several, and often a large number of, behavioural preconditions some of which will have originated many years prior to the actual event. For example, in the case of the Space Shuttle *Challenger* disaster, the initial faulty booster design decisions were made some 13 years before the flight that was destroyed. Finally, and as a direct consequence of the complexity inherent in modern socio-technical systems, it is often a chain of unanticipated interactions between the sets of contributory causes that will lead to a full-scale disaster (Perrow, 1984).

The first comprehensive analysis of the social and organizational preconditions to disaster in large-scale technological systems was by Turner in his 1978 book *Man-Made Disasters*. Turner's model is highly relevant to many of the more recent and prominent large-scale accidents, providing a theoretical framework for understanding their background systemic causes. Turner conducted a detailed sociological analysis of 84 major accidents over a ten year period in the United Kingdom. He concluded that prior to any disaster it is typical to find that a number of undesirable events accumulate, unnoticed or not fully understood, often over a considerable number of years. Turner defined this gradual development of preconditions as the *disaster incubation period*.

The incubation period is brought to a conclusion either by the taking of preventative action to remove one or more of the dangerous conditions where these are noticed, or by a *trigger event*, which might be a final critical error, or a slightly abnormal operating condition. This distinction has since been taken up by Reason (1990) in his discussion of latent and active errors, and also corresponds well with Adams and Payne's (1992) classification of accident causes into system-generated and pilot-generated factors. Both of these writers emphasize the fact that line-operators such as air crew often inherit faulty systems (as a function of particular equipment, procedures or working practices) directly as a result of decisions made elsewhere in the organization.

The trigger results in the catastrophic event and also presents an opportunity for the previously unseen background factors to come to light, as the reasons for a failure, and the lessons to be learned (see Toft,

27

1992a; Turner and Toft, 1988) are sought. However, one problem after the event is that the immediate trigger may be confused with the more systemic background causes to a disaster, or may even be taken to be the sole cause. A further implication of the incubation model is that many incidents are distinguished from full-scale accidents only by the absence of a suitable trigger event, perhaps due to the intervention of chance factors; the background, latent factors remaining common to both.

This in turn points to the value of comprehensive incident analysis, with a view to identifying a developing incubation period, although the full benefits of this can only be realized if the investigator adopts an investigative reality which encompasses the background systemic, as well as the more immediate preconditions to failure. Here Turner's model is of greatest practical value, since it can be used by incident and accident investigators to sensitize themselves to the more subtle organizational preconditions to failure.

Turner's model focuses in particular upon the information difficulties associated with the attempts of both individuals and organizations to deal with uncertain and ill-structured safety management problems. A number of specific types of information difficulty can be found, in retrospect, within the incubation period. First, events may be unnoticed or misunderstood because of wrong assumptions about their significance. Those dealing with them may have an unduly rigid outlook, a particular problem being *organizational* rigidity of perception and belief, leading to the brushing aside of complaints and warnings both from outside and within the organization. Alternatively, personnel may be distracted or misled by 'decoy' events.

Second, dangerous preconditions may go unnoticed because of the difficulties of handling information in complex situations. Poor communications (perhaps between different departments of one organization, or between two different organizations), ambiguous orders and the difficulty of detecting important signals in a mass of surrounding noise may all be important here.

Third, there may be uncertainty about how to deal with formal violations of safety regulations which are thought to be discredited or outdated because of technological advance.

Fourth, when things do start to go wrong, the outcomes are typically worse because people tend to minimize danger as it emerges, or to deny that the failure will happen.

Such information difficulties often compound mechanical malfunctions and human errors, often in ways that conceal their full significance. The result of this is an unnoticed situation that runs counter to the accepted beliefs about hazards and to the established safety norms and procedures. The Embraer 120 case, discussed earlier, is a graphic illustration of how a hazardous situation which should have been avoided can nevertheless develop unnoticed, and MacGregor and Höpfl (in press) argue that all of Turner's categories of information failure are readily found within civilian aviation systems.

A second, related account of failures in socio-technical systems is provided by Perrow (1984), in his influential book *Normal Accidents*. Perrow commences from an analysis of the multiple causes underlying the Three-Mile Island accident, and identifies two general characteristics of large-scale hazardous systems related to accident causation. Adopting terms from systems engineering, Perrow labels these as complexity (a system can be either complex or linear) and coupling (system coupling can be either tight or loose).

Perrow's definition of high system complexity describes a state of interactivity and opacity. This is typically associated with systems comprising a large numbers of elements, where there are multiple potential interactions between the elements, and where only partial or distributed control knowledge of the component parts and their inter-relationships is available. Hence, when things do start to go seriously wrong in highly complex systems they are likely to take the form of unexpected and ambiguously signalled interactions between apparently unrelated system components. As a consequence, the system's behaviour may not be fully comprehended by the operators, with the risk that control actions might exacerbate rather than recover the situation. Nor does the replacement of line-operators with computer-based control systems provide a complete answer, since this merely substitutes the operators' mental models of the system with the software designer's model, which in certain operational contexts may be even more limited (see also Bainbridge, 1987).

The second characteristic, that of tight system coupling, refers to processes such as the rapid onset of and inability to delay events, an absence of built-in resources such as slack and buffering, limited control options to cope with unanticipated events, and few opportunities to deviate from pre-designed safety measures. All of these restrict an operators' opportunities to respond flexibly to unfolding and unexpected events in a 'fire fighting' role.

According to Perrow, some systems, such as nuclear energy and chemical processing, exhibit both high complexity and tight coupling, and their co-occurrence sets particular control difficulties. Such systems will, according to Perrow, be subject almost inevitably to 'normal' accidents. Modern military and civilian 'fly-by-wire' aircraft with computerized flight control systems exhibit high complexity in Perrow's sense. However, the design of these systems may well avoid tight coupling by means of in-built redundancy.

Perrow's account is a good conceptualization of the dangers inherent in large-scale highly integrated systems. However, his model can be criticized, in particular, because his two dimensions confound aspects of system complexity and hence may not be easily distinguished in practice (Turner, 1992). Furthermore, his analysis is based upon an overly mechanistic 'engineering' approach to systems. While this applies to some circumstances, it may not be sufficient to capture all of the more subtle social aspects of socio-technical systems. As we go on to illustrate in the discussion of safety culture in the next section, the overall social and organizational context may be crucial to safety.

## Culture and safety

Current interest in the term 'safety culture' can be traced directly to the accident at the Chernobyl nuclear plant in the former Soviet Union and the response of the Western nuclear industries to the human and organizational causes of the disaster (see OECD, 1987; Pidgeon, 1991). The closely related notion of 'climate of safety' has had a somewhat longer pedigree however (Pugsley, 1972; Zohar, 1980).

The Chernobyl disaster in 1986 was the world's worst nuclear accident. It led to 31 immediate deaths, many more delayed deaths, and spread radiation across a wide area of Europe. The accident was caused by a combination of poor reactor design and operator 'errors' (see Reason, 1987). The Soviet RBMK reactor type had an inherent design defect, which meant that the energy output could become unstable, and the reactor possibly explode, under certain conditions of low power operation. Such a situation had been foreseen by the reactor's designers and operating procedures specified that the reactor should not ordinarily be operated at the low power level. However, to conduct an experiment on the reactor, ironically with the goal of improving safety in the future, the operators contravened Standard Operating Procedures by allowing the reactor power level to fall into the critical low power region. The experiment also required the disengagement of back-up safety systems which might have protected the reactor, allowing the first explosion to occur.

In a number of analyses of the implications of the Chernobyl disaster for the Western nuclear industry the serious human errors and violations of procedures were interpreted as evidence of a poor safety culture at the plant (OECD, 1987). These initial arguments could be taken (rhetorically) to imply that such conditions would not be present in Western nuclear organizations, and therefore that a second Chernobyl could not happen here!

In the United Kingdom, and since Chernobyl, a number of accident inquiries, most notably those that followed the Clapham railway accident (Hidden, 1989) and the King's Cross Underground fire (Fennell, 1988) have identified aspects of the general organizational climate as significant factors. The development of an 'appropriate' safety culture for nuclear operators is now seen, within the Western nuclear industry, as one important goal of reactor operator training (ACSNI, 1990, 1993; Embrey, 1991; INSAG, 1991). And a more general interest in the topic has ensued as attempts to both define and explore the relevance of the concept have been made (see, amongst others, Booth, in press; CBI, 1990; Horbury, in press; ICAO, in press; Maurino, 1992; Pidgeon *et al.*, 1992b; Toft, 1992b; Turner, 1991; Turner *et al.*, 1989; Williams, 1991).

From the perspective of the social sciences safety culture is an important theoretical concept since it explicitly addresses the wider social causes of disasters, and thus represents a significant departure from the traditional human factors approach to safety. With respect to the account of disasters outlined in the previous section, safety culture provides a global characterization for some of the more elusive contributions to the disaster incubation period. It also suggests ways in which reliability and safety questions might be linked to more general social science concepts and findings (to the literature on corporate culture, for example), and thereby to established means of empirical investigation. In addition, Roberts (1989; also Weick, 1987) points out that linking culture with reliability and risk in turn raises novel theoretical issues, since concerns for safety, responsibility and accountability do not appear in traditional organizational culture accounts in social science. Finally, discussion of safety culture raises two interrelated questions for practitioners. First, are there are 'good' or 'poor' safety cultures, which vary across organizations and which influence safety performance and reliability? And second, more controversially, how might interventions be designed to modify existing safety cultures? As we shall go on to see, neither of these questions are simple ones. Nor are they likely to permit of easy answers.

In commenting on the initial OECD (1987) analysis of Chernobyl, Turner *et al.* (1989) point out that here 'the notion of safety culture is reduced, on the one hand, to sets of administrative procedures for training, emergency plans and so on, and on the other to individual attitudes to safety which, it is thought, cannot be regulated' (Turner *et al.*, 1989, p. 6). However, the concept of culture as widely used in the social sciences, as well as in common parlance, refers primarily to a *shared* characteristic or characteristics of a particular social group, organization or society. Therefore, one criticism of the OECD analysis is that it crucially fails to address the shared property that is the defining characteristic of any culture.

31

Within the social sciences literature, however, there are very many views on precisely what is shared by, and defines the culture of, any social group (is it language, attitudes and beliefs, artefacts and technologies, behaviours, or some intangible amalgam of all of these?). In anthropology, sociology and cross-cultural psychology two principal models of culture are to be found (Rohner, 1984). The first of these views culture as a pattern of observable behaviours: that is, the regularly occurring, organized modes of behaviour in technological, economic, religious, political, family and other social groups. The second model holds that culture is a *system of symbols or meanings*: that is a shared cognitive system, or system of meanings in the heads of multiple individuals in a population or group. Researchers who advocate the second model of culture argue that we should not expect to discover any simple one-to-one relationship between culture and observable behaviour.

In our own approach to this issue we view culture in terms of the second model outlined above; that is as principally involving the *exploration of meaning* and the *systems of meaning* through which a given social group understands the world (see also Czarniawska-Joerges, 1992). Accordingly, we have broadly defined safety culture as the set of beliefs, norms, attitudes, roles and social and technical practices within an organization which are concerned with minimizing the exposure of individuals, both within and outside an organization, to conditions considered to be dangerous (Turner *et al.*, 1989). Such a system specifies what is important to individuals and groups, and explains their relationship to matters of life and death, work and danger. A safety culture is created and recreated as members of it repeatedly behave in ways which seem to them to be 'natural', obvious and unquestionable, and as such will serve to construct a particular version of risk, danger and safety. A safety culture also provides a set of assumptions and practices which permit new beliefs about danger to be constructed.

The principal cultural unit within which a safety culture is assumed to be located is that of the organization. The management and regulation of large-scale hazards is typically entrusted to organizations of varying size (both public and private), and it is within such organizations that many practical problems of risk management have to be resolved. In the case of aviation, for example, there are traditional and strong individual responsibilities laid upon flight crew for the safety of their aircraft. However, individual flight crew will all be selected, trained, and work within a corporate setting that both shapes beliefs and regulates behaviour. It is worth noting in addition that the unit of cultural analysis need not always be the full organization; for it is possible to think of the safety cultures of small groups such as particular air crew, of departments, and of divisions, being both nested within, and sometimes overlapping one another.

## Elements of a 'good' safety culture

The discussion now turns to the difficult question of what general elements might be desirable as part of, or contributing to, a safety culture. The definition of culture advanced above implies that to improve reliability in complex socio-technical systems we require more than exhortations to air crew and maintenance personnel to 'change attitudes towards safety', or indeed to 'fly or work more safely'. The following theoretical account (see also Turner *et al.*, 1989) was developed initially from our understanding of the nature of accident causation in socio-technical systems and a consideration of the types of thing to avoid!

Our discussion elaborates these ideas under four principal headings: location of responsibility for safety at *strategic management* level; *distributed attitudes of care and concern* throughout an organization; appropriate *norms and rules* for handling hazards; and on-going *reflection* upon safety practice. Although discussed under separate headings these represent, of course, closely interrelated facets, which in part depend upon and reinforce each other, to generate the somewhat elusive quality of a safety climate or culture. This group of factors, taken as a whole, also map quite well onto a number of other findings in safety research looking at the determinants of *safe* performance in industrial settings. In particular they correspond to Cohen's (1977) and Zohar's (1980) accounts of organizational factors that influence occupational safety, as well as to aspects of more recent findings from a survey of a group of Boeing operators with exceptional safety records (Lautman and Gallimore, 1987).

The first necessary condition for the development of a safety culture is that responsibility for safety should not only reside at a purely operational level, but be an issue for *strategic management* as well. Williams (1991) in his recent review of the safety culture literature notes that most writers agree on the importance of this (see also Cohen, 1977), and Lautman and Gallimore (1987) conclude that the high performing Boeing operators 'characterize safety as beginning at the top of the organization with a strong emphasis on safety [which] permeates the entire organization' (p. 2). Such top management commitment is essential for a number of reasons. It is important because attempts to promote enduring organizational change are unlikely to succeed if senior management are not seen to be closely involved and committed to the initiative. Employees will quickly sense where management's true priorities lie (e.g. optimizing flight operations), and may conform to these even when they conflict with explicit policy statements (always running a safe airline). This issue becomes particularly critical, as pointed out by Maurino (1992), when marginal decisions to go or not are required.

According to Zohar (1980), one indirect sign to employees of management commitment will be the perceived status within the organization of the personnel directly dealing with safety. And, as the ICAO Human Factors Digest no. 9 (in press) points out, a lax culture can be brought about merely by *lack* of disapproval by management of safety transgressions, rather than any positive action on their behalf. In addition Adams and Payne (1992) argue that in the competitive world of contract aviation a strong management commitment to safety is critical to support pilot decision making in the face of external pressures brought by clients. A final factor pointing to the importance of senior management commitment is that the consequences of large-scale accidents are now so serious (both direct, such as deaths and injuries, and indirect, such as from law suits or future sales losses) that safety has become a matter of strategic planning (CBI, 1990), and often requires decisions about the deployment of significant resources. Pauchan *et al.* (1991) describe the shift in corporate philosophy needed here as involving 'an understanding that a corporation can potentially become a *destructive* system in addition to being a productive system' (p. 223).

Senior management commitment to safety is necessary, but is certainly not a sufficient condition for safe operations. A further requirement is for *concern* about safety to be distributed, supported, and endorsed, *throughout* an organization. Distributed concern for safety needs to be 'representative' of organization members, and not imposed in a punitive manner by one group on another. Only in this way is it possible to move towards a state in which the recognition of the necessity and desirability of safety rules provides a motivation to conform to them in their spirit as well as according to their letter. Under such circumstances everybody in the organization would regard the policing of hazards as a personal as well as a collective goal. In this respect, Zohar (1980) describes an appropriate management philosophy as 'not strictly production oriented but also people oriented' (pp. 97–98). Turner (1991) recommends that formal safety directives should be complemented with more subtle approaches aimed at promoting *caring* on the part of employees and the organization, expressed in terms of both concern for the outcome of dealing with risks, and solicitude for the effects of their activities upon people.

The specific *norms and rules* governing safety within an organization, many of them tacit, will also be at the heart of a safety culture. As corporate guidelines for action, these will shape the perceptions and actions of the individuals in the organization in particular ways, defining what is and is not to be regarded as a significant risk, and what will represent an appropriate response. In an ideal world one might attempt to specify a set of complete, up to date, and practical contingencies that anticipate *all* foreseeable hazards. Being alert to well-defined hazards requires appreciation of the individual and organizational difficulties that tend to conceal and distort significant available information (as noted earlier, such things as communication problems across departments, and rigidities of belief and perception).

However, there is always a tension between the need to cope both with hazards which are well-defined in advance, and those that are ill-defined or unexpected, perhaps because they arise only infrequently in periods of crisis or because they are completely beyond the boundary of current operational experience (Collingridge, 1980; Pidgeon, 1988). Being alert to both well-defined and ill-defined or unseen hazards sets a demanding task, since the inflexible or ritual application of existing rules and Standard Operating Procedures to guard against anticipated hazards might lead to crucial oversights as a result of cognitive mind-set, or Groupthink (Janis, 1982). Guarding against mind-set, both individually and collectively, involves a willingness to monitor ongoing technologies in diverse ways; to research, and to accept uncertainty and incompleteness of knowledge as facts of life; to be prepared both to solicit opinions about risk from outsiders, as well as to reward rather than ignore or punish internal 'whistle-blowers' (see Pauchant *et al.*, 1991); and finally, to exercise creativity and 'safety imagination' (Ramsden, 1984) as aids in assessing the available information about hazards.

The onset of ill-defined hazards is particularly characteristic of complex and tightly coupled systems, such as nuclear aircraft carrier operations. Rochlin (1989), as part of the 'high-reliability organizations' project, describes how the complex networks of organizational arrangements on US carriers cope with the crises that frequently arise during flight operations. First, personnel on the carrier constantly monitor communications and their immediate environment to detect any variations from normal procedures and operating conditions. And second, when significant disturbances are detected *ad hoc* problem-solving groups and communication networks are rapidly set up. Significantly for such a hierarchical context as the military, Rochlin claims that these 'phantom' networks often cut completely across the traditional rank hierarchies and formal organizational sub-divisions on board the carrier. Rather, precedence is given to individuals who hold knowledge and experience sufficient to solve the problem at hand over those with status or particular formal roles. Once the crisis has receded such networks disperse rapidly. However, it is worth noting that organizations such as nuclear carriers experience operational crises on a relatively regular, if perhaps not entirely 'routine' basis, and that this may restrict generalization of these findings to other organizational contexts.

*Ongoing reflection* about current practices and beliefs involves the search for, and generation of, meaning and new knowledge in the face of initial ambiguity and uncertainty about what might prove a significant hazard. This process is crucial if a group or organization is to learn, as well as to adapt to changing circumstances. As noted above, one function of reflection is to act as a precaution against the over-rigid application of existing safety rules and procedures. In aviation such reflection has traditionally been facilitated by institutionalized mechanisms of accident investigation, together with proactive incident reporting and feedback. For example, Lautman and Gallimore (1987) note that in the high performing Boeing operators

> there is an acute awareness of the factors that result in accidents, and management reviews accidents and incidents in their own airline and other airlines and alters their policies and procedures to best guard against recurrence. There is a method for getting information to the flight crews expeditiously and a policy that encourages confidential feedback from pilots to management (p. 2).

Open communication links between management and personnel have also been found to be associated with safe climates in industrial organizations (Cohen, 1977), and in the US studies of high reliability organizations it has been found, perhaps not surprisingly, that this is fostered where organizations actively avoid laying blame for mistakes and errors (e.g. see Westrum, 1987). This latter consideration sets special responsibilities, once more, on senior management for setting the corporate framework within which safety can gain suitable priority. In very general terms the processes of reviewing accidents *and* acting accordingly have been characterized by Toft (1992a) as *active* learning. He distinguishes this from the maladaptive, but nevertheless only too common passive learning, which occurs when incidents are reviewed but there is then a failure to put the knowledge gained into practice.

## From safety culture to institutional design? Some remaining dilemmas and challenges

Having broadened the focus for the investigation of the causes of aviation accidents we can now ask whether the concept of a safety culture can be of utility as part of a proactive process of institutional design for safety (see Maurino, 1992)? By institutional design we mean, quite simply, the specification of appropriate organizational arrangements for managing hazards. Of course, many of the considerations discussed above suggest possible design strategies; some of which, like incident reporting, are tried and tested in aviation practice already.

However, as we have stated in the introduction, it is important at the same time to pass a critical eye over the concept of safety culture, and to stress the fact that it will be no simple matter to translate the many theoretical treatments of this concept into practical action. Nor will it be easy to resolve a number of the generic dilemmas which are associated with any attempt at institutional design. There are very clear reasons why we should not view institutional design for safety as a simple, un-problematic issue, merely to be solved by a form of applied or 'consultancy' social science.

A basic problem stems from the fact that there is evidence within the management science literature to indicate that recent attempts to manipulate corporate behaviour by changing organizational cultures have met with only limited success (e.g. see Nord, 1985); organizational cultures are notoriously resistant to change, and there is no reason to suppose that a safety culture will be any different in this respect (Johnson, 1991)! Elaborating on this observation, Toft (1992b) argues that attempts to change safety cultures solely by management edict (decree) or by imposition of external regulation (prescription) will meet with limited success. According to Toft, permanent change is best addressed through long-term 'organizational-learning'. One way in which this arises is through an organization having to respond to its own involvement in an accident or series of accidents, and another might be through the use of crisis simulation exercises. This accords well with Rochlin's (1989) observation that high reliability organizations probably evolve many of their crisis management strategies not through imposition from above or outside, but by a process of institutional self-design.

Olson (1987) points out that the further back in an accident causation chain one seeks to investigate, such as at production pressures on safety, or at corporate attitudes towards risk, the more difficult it is to say with certainty how specific variables identified relate to safety performance in general. A corollary of this, noted by Williams (1991), is that even if corporate culture can be shown to have changed at the same time as safety performance has improved, it will not be easy to demonstrate un-ambiguously that the two are directly linked. Williams goes on to argue that it might be easier to gauge the impact of culture change on an organization through a wider range of measures, including those of quality, reliability and competitiveness. He concludes, from his review of available studies, that a 'good' safety culture may have an influence on safety performance only indirectly, through its relationship with raised quality standards within an organization more generally.

The above arguments support what at first may seem a surprising claim: that safety and goals such as quality or production may not always need to be traded-off, but may sometimes go hand-in-hand (see also Blockley, 1992; CBI, 1990). This is an important argument, not least because it provides a powerful economic justification for the allocation of resources to safety programmes.

A number of institutional design questions are raised when we consider the fact, noted earlier, that any commercial aviation organization will comprise many different, and sometimes overlapping, social sub-groupings (e.g. pilots and cabin crew, air crew across different fleets, maintenance and ground handling personnel, different tiers of management, etc.), perhaps with their own distinctive sub-cultures. For example, it could be argued that different styles of management evolve in flight and ground operations departments, promoting different levels of concern for safety issues amongst flight and ground personnel. A second sub-culture problem arises where an incident is deemed serious by operational personnel, but the management to whom it is reported do not allocate a similar priority to investigating the aetiology of incidents, perhaps merely seeking to lay blame for the event. In both of these examples, the question arises as to whether a unified programme of culture change can be designed, and at minimum institutional change in these circumstances must start from a clear understanding of the existing sub-cultures.

On a much wider cross-cultural level one can also ask whether the concept of a corporate culture, which is primarily a Western business concept, will prove useful if applied to airlines from non-Western national cultures. Johnston (1993) presents some related arguments concerning the cross-cultural transferability of Crew Resource Management programmes. And in this respect attempts to change culture may only serve to divert attention away from more pressing issues that undermine safety, such as poor infrastructure and lack of resources. Culture cannot be a panacea for lack in any of these areas.

The final issue that we consider here is the role of blame in the risk management process, and it is one which is particularly difficult to resolve. A number of generic dilemmas of risk management and institutional design are discussed by Hood *et al.*, (1992). Blame sets a dilemma, according to Hood *et al.*, because the call for society to place strong sanctions upon individuals and organizations who act unsafely must be balanced against the need to learn from past events. At a theoretical level, the anthropologist Mary Douglas (1992) reminds us that processes of blaming may be inherent to most social settings, and will always surround the topics of safety, danger and risk. However, Johnston (in press) argues strongly that the social functions of blaming should be clearly seen for what they are, and be separated from the exercise of risk management.

In an ideal world the view of accidents as socio-technical phenomena would be accompanied by an investigative emphasis which seeks for *what* is wrong rather than *who* made the mistake. In this respect the ICAO digest argues that 'Blame and punishment do not have, in themselves, any prevention value' (in press, p. 6). Thus blame and punishment should be avoided because the knowledge that a 'culprit' has to be found whenever an error has occurred will invariably prevent the full and candid reporting of incidents and unsafe events to the detriment of opportunities for learning about the system.

However, even assuming that a blame-free safety culture can be established within an organization, there is no ultimate guarantee that knowledge gained about system deficiencies will be translated into active learning. At a corporate level, therefore, legal sanction may always be necessary to ensure that organizations set up and maintain effective safety systems. In this respect there have been attempts in the UK, principally by relatives of disaster victims, to establish a legal principle of corporate manslaughter (Crainer, 1993; Wells, 1993). The balance to be struck between the issues of blame and sanction at the individual and corporate level cannot be ignored when questions of safety and institutional design are considered.


## Conclusion

We have argued for a widening of perspectives within aviation practice to incorporate a socio-technical systems approach to safety. This involves adopting an understanding of unsafe technical events and individual actions within their social and organizational contexts. An immediate and tangible benefit of viewing safety in socio-technical terms is that the potential of accident and incident analysis is greatly enhanced. We have also argued that a significant part of the social and organizational context is captured by the notion of a safety culture, which we see as an influential background factor intertwined with ultimate safety performance. However, we recognize that one drawback to the characterization of safety culture presented here (i.e. fundamentally in terms of a shared system of meanings or understandings about risk and danger) is that a clear link with risk management practice does not become immediately visible. This point is further reinforced by our discussion, in the preceding section, of the very many difficulties associated with attempts at institutional design for safety. And in this respect, the chapter should not be taken to imply that institutional design and risk management efforts should now be solely directed towards programmes for safety culture change. Lasting culture change may be possible in only some circumstances, and although our discussion in the penultimate section of the chapter points to ways in which this task might be approached, empirical studies are now needed to explore further the feasibility of this.

However, it is equally clear to us that all efforts at risk management through institutional design (and many of the chapters in the current volume are concerned with just this issue) will have to be conducted against the backdrop of the existing safety culture(s) within and across the organizations involved. A central message should be, therefore, that at least some understanding of existing cultures must be gained *before* risk management efforts such as new training, reporting systems, procedural frameworks and resource management programmes are designed and initiated. This is important both because existing cultures may generate unintended consequences that subvert the intended outcomes of an otherwise well-designed risk management programme, and conversely because the introduction of a new programme always has the potential to change the existing culture. Regarding the latter point, it is important to recognize that safe cultures can be harmed as well as encouraged!

To conclude, the question of whether we can design safe organizations is perhaps the most important issue in risk management today, both for aviation and other high-risk systems. The discussion here should caution practitioners against approaching this difficult problem with an overly optimistic view, while at the same time indicating some of the ways in which our knowledge about this question can move forward. Hopefully, the result of this will be a more contextually grounded understanding, both in theoretical and practical terms, of risk and safety issues in the aviation arena.

## Acknowledgements

We wish to thank the editors for both reading and making useful comments on earlier drafts of this chapter, and in particular to Neil Johnston for drawing our attention to the Embraer 120 accident report; David Blockley, John Stone and Brian Toft commented on the manuscript; our colleague Barry Turner contributed to the development of many of the ideas on safety culture presented in this chapter.

## References

ACSNI (1990), *Advisory committee on the safety of nuclear installations: Study group on human factors. First report on training and related matters*, London, HMSO (Health and Safety Commission).

ACSNI (1993), *Advisory committee on the safety of nuclear installations: Study group on human factors. Organising for safety*, London, HMSO (Health and Safety Commission).

Adams, R. J. and Payne, B. (1992), 'Administrative risk management for helicopter operators', *The International Journal of Aviation Psychology*, **2**(1), 39–52.

Bainbridge, L. (1987), 'The ironies of automation', in J. Rasmussen, K. Duncan and J. Leplat (eds), *New Technology and Human Error*, London, Wiley.

Bignell, V., Peters, G. and Pym, C. (1977), *Catastrophic Failures*, Milton Keynes, Open University Press.

Blockley, D. I. (1992), 'Concluding reflections', in D.I. Blockley (ed.), *Engineering Safety*, Maidenhead, McGraw-Hill.

Booth, R. (in press), 'Safety culture: Concept, measurement and training implications', *Journal of Health and Safety*.

CBI (1990), *Developing a Safety Culture: Business for safety, London*, Confederation of British Industry.

Cohen, A. (1977), 'Factors in successful occupational safety programs', *Journal of Safety Research*, **9**(4), 168–178.

Collingridge, D. (1980), *The Social Control of Technology*, Milton Keynes, Open University Press.

Crainer, S. (1993), *Zeebrugge: Learning from Disaster – Lessons in corporate responsibility*, Wisbech, The Herald Charitable Trust.

Czarniawska-Joerges, (1992), *Exploring Complex Organizations: A cultural perspective*, London, Sage.

Douglas, M. (1992), *Risk and Blame: Essays in cultural theory*, London, Routledge.

Embrey, D. (1991), 'Bringing organizational factors to the fore of human error management', *Nuclear Engineering International*, October, 50–53.

Enders, J. H. (1992), 'Management inattention greatest accident cause, not pilots', *Flight Safety Foundation Newsletter*, February/March/April, **33**(2), 1.

Fennell, D. (1988), *Investigation into the King's Cross Underground Fire*, London, HMSO (Department of Transport).

Hawkins, F. H. (1987), *Human Factors in Flight*, Aldershot, Gower.

Hidden, A. (1989), *Investigation into the Clapham Junction Railway Accident*, London, HMSO (Department of Transport).

Hood, C., Jones, D. K., Pidgeon, N. F., Turner, B. A. and Gibson, R. (1992), 'Risk management', in *Risk: Analysis, Perception and Management*, London, The Royal Society.

Horbury, C. (in press), 'Safety culture: Future dimensions in the light of new knowledge', *Journal of Health and Safety*.

Horlick-Jones, T. (1990), *Acts of God? An Investigation into Disasters*, London, Emergency Planning and Information (EPI) Centre.

ICAO (in press), *Human Factors Digest No. 9: Human factors in management and organization*, Montreal, International Civil Aviation Organization.

INSAG (1991), *Safety culture: A report by the International Nuclear Safety Advisory Group*, (safety series No. 75-INSAG-4), Vienna, International Atomic Energy Agency.

Janis, I. L. (1982), *Groupthink*, 2nd edn, Boston, MA, Houghton-Mifflin.

Johnson, B. B. (1991), 'Risk and culture research: some cautions', *Journal of Cross-Cultural Psychology*, **22**(1), 141–149.

Johnston, A. N. (1991), 'Organizational factors in human factors accident investigation', *Proceedings of the 6th Symposium on Aviation Psychology*, Ohio, May.

Johnston, A. N. (1993), 'CRM: Cross-cultural perspectives', in E. Wiener, B. Kanki and R. Helmreich (eds), *Cockpit Resource Management*, San Diego, CA, Academic Press.

Johnston, A. N. (in press), 'Blame, punishment and risk management', in D. Jones, C. Hood, N. F. Pidgeon and B. A. Turner (eds), *Accident and Design: Contemporary Debates in Risk Management*, London, University College Press.

Lautman, L. G. and Gallimore, P. L. (1987), 'Control of the crew caused accident', *Airliner*, April–June, 1–6.

MacGregor, C. and Höpfl, H. (in press), 'Integrating safety and systems: The implications for organizational learning', *Proceedings of the 45th Annual International Air Safety Seminar*, Arlington, VA, Flight Safety Foundation.

Maurino, D. (1992), 'Corporate culture imposes significant influence on safety', *ICAO Journal*, April, 16–17.

Murphy, T. (1992), 'Risk assessment in civil air transport', *International Risk Assessment Conference*, London, UK, 5–9 October.

NTSB (1992), *Aircraft accident report: Continental Express flight 2574 in-flight structural breakup, EMB-120RT, N33701, Eagle Lake, Texas*, (NTSB/AAR-92/04), Washington, DC, National Transportation Safety Board.

Nord, W. R. (1985), 'Can organizational culture be managed?', in P. J. Frost, L. F. Moore, M. R. Louis, C. C. Lundberg and J. Martin (eds), *Organizational Culture*, London, Sage.

OECD (1987), *Chernobyl and the safety of nuclear reactors in OECD countries*, Paris, Organization for Economic Cooperation and Development.

Olson, J. (1987), 'Measuring safety performance of potentially dangerous technologies', *Industrial Crisis Quarterly*, 1(4), 44–53.

Pauchant, T. C., Mitroff, I. I. and Lagadec, P. (1991), 'Toward a systemic crisis management strategy: Learning from the best examples in the U.S., Canada and France', *Industrial Crisis Quarterly*, 5, 209–232.

Perrow, C. (1984), *Normal Accidents*, New York, Basic Books.

Pidgeon, N. F. (1988), 'Risk assessment and accident analysis', *Acta Psychologica*, 68, 355–368.

Pidgeon, N. F. (1991), 'Safety culture and risk management in organizations', *Journal of Cross-Cultural Psychology*, 22(1), 129–140.

Pidgeon, N. F., Hood, C., Jones, D. K., Turner, B. A. and Gibson, R. (1992a), 'Risk perception', in *Risk: Analysis, Perception and Management*, London, The Royal Society.

Pidgeon, N. F. and Turner, B. A. (1986), 'Human error and socio-technical system failure', in A. S. Nowak (ed.), *Modeling Human Error in Structural Design and Construction*, New York, American Society of Civil Engineers.

Pidgeon, N. F., Turner, B. A., Toft, B. and Blockley, D. I. (1992b), 'Hazard management and safety culture', in D. J. Parker and J. Handmer (eds), *Hazard Management and Emergency Planning: Perspectives on Britain*, London, James and James.

Pugsley, A. (1972), 'The engineering climatology of structural accidents', in A. M. Freudenthal (ed.), *Proceedings of ICOSSAR, International Conference on Structural Safety and Reliability*, New York, Pergamon.

Ramsden, J. M. (1984), 'Shall safety be maintained?' *Flight International*, 17 November, 1328–1331.

Reason, J. T. (1987), 'The Chernobyl errors', *Bulletin of the British Psychological Society*, 40, 201–206.

Reason, J. T. (1990), *Human Error*, Cambridge, Cambridge University Press.

Roberts, K. H. (1989), 'New challenges in organizational research: High reliability organizations', *Industrial Crisis Quarterly*, 3, 111–125.

Rochlin, G. I. (1989), 'Informal organizational networking as a crisis-avoidance strategy: US naval flight operations as a case study', *Industrial Crisis Quarterly*, 3, 159–176.

Rohner, R. P. (1984), 'Towards a conception of culture for cross-cultural psychology', *Journal of Cross-Cultural Psychology*, 15, 111–138.

Toft, B. (1992a), 'The failure of hindsight', *Disaster Prevention and Management*, 1(3), 48–60.

Toft, B. (1992b), 'Changing a safety culture: Decree, prescription or learning?' *IRS Risk, Management and Safety Culture Conference*, 9 April, London Business School.

Turner, B. A. (1978), *Man-made Disasters*. London, Wykeham.

Turner, B. A. (1991), 'The development of a safety culture', *Chemistry and Industry*, 1 April, 241–243.

Turner, B. A. (1992), 'The sociology of safety', in D.I. Blockley (ed.), *Engineering Safety*, Maidenhead, McGraw-Hill.

Turner, B. A., Pidgeon, N. F., Blockley, D. I. and Toft, B. (1989), 'Safety culture: Its importance in future risk management', *Second World Bank Workshop on Safety Control and Risk Management*, 6–9 November, Karlstad, Sweden.

Turner, B. A. and Toft, B. (1988), 'Organizational learning from disasters', in H. B. F. Gow and R. W. Kay (eds), *Emergency Planning for Industrial Hazards*, London, Elsevier.

Weick, K. E. (1987), 'Organizational culture as a source of high reliability', *California Management Review*, **29**(2), 112–127.

Wells, C. (1993), *Corporations and Criminal Responsibility*. Oxford, Oxford University Press.

Westrum, R. (1987), 'Management strategies and information failure', in J. A. Wise and A. Debons (eds), *NATO ASI series F, Computer and Systems Science, Vol. 3*. Berlin, Springer-Verlag.

Williams, J. C. (1991), 'Safety cultures: Their impact on quality, reliability, competitiveness and profitability', in R. H. Matthews (ed.), *Reliability '91*, London, Elsevier Applied Science.

Zohar, D. (1980), 'Safety climate in industrial organizations: Theoretical and applied implications', *Journal of Applied Psychology*, **65**(1), 96–102.

# 3 Philosophy, policies, procedures and practices: The four 'P's of flight deck operations

*Asaf Degani and Earl L. Wiener*

## Introduction

### Background

A complex human–machine system is more than merely one or more operators and a collection of hardware components. To operate a complex system successfully, the human–machine system must be supported by an organizational infrastructure of operating concepts, rules, guidelines and documents. The coherency, in terms of consistency and logic, of such operating concepts is vitally important for the efficiency and safety aspect of any complex system.

In high-risk endeavours such as aircraft operations, space flight, nuclear power, chemical production and military operations, it is essential that such support be flawless, as the price of deviations can be high. When operating rules are not adhered to, or the rules are inadequate for the task at hand, not only will the system's goals be thwarted, but there may be tragic human and material consequences. Even a cursory examination of accident and incident reports from any domain of operations will confirm this.

To ensure safe and predictable operations, support to the operators often comes in the form of Standard Operating Procedures (SOP). These provide the crew with step-by-step guidance for carrying out their operations. SOPs do indeed promote uniformity, but they do it at the risk of reducing the role of the human operators to a lower level. Furthermore, an exhaustive set of procedures do not absolutely ensure flawless system behaviour: deviations from SOP have occurred even in highly procedurized organizations.

The system designers and operational management must occupy a middle ground: operations of high-risk systems cannot be left to the whim of the individual. But they likewise must recognize the danger of overprocedurization, which fails to exploit one of the most valuable assets in the system, the operator who is close to the actual operation. Furthermore, the alert system designer and operations manager recognize that there cannot be a procedure for everything, and the time will come when the operators of a complex system face a unique situation for which there is no procedure. It is at this point that we recognize the reason for keeping humans in the system; since automation, with all its advantages, is merely a set of coded procedures executed by the machine. Procedures, whether executed by humans or machines have their place, but so does human cognition.

A dramatic example was provided by the Sioux City accident in which a United Airlines DC-10 suffered a total loss of hydraulic systems, and hence aircraft control, due to a disintegration of the centre engine fan disc (NTSB, 1990a). When he had sized up the situation, the captain turned to the flight engineer and asked what the procedure was for controlling the aircraft. The reply is worth remembering: 'There is none.' Human ingenuity and resource management were required: the crew used unorthodox methods to control the aircraft. This resulted in a crash landing, which well over half of the passengers and crew survived.

This chapter is a continuation of our previous work on the human factors of aircraft checklists in air carrier operations (Degani and Wiener, 1990). Our work in this area was undertaken largely as a result of the discovery, during the investigation of the Northwest 255 crash, that checklists, for all their importance to safe operation, had somehow escaped the scrutiny of the human factors profession. The same, we found out, can be said of most flight deck procedures. We felt that our work in checklist design and usage would not be complete until we gave equal consideration to cockpit procedures.

## Procedural deviation: Its influence on safety

Problems within the human-procedure context usually manifest themselves in procedural deviation. If all goes well, these deviations are not apparent to the operational management, and in most cases are left unresolved. They do become apparent, however, following an incident or an accident. Lautman and Gallimore (1988) conducted a study of jettransport aircraft accident reports to 'better understand accident cause

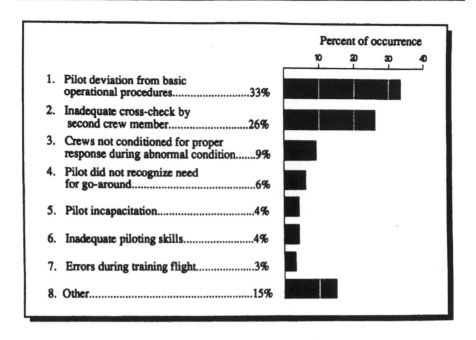

Percent of occurrence

1. Pilot deviation from basic
   operational procedures....................33%
2. Inadequate cross-check by
   second crew member....................26%
3. Crews not conditioned for proper
   response during abnormal condition.......9%
4. Pilot did not recognize need
   for go-around.............................6%
5. Pilot incapacitation.........................4%
6. Inadequate piloting skills......................4%
7. Errors during training flight...................3%
8. Other.......................................15%

**Figure 3.1    Significant crew-caused factors in 93 hull-loss accidents**
(Source: Adapted from Lautman and Gallimore, 1988)

factors' in commercial airline operations. They analysed 93 turbojet hull-loss accidents that occurred between 1977–1984.

The leading crew-caused factor in their sample was 'pilot deviation from basic operational procedures' (Figure 3.1). These findings are clearly supported by three airline accidents that occurred in the last five years: in the first, Northwest Airlines Flight 255, an MD-82, crashed shortly after take-off from Detroit Metro Airport following a no-flap/no-slat take-off (NTSB, 1988). In the second, Delta Air Lines Flight 1141, a B-727, crashed shortly after lifting off from Dallas–Fort Worth International Airport, following a no-flap/no-slat take-off (NTSB, 1989). In the third, USAir Flight 5050, a B-737, ran off the runway at LaGuardia Airport and dropped into adjacent waters, following a mis-set rudder trim and several other problems (NTSB, 1990b).

We submit that the classification of 'pilot deviation from basic operational procedures' may be somewhat misleading. One should first ask whether the procedures (from which the flight crews deviated) were adequate for the task. Were the procedures compatible with the operating environment? Were they part of a consistent and logical set of procedures? Most important, was there something in the design or the manner in which the procedure was taught that led a responsible flight crew member to deviate from it?

We argue that if we wish to understand how operators conduct flight deck procedures, we cannot look only at the aggregate level, i.e. procedures, but we also must examine the infrastructure, i.e. the policies and concepts of operation, that are the basis on which procedures are developed, taught and used.

## Theory of the three 'P's: Philosophy, Policy and Procedures

*Procedure development*

Procedures do not fall from the sky, nor are they inherent in the equipment. Procedures must be based on a broad concept of the user's operation. These operating concepts lend themselves into a set of work policies and procedures that specify how to operate the equipment efficiently. There is a link between procedures and the concepts of operations. We call that link 'The three 'P's of cockpit operations': philosophy, policies and procedures. In this chapter we shall explore the nature of these links, and how an orderly, consistent path can be constructed from the company's most basic philosophy of operation to the actual conduct of any given procedure.

*Procedures: What and Why?* In general, procedures exist to specify unambiguously six things:

1   What the task is.
2   When the task is conducted (time and sequence).
3   By whom it is conducted.
4   How the task is done (actions).
5   The sequence of actions.
6   What type of feedback is given (callout, actions etc.).

The function of a well-designed procedure is to aid the operators by dictating and specifying a progression of sub-tasks and actions to ensure that the primary task at hand will be carried out in a manner that is efficient, logical and also error resistant. Another important function of a cockpit procedure is that it should enhance coordination between agents in the system, be they cockpit crew, cabin crew, ground crew, or others. Procedures are also a form of quality control by management and regulating agencies over the operators.

*Standard operating procedures*  A set of procedures that apart from being merely a specification of a task, also serve to provide a common ground for two or three individuals (comprising a flight crew) that at times may be totally unfamiliar with each other's experience and technical capabilities. So strong is the airline industry's belief in SOPs that it is believed that in a well standardized operation, in mid-flight a cockpit crew member could be plucked from the cockpit and replaced with another and the operation would continue safely and smoothly.

As mergers and acquisitions create 'mega-carriers', the process of standardization, and the need to render consistent and unambiguous manuals, procedures, policies, and philosophies becomes increasingly important, costly and difficult to achieve. This is because not all flight crews equally share the corporate history and culture that lead to a certain concept of operation. Nevertheless, any operator knows that adherence to SOPs is not the only way to operate a piece of equipment; and that there are several other ways of doing the same task with reasonable level of safety (Orlady, 1989). For example, most carriers require that crews enter the magnetic course of the runway into the heading select window on the Mode Control Panel (MCP) before take-off. One company requires that the first published or expected heading after take-off be entered. Good reasons exist for both procedures.

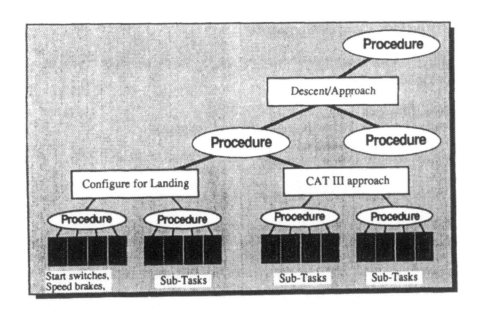

**Figure 3.2  The task-procedure framework**

*Tasks and procedures* As mentioned earlier, it is common in all high-risk systems that critical tasks that affect the objectives of the system are always accompanied with a set of procedures. Procedures, in turn, specify a set of sub-tasks or actions to be completed; that is, each procedure can be shown to lie between a higher level task and lower level sub-tasks. Figure 3.2 shows this structure. The structure, or pyramid, ends with the system goal, e.g. flying passengers from point A to point B.

The task-procedure hierarchy allows one to make the distinction between a checklist and a procedure (as the two are often confused). A checklist is a device (paper, mechanical, audio or electronic format), that exists to ensure that certain actions are carried out. A checklist is not, however, a procedure *per se*. The confusion may arise from the fact that conducting the checklist procedure is a task which is specified by a higher level procedure (e.g. 'the *taxi checklist* shall be conducted once the aircraft starts to move on its own power').

## Philosophy and policies

*Philosophy* The cornerstone of our approach to the concepts of cockpit procedures is philosophy. By philosophy we mean that the airline management determines an over-arching view of how they will conduct the business of the airline, including flight operations. A company's philosophy is largely influenced by the individual philosophies of the top decision makers, but also by the company's culture, a term that has come into favour in recent years in explaining broad-scale differences between corporations. The corporate culture permeates the company, and a philosophy of flight operations emerges. (For a discussion of cultural differences between carriers, see various chapters in Wiener *et al.*, 1993.)

Although most airline managers, when asked, cannot clearly state their philosophy, such philosophies of operation do indeed exist within airlines. They can be inferred from procedures, policies, training, punitive actions, etc. For example, one company that we surveyed had a flight operation philosophy of granting considerable discretion (they called it 'wide latitude') to the individual pilot. Pilots are schooled under the concept that they are both qualified and trained to perform all tasks. Consistent with this philosophy, the company until recently allowed the first officer to call for as well as conduct the rejected take-off (RTO) manœuvre (a manœuvre which is only at the captain's discretion at most carriers).

The emergence of flight deck automation as an operational problem has recently generated an interest in the philosophy of operations, partly due to lack of agreement about how and when automatic features are to be used, and who may make that decision (Wiener, 1989). This led one carrier, Delta Air Lines, to develop a one-page formal statement of automation philosophy (see Appendix 1). This philosophy is discussed by Wiener *et al.* (1991). It is the first case that we are aware of where an airline management actually wrote out its philosophy and the consequences of its philosophy on doing business.

*Policies*  The philosophy of operations, in combination with economic factors, public relations campaigns, new generations of aircraft, and major organizational changes, generates policies. Policies are broad specifications of the manner in which management expects operations to be performed (training, flying, maintenance, exercise of authority, personal conduct, etc.). Procedures, then, should be designed to be as consistent as possible with the policies (which are consistent with the philosophy). Figure 3.3 depicts this framework.

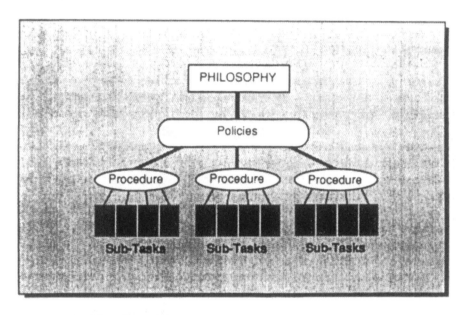

**Figure 3.3   The three 'P's**

The levels in the three 'P's framework are not rigid. For some aspects of flight operations there may be several policies, for others there may be only a philosophy. For example, checklist SOP is a mature aspect of flight operation: there may be an overall checklist philosophy, checklist policies for normal, abnormal and emergency situations. Flight deck automation is still an immature aspect; in this case, there may be only a philosophy and procedures. As the operation becomes more mature, policies are defined and added. Philosophies may also change with time.

To illustrate the three 'P's, let us assume that the task at hand is the configuration of an advanced technology aircraft for a Category-I ILS approach:

1  *Philosophy*: Automation is just another tool to help the pilot.
2  *Policy*: Use or non-use of automatic features (within reason) is at the discretion of the crew.
3  *Procedure*: On a Category-I approach, the flight crew will first decide what level of automation to use (hand-fly with flight director; autopilot and mode control panel; coupled; etc.), which determines what must be done to configure the cockpit.
4  *Sub-tasks (or actions)*: Follow from procedures (e.g. tune and identify localizer and compass locator, set decision height, select autopilot mode, etc.).

Consider the following example of how policies that are actually remote from flight operations can affect procedures. One carrier's new public relations policy called for more interaction between the cockpit crew and the passengers. It was recommended that at each destination the captain stand at the cockpit door and make farewells to the passengers as they departed the cabin. This dictated a change in the procedure that most of the *secure-aircraft* checklist will be done by the first officer. Thus checklist procedures which would normally be run by both pilots, probably as a challenge-and-response, yielded to public relations to be performed by a single pilot. The marketing department considered this particularly important, as they wanted the captain to be in place at the cockpit door in time to greet the disembarking first-class passengers.

To conclude, it is our position that procedures should not (1) come solely from the equipment supplier, or (2) simply be written by the individual fleet manager responsible for the operation of the specific aircraft. They must be based on the operational concept of the organization, and on the organization's examination of its own philosophies. We hypothesize that if procedures are developed in this manner, and a logical and consistent set of cockpit SOPs are thus generated, that there will be a higher degree of conformity during line operations, flight training will improve, and the general quality of flight operations will be enhanced.

If flight management attempts to shortcut the three 'P's process by jumping right into procedure writing, the risk is a set of ill-conceived and inconsistent procedures.

## The fourth 'P': Practices

*An extension of the three 'P's*

In the first two sections of this chapter we focused on the global aspects of flight operations: the philosophy, policies and procedures (Degani and Wiener, 1991). As we progressed in this research, it appeared to us that something was missing. We neglected and ignored the ultimate consumer of the procedure – the pilot– whose decisions and actions determine the 'system outcome'.

To correct this, we have added an additional component – practices. A practice is the activity actually conducted on the flight deck. While procedures must be part of a structured framework, it is the crew members who must carry them out. It is the pilot who will either conform to a procedure or deviate from it. The procedure is specified by management – the practice is conducted by the crew. Ideally they should be the same. The high prevalence of the 'pilot deviation from SOP' classification (Lautman and Gallimore, 1988) indicates that no one can assume that operators will always follow any given procedure dictated by flight management.

The goal of flight management is to promote 'good' practices by specifying coherent procedures. But we must also recognize that this is not

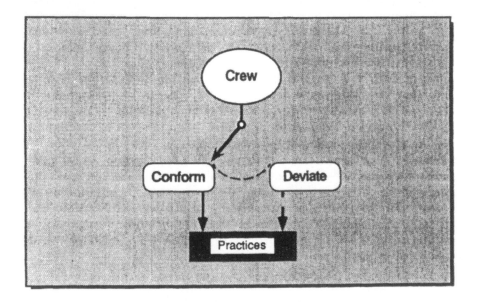

**Figure 3.4  The deviate versus conform 'switch'**

always the case – procedures may be designed poorly. The crew can either conform to a procedure or deviate from it. The deviation may be trivial (e.g. superimposing some non-standard language on a procedural callout), or it may be significant (e.g. not setting the auto-brakes according to the take-off procedures). The alternatives of conformity *versus* deviation can be visualized as a switch (Figure 3.4). This may be somewhat of an over-simplification, but it expresses the choice that the crew member must make: to conform or to deviate. The reasons for and consequences of 'placing the switch' in the 'deviate' position will be explored later in this chapter.

We envision a term '$\triangle$' – delta, or the degree of difference between procedures and practices (Figure 3.5). This '$\triangle$' (not to be interpreted as a quantitative value by any means) expresses the amount of deviation from a specified procedure. This term has two components: (1) the magnitude of deviation from the procedure, and (2) the frequency of such deviations during actual line operations. The goal of flight management is to minimize $\triangle$. When $\triangle$ is large (flight crews constantly deviate from SOP and/or deviate in a gross manner) there is a problem. It may be due to a culprit crew, or in the case where there are frequent violations by many flight crews, a problem in the procedure itself.

The pilot, in this situation, is analogous to a 'filter'. From the above, standards and training departments dictate and teach the way procedures

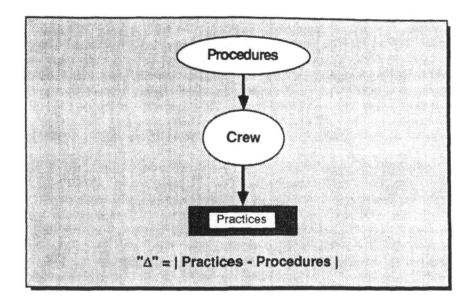

**Figure 3.5   The quantity '$\triangle$' – deviation from procedures**

should be performed. However, in daily line operations (and not under the watchful eye of a check-airman), the individual pilot may adjust the gauge of the filter. This gauge will determine just how much of the SOPs will be actually used, modified, misused or completely unused. The process and the purpose of standardization is to bias the filter towards prevention of deviations.

The consequences of the failure to conform to a procedure can be seen in the following report from NASA's Aviation Safety Reporting System (ASRS) (note that ASRS reports included in this chapter are quoted verbatim):

> The flight was delayed approximately 2 hours due to late arrival of outbound aircraft. F/O arrived and observed how 'tired' he was since he had done 'yard work' all day at home. Our flight finally departed late PM local time for the 4:30 plus flight to SFO. F/O was PF. En route discussed necessity to request lower altitudes with both OAK CTR and Bay Approach when approaching SFO due to tendency to be 'caught high' on arrival in this aircraft type. Area arrival progressed smoothly and we were cleared for the QUIET BRIDGE visual to 28R. When changing radio frequency from approach to tower (head down), F/O selected 'open descent' to 400 feet MSL. Autopilot was off, both flight directors were engaged, and autothrust was on. While contacting SFO tower I became aware that we were below the glideslope, that airspeed was decaying, and that we were in an 'open descent'. Instructed the F/O to engage the V/S mode in order to stop our descent, restore the speed mode for the autothrust, and continue the approach visually once above the 28R ILS glideslope. Company procedures explicitly prohibit selecting an altitude below 1500 feet AGL for an open descent, since this places the aircraft close to the ground with engines at idle. I attempted to explain to the F/O when we were parked at the gate that he had configured the aircraft improperly. Lack of adherence to SOP: 'highly automated' aircraft demands explicit following of established procedures. Unfortunately, it is possible to fly the aircraft numerous ways that will degrade your safety margin rapidly. Adherence to procedures would have prevented this incident (ASRS Report No. 149672).

To summarize, the ultimate factor that determines the quality of the system outcome is the actual practices. These may be governed by procedures, but they are not the procedures themselves. Management's role does not end with the design of the procedure. Management must maintain an active involvement as the procedures move from the flight management offices to the line, remaining concerned with practices, and committed to management of quality through reduction of $\Delta$. This is generally approached as a 'standardization', a form of quality management aimed at ensuring compliance. Standardization is also a check on the quality of the procedures themselves, as well as on the training function.

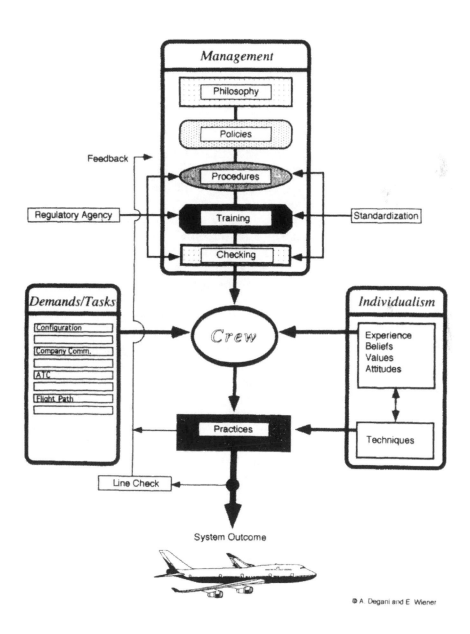

© A. Degani and E. Wiener

**Figure 3.6    Schematic linking all four 'P's**

## Features of the four 'P's framework

Our four 'P' framework is an extension of the three 'P' framework, taking into account the following: tasks, crews, practices, quality assurance and the system outcome. Figure 3.6 is a global depiction of the inter-relationship of these elements. The top half of the chart is essentially the same as Figure 3.3. But when we get to the circle 'crew' we open the door to *practices* (and △).

*Deviant behaviour*   In this section we shall examine several reasons why △ exists. Why would a well-trained, and presumably well-motivated pilot, purposely deviate from the company's published procedures? Listed below are a few of the reasons.

- *Individualism.* △ arises primarily due to the fact that pilots are individuals, and in spite of training, loyalty and generally a devotion to safe practices, they will impose their individuality on a procedure. This may or may not adversely affect the system. We also recognize that there is a positive side to individualism: it is one of the differences between humans and computers. Individualism makes life interesting and provides us with an incentive to achieve. Pilots are not 'procedure executers': they are individuals who bring to their job certain biases, prejudices, opinions and self-concepts. Furthermore, humans possess brains that allow great flexibility, and this can become critically important in extreme cases where no procedure is available, e.g. United's Sioux City accident, previously mentioned (NTSB, 1990a). The problem is the potential conflict between individualism and standardization in high-risk enterprises.

  We once observed a captain making a altitude change (FL370 to FL 260). He abandoned the already programmed VNAV mode in favour of the V/S mode. Asked why he preferred to disengage VNAV he said: 'just because its fun to have "manual" control over the aircraft.'

- *Complacency.* It is well established in aviation that a pilot's vigilance may not always remain at its highest, or even an acceptable level at all time. This phenomenon, of dropping one's guard, is generally labeled 'complacency'. Wiener (1981) has questioned whether the term has any real meaning, and whether its use makes any real contribution to understanding safety. Pending an answer to this question, it seems safe to say that complacency, as the term is used, may be responsible for many departures from SOP.

  It is the very safety of the system that may generate complacency and non-adherence to SOPs. If day after day, year after year, pilots encounter few threats, and few genuine emergency situations, the temptation to ease up and accept less than standard performance is understandable. Recent work by Parasuraman and his collaborators have examined what they call 'automation complacency', the tendency to become overly trustful and over-dependent on various automatic devices in the cockpit (Parasuraman *et al.*, 1991). Again, it is the unerring quality, the high reliability of these devices, that may induce pilot complacency.

• *Humour*. Humour is closely related to individualism, while its negative consequence may be related to complacency. Humour in the cockpit represents the desire to inject some variety and stimulation into an otherwise humourless situation. Humour, like individualism, has its place. It makes life enjoyable, overcomes the tedium of a highly precise job, and establishes a form of communication between crew members. It also carries potential hazards – it can be at odds with standardization.

We have observed in checklist reading behaviour, for example, when the pilot reads 'gasoline' where the checklist requires a challenge of 'fuel', or the use of the Spanish term 'unomas' instead of 1000 ft to level-off callout. These departures are inevitable, as they break the monotony of a highly standardized and procedurized situation. The meanings are *assumed* to be clear, so the departure from SOP is in most cases harmless. However, that is exactly what cockpit standardization is all about – trying to eliminate the need to make unnecessary assumptions during high-risk operations. The difficult question, of course, is where to draw the line. Unfortunately, the absolute distinction between what is humour and what is a deviation from SOP depends on the outcome. If this humour caused a breakdown in communication that led to an incident, then it would be labelled 'deviation from SOP'. If it did not result in an untoward consequence, it could be regarded as humour.

Nonetheless, we take the position that these are still breaches in cockpit discipline that should not be taken lightly, as the following example illustrates. We once observed a take-off in which the captain was the pilot flying. The first officer was supposed to make standard airspeed calls of V-1, V-r and V-2. Instead, he combined the first two into a non-standard call of 'V-one-r', and at V-2 called 'two of 'em.' Obviously, the captain knew what was meant by these strange calls, and while one cannot say that this was a dangerous compromise with safety, it did represent a serious departure from SOPs. Perhaps worse, it established an atmosphere of tolerance on the part of the captain of sub-standard performance which may lay the groundwork for more serious SOP departures later.

We may question what moved the first officer to depart from standard procedures and utter nonsensical callouts. This example could of course be attributed as well to complacency, and as explained above. The link may be that complacency induces the introduction of 'humour' in place of standardization. But during a critical phase of flight operations is probably not an appropriate arena for too much humour.

Perhaps the appropriateness of humour on the flight deck is an area addressable by Crew Resource Management (CRM). It might be a relatively simple matter for the captain, during his initial briefing, to advise the subordinate crew member(s) on how he feels about humour in the performance of duties. It might be somewhat more difficult, in spite of the fact that CRM training stresses free two-way communication in the cockpit, for the other crew persons to do the same for a 'humorous' captain.

- *Frustration.* Pilots may feel that they have been driven to non-conformity by frustrating forces beyond their control. An example would be the failure to use the oxygen mask (above FL 250) when one pilot leaves the cockpit of a two-pilot aircraft. Our impression is that there is low conformity to this regulation, and the reason is equipment-induced. First, it is not comfortable to wear any mask, and furthermore, in some modern aircraft, it is difficult to replace the new inflatable masks in their receptacles. Pilots find it a frustrating task and avoid it by simply not conforming to the regulation.

  We observed an interesting ploy to overcome the mask while still obeying the regulation. In a two-pilot aircraft with the inflatable mask, the captain left the cockpit briefly while the aircraft was climbing unrestricted to FL 330. At about FL 200 the first officer called ATC and requested level off at FL 250, which he maintained until the captain returned, and then requested continuation of his climb. The mask is not required at this altitude. In this case the pilot conformed to the regulation and procedure, but possibly at some cost to the company (wasted fuel resulting from sub-optimal climb profile) and possibly also to the ATC system.

*Technique*    The use of technique in the cockpit allows the pilot to express individualism and creativity without violating procedural constraints. A technique may be defined as a personal method for carrying out a task. If the technique is consistent with the procedure, then the task is conducted correctly, and $\Delta$ is zero.

Techniques have been developed by pilots over their years of experience of flying various aircraft. Every pilot carries with him a virtual catalogue of techniques. They are often fine points which pilots have discovered for themselves, experimented with, or learned from other pilots. Consider the following technique: the Quiet Bridge visual approach to runway 28R at San Francisco (SFO) requires a profile descent with fixes at 6000 ft 18 DME, 4000 ft 13 DME, and recommended 1900 ft 6 DME from the SFO VOR. We once observed a crew that, in preparation for this approach, built these fixes into the FMC and named them '6000', '4000', and '1900' (in an A-320 FMC it is possible to give any name for a 'man-made' waypoint, as opposed to 'SFO01', etc. in other FMS aircraft). As they flew this approach using only the autopilot and manual flying, the depiction of these fixes and their associated names (altitudes) on the map display provided an enhanced situation display.

Why does the procedure writer not include the techniques as part of the procedure? Generally this is not advisable: the techniques are too fine-grained. If SOPs were replaced with the detailed descriptions necessary for one to carry them out, the flight operations manuals would be many times their present size. The company should be happy to specify the procedure, and leave it to the individual pilot to apply what he/she considers the best technique.

To the credit of the flying profession, pilots are always looking for a better technique. The motivations are various: professional pride, overcoming boredom, expression of individuality, the comfort of the passengers, and perhaps most salient, a feeling that they can find a better way. Note that some of the motivations are the same as those that led to deviant behaviour and $\Delta$, but in the case of techniques, they led to a more favourable result. Thus we see that there is room for flexibility and individualism even within a rigid procedurization and standardization.

We would add one caution regarding technique. Any given technique may indeed conform to the written procedure and thus not add to $\Delta$, but would still entail an unnecessary risk. Although $\Delta$ is not increased time after time when this technique is employed, the seeds of a latent error may be planted. Therefore, within the context of our four 'P' framework, it is not enough for the technique to conform to the procedure; it must also be consistent with policies. If the technique is not consistent with both the published procedure and the published policy, calling it a technique means nothing – it is simply a deviation from SOP.

- *Automation and Technique.* The introduction of cockpit automation has brought a plethora of techniques, largely consisting of ways in which the pilots chose to employ the automatic devices and modes. These techniques are the result of the great variety of ways that a task can be accomplished in a high-technology aircraft, due to its many modes and options.

A common example is the automatic level-off manœuvre. Many pilots feel that left to its own, the auto-levelling produces flight manœuvres that are safe and satisfactory, but could be smoother and more comfortable for the passengers. Pilots also believe that in the auto-level-off manœuvre the autothrottles are too aggressive. As a result of this, many have developed techniques to smooth these actions; most of these techniques involve switching autopilot modes during the level-off. We emphasize that these are *techniques* and not procedures. They represent the superimposition of the pilot's own way of doing things upon a standard procedure, and as long as the SOP is not violated, it may be all to the good.

Pilot technique is actually accommodated by some modern flight guidance systems. The bank angle limiter, for example, invites the crew to express their preference for maximum bank angles and rates of turn, consistent with the demands of ATC, and the comfort of their passengers.

Other techniques have been developed to 'trick the computer', as Wiener discussed in his 1989 report on glass cockpit human factors. For example, the pilot of a glass cockpit aircraft, wishing to start a descent on VNAV path earlier than the displayed Top of Descent (TOD) point, could either enter a fictitious tailwind into the flight guidance computer, or could enter an altitude for turning on thermal anti-ice protection (which he had no intention of actually doing). Both methods would result in a recomputation of the TOD and VNAV path, with an earlier descent. Why would the pilots do it? Because experience had taught them that the correctly computed VNAV path would result in speeds that would require the use of spoilers, which pilots consider as unprofessional, as well as creating vibration that would discomfort the passengers.

Perhaps the most unusual technique we have observed was demonstrated by a captain of a B-757. Acting as Pilot Not Flying (PNF), he tuned the arrival ATIS on the VHF radio, listened to it, and then rather than writing it on a pad or a form, he proceeded to encrypt it into the scratch line of the Control Display Unit (CDU). He then read it from the CDU to the pilot flying. This was a captain who obviously wanted to make maximum use of his automated devices. Of course, this method of recording the ATIS has its limitations, the most severe being that only one person in the world could decode the message as recorded. Still we must presume that this was a technique and not a procedural deviation, unless the company's manual said that the ATIS was to be written out by 'pen and pencil'.

Regardless of whether this was or was not a technical violation of company policy, it did seem to violate common sense, that ATIS information should be available to all pilots. What would we have thought if the captain had scribbled the ATIS on a form so illegibly that the first officer could not read it? Furthermore, had a message come into the scratch line, the ATIS message would have been lost. As we have previously said about individuality and humour, technique has its place. It may also have its price.

- *Management's view of technique.* What view should management take of pilots developing their own techniques? Does the superimposition of 'personal' technique on SOPs represent a compromise with standardization? Once again the answer is to be found in the four 'P's. Management must develop a philosophy that governs the freedom of the pilot to improvise, and from this philosophy will flow company policies that will state exactly what the company expects on the line. Our own view is, of course, to return to the definition of $\Delta$. If the techniques employed on the line lead to practices that are consistent with the procedure and the policy, then $\Delta$ is zero and management should take little notice of the techniques employed.

  If management discovers, through standardization and quality management techniques, or the feedback loop (to be discussed next) that certain techniques may have potential for procedural deviation, then this can be dealt with through the normal quality assurance processes. It is entirely possible that the opposite could occur, that the quality management or feedback processes could discover superior techniques that should become procedures. Check-airmen play a vital role here. While their job is generally quality assurance and standardization, they should be watchful for line-generated techniques that could and should be incorporated into the company's SOPs.

- *Technique and CRM.* Our discussion of technique has centred on the means of executing company-generated cockpit procedures. The same principles apply to the vast and ill-defined area known as cockpit resource management (Wiener *et al.*, 1993). Pilots develop communication, team-building, stress management and other mechanisms for getting the job done effectively. These can also be viewed as techniques, personalized ways of carrying out procedures. CRM training programmes attempt to teach principles of communication, not techniques. Just as in cockpit techniques, these are developed, largely by trial and error, as well as observations of others, during a pilot's experience. We have all seen examples of good and bad communications techniques in the cockpit and elsewhere. We can again apply the definition of $\Delta$. If one's personal CRM techniques lead to congruence between policies, procedures and practices, they should be considered adaptive. If not, they generate $\Delta$ and must be dealt with through the same quality management mechanisms that are invoked by unsatisfactory piloting. We extend the comments of the last section to CRM as well: check-airmen should be vigilant in observing adaptive and maladaptive CRM techniques on the line and in training.

The hazards of tricking the computer combined with poor crew coordination can be seen in the following ASRS report.

> We were cleared to cross 40 MN west of LINDEN VOR to maintain FL 270. The captain and I began discussing the best method to program the CDU to allow the performance management system to descend the aircraft. We had a difference of opinion on how best to accomplish this task (since we are trained to use all possible on-board performance systems). We wanted to use the aircraft's capabilities to its fullest. As a result, a late descent was started using conventional autopilot capabilities (*vertical speed, maximum indicated mach/airspeed* and *speed brakes*). Near the end of the descent, the aircraft was descending at 340 knots and 6000 fpm. The aircraft crossed the fix approx. 250–500' high .... This possible altitude excursion resulted because: 1) captain and F/O had differences of opinion on how to program the [FMC for] descent. Both thought their method was best: the captain's of programming (fooling) the computer to believe that anti-ice would be used during descent, which starts the descent earlier. The F/O's of subtracting 5 mile from the nav fix and programming the computer to cross 5 mile prior to LINDEN at FL 270. 2) Minor personality clash between captain and F/O brought about by differences of opinion on general flying duties, techniques of flying, and checklist discipline. 3) Time wasted by both captain and F/O (especially F/O) in incorrectly programming CDU and FMS for descent, which obviously wasted time at level flight, which should have been used for descent (ASRS Report No. 122778).

## Standardization

Standardization is the palace guard of procedures. It is a management function which begins with the writing of procedures, to ensure that they are consistent with the first two P's (philosophy and policies), are technically correct, and are published in a manner that will be clear to the line pilots. Standardization also extends to the various quality assurance methods that allow flight management to monitor line performance, training performance (of both instructors and trainees), and to guarantee conformity to SOPs (low values of $\triangle$). These methods include recurrent training, LOFT, line checks, and simulator checkrides. We have stated previously in this chapter the vital role of standardization personnel as elements in the feedback loop which links the line to flight management, this we shall now discuss.

*Feedback*

We stress the importance of the feedback process. Feedback is an essential process because this is not a perfect world, and some procedures are not perfectly designed and this may lead to deviations on the part of the operators.

The frustration of a crew member who feels that management is unresponsive to feedback from the line can be seen in the following ASRS report:

> I am very concerned about the safety of the company's new checklist policy. The climb checklist has three different segments and it is not completed until about 18 000 feet. The approach checklist has a descent check that precedes it. The landing check has four segments. The 'landing check' is called after the approach flap settings. The 'landing gear' call stands alone, and the final segments are completed after the final flaps are set. The last segment requires both pilots to watch the flaps come down, no matter how busy an approach [we are flying]. We have had several major checklist changes over the last two years. This latest one is the most radical. Having flown with 20–30 F/Os using it, I find that about 30 to 40 percent of the time we are able to do all the checklists correctly. Since it takes so long to complete all the segments of the list, something usually gets left out. Many times it's the 'gear down' check, since it no longer is 'gear down/landing check' as we have all done since day-one in our careers. Also much of the check is done with a flow that does not match the checklist. After time off you have to re-memorize the flow since it's so different from the list. I note that when a situation is tight we are all, at times, reverting back to some of the calls from the previous procedures. Even the new hires who never used another checklist are not able to remember all the steps. The company imposed the procedures without input from the line, and is not interested in our input. Please help us convince the company that these procedures are not user friendly before someone makes a serious mistake (ASRS Report No. 155183).

As we have noted, one of the reasons why pilots deviate from accepted procedures (create positive $\Delta$) is that they think that they have a better way. In some cases they might. This view of $\Delta$ portrays it as a negative feedback signal in a closed-loop system. If a corrective path is available, ideally $\Delta$ will be a self-eliminating quantity.

One way of promoting conformity to procedures is by providing a formalized feedback between the operational world and flight management. Some may argue that this is not necessary and that flight management *is* part of the operational world. On the other hand, the performance of line pilots is the ultimate measure of the adequacy of procedures because of their daily interaction (and sometimes confront-ation) with procedures. When written procedures are incompatible with the operational environment, or have technical deficiencies, or increase workload, or create conflicts in time management, etc., flight crews may react by resisting and deviating from SOP. This can be minimized by establishing a clear feedback path that will provide a channel of communication between the line and management. If the line pilot is resisting certain procedures, or if s/he feels that there is a better way, clearly this information should be brought to the attention of the procedure writers in management for reevaluation.

One value of a well managed feedback loop is to eliminate the difference between what is taught in the training centre and what is expected on the line. The oft-expressed instruction 'I don't care what they are teaching you in the simulator and ground school, this is how we do it on the line' reveals not a minor quirk, but a serious management and training failure.

We have used the word 'formal' to describe the desired feedback path. By this we mean that a clear mechanism be established for movement of information and suggestions from the line to management. Bland statements from management such as 'my door is always open,' or 'you can always go to your chief pilot' do not constitute a sufficient feedback path. The line pilot must feel that his input is desired, and will be taken seriously. Offhand comments given in passing in the corridors and coffee shops do not qualify as effective feedback mechanisms.

Discussing the feedback path from line to management forces us to consider briefly labour–management relations at airlines. To be success-ful, the feedback process must involve the participation of the appropriate pilots' representative group. At most carriers this would be the Air Line Pilots Association Safety Committee, or perhaps other committees such as Training or Professional Standards. The feedback path then would consist of a communication from the line to the representative group, and thence to management. This has some advantages over direct pilot-to-management communication, in that the pilot may wish for various reasons to be insulated from his managers. Also, by working through a committee, patterns can be noted by the committee members.

For this system to be effective, it is essential that a cooperative, non-adversarial relationship exist between management and the representative group. This is sometimes difficult when either contract negotiations are underway, or for whatever reason tensions exist between pilots and management. The feedback process can be effective only if management makes it clear that they are eager to receive input from the pilots' representative group on a non-adversarial basis, and the pilots' group in turn must resolve to stick to its safety mandate, and not be tempted to use safety as a smokescreen for contractual/industrial matters. It is a measure of the maturity of the management of both the company and the union if both sides can transcend 'politics as usual' for the sake of promoting safety.

We recommend that a clear, well-defined feedback loop be established and supported so as to provide an effective channel of communication between line and management. To be effective, the feedback process must be easy to use, non-threatening, and above all must engender at least the promise that the line pilot can affect something.

## Conclusions

We believe that the four 'P's concept detailed in this chapter forms the foundation for writing and enforcing flight deck procedures in a consistent and logical manner, within and across fleets. Consistent and technically correct procedures in turn ensure both the economical utilization of humans and equipment and the safe conduct of flight. Any procedure, even the best one, cannot be 'bullet-proof'. It can only be a baseline. The role of management should be to provide the best possible baseline for its crews, and then train and standardize to this baseline. No procedure is a substitute for an intelligent operator.

## Acknowledgements

This research was conducted under two research grants from the NASA Ames Research Center: NCC2–327 to the San Jose State University Foundation, and NCC2–581 to the University of Miami. The University of Miami grant was jointly supported by NASA (the Office of Space Science and Applications, and the Office of Aeronautics, Exploration, and Technology), and the Federal Aviation Administration. The contract technical monitors were Drs Barbara G. Kanki and Everett A. Palmer. A more complete report on the design of flight deck procedures is currently being written by the authors, and will be available as a NASA report.

# References

Degani, A. and Wiener, E. L. (1990), *The human factors of flight-deck checklists: The normal checklist*, (NASA Contractor Report 177549). Moffett Field, CA, NASA-Ames Research Center.

Degani, A. and Wiener, E. L. (1991), 'Philosophy, policies, and procedures: The three P's of flight-deck operations', in R. S. Jensen (ed.), *Proceedings of the Sixth International Symposium on Aviation Psychology Conference*, Columbus, OH, The Ohio State University, pp. 184–191.

Lautman, L. G. and Gallimore, P. L. (1988), *Control of the crew caused accidents*, Seattle, WA, Boeing Commercial Airplane Company.

NTSB (1988), *Northwest Airlines. DC-9–82 N312RC, Detroit Metropolitan Wayne County Airport. Romulus, Michigan. August 16, 1987*, (Aircraft Accident Report, NTSB/AAR-88/05), Washington, DC, National Transportation Safety Board.

NTSB (1989), *Delta Air Lines, Boeing 727–232, N473DA. Dallas-Fort Worth International Airport, Texas. August 31, 1988*, (Aircraft Accident Report, NTSB/AAR-89/04), Washington, DC, National Transportation Safety Board.

NTSB (1990a), *United Airlines Flight 232, McDonnell Douglas DC-10–10, Sioux Gateway Airport, Sioux City, Iowa. July 19, 1989*, (Aircraft Accident Report, NTSB/AAR-90/06), Washington, DC, National Transportation Safety Board.

NTSB (1990b), *USAir, Inc., Boeing 737–400, N416US. LaGuardia Airport. Flushing, New York. September 20, 1989*, (Aircraft Accident Report, NTSB/AAR-90/03), Washington, DC, National Transportation Safety Board.

Orlady, H. W. (1989, January), 'The professional airline pilot of today: all the old skills – and more', *Proceedings of the International Airline Pilot Training Seminar conducted by VIASA Airlines and the Flight Safety Foundation*, Caracas, Venezuela.

Parasuraman, R., Molloy, R. and Singh, I. L. (1991), *Performance consequences of automation-induced 'complacency'*, (Technical Report No. CSL-A-91–2), Cognitive Science Laboratory, Catholic University, Washington.

Wiener, E. L. (1981), 'Complacency: Is the term useful for air safety?' *Proceedings of the Flight Safety Foundation Seminar on Human Factors in Corporate Aviation*, Denver, CO.

Wiener, E. L. (1989), *The human factors of advanced technology ('glass cockpit') transport aircraft*, (NASA Contractor Report 177528), Moffett Field, CA, NASA-Ames Research Center.

Wiener, E. L., Chidester, T. R., Kanki, B. G., Palmer, E. A., Curry, R. E. and Gregorich, S. E. (1991), *The impact of cockpit automation on crew coordination and communication: I. Overview, LOFT evaluations, error severity, and questionnaire data*, (NASA Contractor Report 177587), Moffett Field, CA, NASA-Ames Research Center.

Wiener, E. L., Kanki, B. G., and Helmreich, R. L. (eds) (1993), *Cockpit Resource Management*, San Diego, CA, Academic Press.

# Appendix 1 – Delta Air Lines automation philosophy statement

The word 'Automation', where it appears in this statement, shall mean the replacement of human function, either manual or cognitive, with a machine function. This definition applies to all levels of automation in all airplanes flown by this airline. The purpose of automation is to aid the pilot in doing his or her job.

The pilot is the most complex, capable and flexible component of the air transport system, and as such is best suited to determine the optimal use of resources in any given situation. Pilots must be proficient in operating their airplanes in all levels of automation. They must be knowledgeable in the selection of the appropriate degree of automation, and must have the skills needed to move from one level of automation to another.

Automation should be used at the level most appropriate to enhance the priorities of Safety, Passenger Comfort, Public Relations, Schedule, and Economy, as stated in the Flight Operations Policy Manual.

In order to achieve the above priorities, all Delta Air Lines training programs, training devices, procedures, checklists, aircraft and equipment acquisitions, manuals, quality control programs, standardization, supporting documents, and the day-to-day operations of Delta aircraft shall be in accordance with this statement of philosophy.

(Reprinted from Wiener *et al.*, 1991, p. 146.)

# 4 The management of safety on the airport ramp

*Nick McDonald and Ray Fuller*

## Introduction

Aircraft ground handling comprises all those operations servicing an aircraft during a normal turnaround between landing and departure, including marshalling, chocking, refuelling, servicing water, toilets and catering, passenger embarkation/disembarkation, the loading and unloading of baggage and freight and aircraft towing and pushback. It is an integral part of the aviation flight cycle upon which both the safety and efficiency of aviation operations crucially depend.

This chapter will begin by exploring goals and motivation for safety management before addressing a number of related organizational and technical issues. It will then turn to the focal issue of the human factor in ramp accidents and the necessity of designing compatibility in the interface between the human component and the entire working environment. The significance of the systematic gathering of ramp accident data for the development of our understanding of accident causes and the design of preventative measures will then be reviewed. The chapter concludes with a broad analysis of the management of ramp safety through consideration of where the responsibility for safety lies, the role of accident reporting systems, the use of accident information and the development of countermeasures.

## Safety management goals

Why should ground handling management be concerned with safety?

Clearly there is an ever-present potential for unreported ground handling damage to aircraft to cause a major flight disaster. But this is only one of a number of considerations concerning safety in ground handling. Costs are a major concern, not only because aircraft are both easy to damage and expensive to repair, but also because of the indirect costs associated with schedule disruption which follow from accidental damage. The ramp is also a dangerous environment in which to work given the risk of death or disabling injury to those who work in it. Thus the focus of interest in ramp safety is as much on the relatively frequent small-scale accident as it is on the prevention of large-scale disasters.

One can look at the motivation for safety from three points of view, namely those of the organization, the operative and of the individual manager.

*Protection of the organization*

The organization has a direct interest in safety from a variety of points of view; in particular the financial and operational savings which can accrue from reducing the direct and indirect costs of accidental damage. In the approach of one airline (SAS), Busk (1992) describes the use of setting cost-minimizing targets in diminishing the frequency of damage to aircraft, documenting costs under three headings, direct costs, production costs and capital costs.

Good safety management practice has also been linked to:

- efficient systems of monitoring, training and performance;
- better planning; and
- the protection of the organization in cases of third party or employers' liability claims (see, for example, HSE, 1991).

Safety has been built into some quality management programmes in ground handling (Hanus, 1992; de Boek, 1992; Kerkloh, 1992). On the other hand, we are aware of other quality management programmes which do not appear to have taken account of safety issues.

*Protection and enhancement of the workforce*

A core objective of ramp safety management is the protection of the worker against injury or ill-health arising out of work. Increasingly, the promotion of health and the enhancement of the organization's human resources have become an explicit goal justifying significant organizational effort and resources (Wynne, 1989; Kerkloh, 1992). Motivation towards safety can also be a powerful springboard towards a higher quality all-round performance at work.

69

*Management development*

Modern line management requires its participants to be multiskilled. Best practice in safety management provides the opportunity for management to develop their own skills and competencies with expertise as much in the management of people and processes (quality, safety, etc.) as in the delivery of a particular service or product.

Safety management thus spans both operational safety and occupational safety. The difference between these two is not always clear but is worth elaborating because it can lead to different emphases in strategy and approach. Operational safety starts from the point of view of the safety of the operation, be it the cycle of flight and ground handling operations or whatever. The important goal is the smooth and effective completion of the operation without disastrous consequences for the user or consumer. Occupational safety adopts the viewpoint of the operative, in particular the impact of work on personal safety, health and well-being. In many organizations the administration of operational and occupational safety follow quite distinct channels. For example, a staff role in occupational safety may be the responsibility of personnel, health or welfare sections of the organization while operational safety remains the main focus of operations management and the safety department. An airline's safety management system may tend to have a strong focus on the operations requirements of flight safety, which may in turn influence the orientation of and resources available to its ground handling department.

The organizational roles and responsibilities which provide the framework for any safety management system are set within the context of laws concerning safety of flight operations, the framework of laws concerning occupational safety and health, and specific laws and regulations on a variety of hazards. While in the European Community there is a process of harmonization of this legislative framework, in a broader international framework there is considerable diversity (see Kelman, 1981). Even within the emergent supra-national framework of the EC there is considerable national diversity of institutions, organizational cultures, practices and operational style with respect to occupational safety and health.

## Organizational and technical issues

*Organizations*

Ground handling is done by a variety of different types of organization, and although the service or range of services is similar throughout the world, the commercial and administrative arrangements for the delivery of these services vary widely (Ashford *et al.*, 1984). Perhaps the most comon arrangement is for airlines themselves to have their own internal ground handling section which handles that airline's aircraft. Such companies also frequently handle other airlines' aircraft on a contract basis. Airport authorities are another agency providing aircraft ground handling services and there are independent general ground handling companies who offer a range of similar services. Lastly, there are companies which provide a service around a specialized product such as fuel or catering. Different organizations vary considerably in size and they can be widely dispersed geographically in small units. The commercial environment varies from the monopoly situation enjoyed by most of the airport authorities in Germany to the deregulated competitive environment of the United States and Canada.

The implications of this diversity are that the problems and issues for the management of ground handling organizations in ensuring a safe system of work are not always the same. It is not only a question of scale and of the resources available to organizations, which are of great importance in developing a safety management system, but there are other implications for the way in which safety is managed. Examples include the ability (or inability) to plan personnel requirements in a way that facilitates the development of training, or to control the design of the workplace infrastructure.

*Technology*

The technology of the ground handling process also poses interesting problems for safety. The truism that aircraft are designed to fly carries with it the corollary that they are not optimally designed for ground handling: their shape makes no compromises in interfacing with ground handling equipment and differences between types and models of aircraft make inconsistent and inconvenient demands on equipment operation; for example, avoiding wings when accessing fore- and aft-located doors. The aircraft's size and location in space relative to the operator invite misperceptions and misjudgements of distance and location; the fragile skin and appendages are easily damaged and such damage if undetected can have disastrous consequences in flight. Key safety considerations must include the preservation of the aerodynamic and structural integrity of the aircraft and its pressurized hull. The constraints of space and time in ground handling operations make issues of work organization, perceptual psychology and ergonomics of central concern.

The advanced technological systems for controlling flight operations contrast markedly with the available technology in many ramp

71

operations (to put it crudely – muscle power, conveyor belts, lifting platforms, specialist surface vehicles and a variety of adapted mechanical road vehicles). Apart from the development of the size and sophistication of individual pieces of ground handling equipment to match the size and scale of modern aircraft, three developments are worth a particular mention for the actual or potential way in which they influence the nature of ground handling work.

First is the containerization of goods and baggage. This has reduced the amount of handling of individual items through the introduction of mechanical lifting and transport devices. A second is the development of automated baggage sorting and conveyor systems, interfacing at one end between the check-in desk and the baggage hall. At the other end of the system there may be a 'moving carpet' facility within the aircraft hold itself. A logical development of this process is the experimental 'vehicle-free' ramp which has been developed at Arlanda in Sweden. In this, all services are brought to the aircraft stand and are connected directly to the aircraft (underground conduits for fuel, water, power, waste and baggage conveyor belt; passenger bridges to the front and rear of the aircraft, extending over-wing). The third area of technological development is in computer-based systems for the dispatch and movement control of all ramp vehicles, which has been in place for some time for the control of aircraft movements, but developed for the contol of other ramp vehicles by airports such as Munich.

Apart from these developments, the core tasks of ground handling activities have not changed markedly over the last few decades. However, they continually throw up issues to do with the application of fundamental principles of ergonomics to the design of the interfaces between the operatives working on the ramp, their equipment and vehicles, the aircraft, and the layout of the ramp facilities and structures.

## Human factors on the ramp

The air transport system is socio-technical, involving the integration of organized human labour with a sophisticated technology to deliver people and goods from one location to another. As it currently functions, the human element is an indispensable component of the system, no more so on the ramp than in the air. The ramp worker operates expensive, highly specialized ramp service equipment of various types, each making different demands on the operator. He frequently has to work in limited space in the midst of congestion from other service equipment and vehicles, aircraft and pedestrians, under conditions complicated by time pressure, noise, jet-blast and all types of weather (National Safety

Council, 1975). The cyclical nature of this job often requires immediate and demanding increases of both mental and physical performance from a resting and perhaps fatigued level. At other times there may be a demand for sustained physical and mental work over prolonged periods. The primary human virtue under such conditions is adaptability, the ability to achieve the service goal in spite of enormous variation in task requirements, in spite of failures in other parts of the system, in spite of the unending parade of new problems to be solved. And an important part of this adaptability is the plasticity of human behaviour: the ability to modify what is done and how it is done in the light of experience.

However, human adaptability is a two-edged sword, for the potential for variation in performance which it embodies also represents a potential for unreliability. The performance of the human component in the socio-technical system does not have the reliability of a highly engineered machine component. It also has to be recognized that human capability can rapidly reach its limitations, both in absolute and relative terms. Examples of the former are clearly evident in limits of muscular strength and endurance, in decrements in visual acuity under degraded light conditions and in impaired spatial judgements in the absence of cues to depth and distance. The restricting effects of factors such as fatigue, stress and a lack of motivation are examples of relative limitations in capability.

These considerations lend support to attempts to design the work environment and the way work is organized so as to avoid exceeding the mental and physical capabilities of the ramp worker and to minimize the effects of human variability in performance. The fundamental ergonomic concept here is that human characteristics should be taken into account in the design of all of the interfaces between the operator and the equipment he has to use, the environment in which he has to work and the procedures he has to follow.

Up to the present time, the application of ergonomic principles to design aspects of the ramp worker's task has been typically fragmentary and unsystematic and some applications, such as those relating to the control of exposure to high noise levels, have arisen more in response to issues of occupational health and related legislation. The general picture is one of reliance on the flexibility and adaptability of the operator to cope with the deficiencies in the system which emerge after other considerations have taken precedence; deficiencies such as:

- working on a congested apron and in confined space;
- working under time-pressure to recover schedule delays;
- working in height-restricted holds;
- serially operating equipment and vehicles which differ in how a particular function is controlled (e.g. gear selection; rear-wheel

steering as opposed to front wheel), which display the same information in a different way or place (e.g. brake air-pressure) and which have different handling characteristics (e.g. rates of acceleration and braking).

Furthermore, ergonomists representing ground handling operations are rarely involved, if ever, in the design of service vehicles and equipment or at the design stage of an aircraft fuselage, so that, for example, ergonomically unfavourable working postures in cargo and baggage holds can be minimized (Ruckert, Romhert and Pressel, 1992). Similarly, human factors considerations have not been a common feature either of ramp safety training, of ramp accident reporting (see later), or of ramp safety audits.

This situation is perhaps surprising given the evidence that in the majority of ramp accidents the human factor is partially or completely the element in the system that is the 'final cause' in the chain or pattern of events. For this reason, an ICAO circular has declared that the expansion of human factors awareness presents the international aviation community with the single most significant opportunity to make aviation both safer and more efficient (ICAO, 1989).

A systematic representation of ergonomic (or human factors) interfaces is provided by the SHEL model (Hawkins, 1987) represented in Figure 4.1, where SHEL is an acronym for Software, Hardware, Environment and Liveware. Liveware stands for the human element in the system, with the worker of interest represented in the centre of the figure. In the normal work setting he has to interface with Software, a term used here to describe controlling instructions to which he is subject (rather in the manner of software controlling the operations of computer hardware).

**Figure 4.1    Interfaces of the human factor – the SHEL model**

Examples would be the contents of an instruction manual, checklists for procedures, rules and regulations and the commands given by a supervisor.

Hardware refers to the equipment the worker has to operate and includes vehicles, conveyor belts, service equipment, aircraft hatches, doors and connecting points, on-board devices, and so on. Environment represents physical variables which may impact on the worker such as air temperature, wind speed, humidity, radiation, noise, vibration and light levels. Finally, there is an interface with other Liveware, representing factors of communication, the dynamics of social interaction and the interaction of differing personality types. This element is of particular importance in the ramp situation as the work typically involves small groups operating as teams with possibly frequent shared opportunities to socialize in between servicing flights. Thus there is present the opportunity for powerful social forces (such as of conformity, compliance, coercion and cooperation) to work either in a positive or negative direction on ramp worker attitudes and behaviour. The social group can be a major determinant of norms of behaviour and their implementation. Furthermore, ramp workers usually operate under supervision which makes the identification and application of leadership skills of fundamental importance (see later).

Note in Figure 4.1 that the Liveware in the centre is contained by an irregular line, representing human variability and other limitations and that each interface is shaped so as to accommodate as much as possible to these limitations: a simple graphic representation of the ethos of ergonomics.

This model is useful because it is based on the concept of matching the task to the operator, of ensuring that the demand characteristics of the working environment are compatible with the supply characteristics of the worker. This matching requirement has both a fixed and a dynamic dimension. On the one hand it is necessary that controls and displays are designed in such a way that the range of potential human workers will be able to read and operate them. On the other it is essential that the competence of the operator is stable enough over time so as not to undermine the quality of what he does. Factors such as motivation, fatigue, attention and emotion are all potential candidates for creating a mismatch. The evidence suggests that a significant proportion of ramp accidents occurs, perhaps surprisingly, not when work demands are high and 'the place is buzzing' but when activity levels are relatively low. Under these conditions it appears that the operator's performance simply fails to rise to the level required by the task, rather than the task being too demanding in a more absolute sense.

There are other more subtle conditions which can undermine the stability of the operator's performance. Cumulatively, experience can modify behaviour in undesirable ways. For example, if an operator gets away with a particular 'short-cut' on a number of occasions, this behaviour may become established and habitual, eventually leading to a hazardous situation (see Chapter 9). The effects of rapid transition from one task to another can also undermine competence where, for example, the displays and controls of one vehicle are inadvertently treated as if they were the displays and controls of a vehicle just abandoned.

## Ramp accident and injury data

At present, remarkably little is known about contributing human factors in ramp accidents because such information has not been systematically collected. While many companies have their own internal accident or incident databases there are very few public sources of such data. For nine years (1982–1990) the joint Ramp Safety Group of IATA and the AACC ran an annual ramp safety campaign in the month of November. This invited participating organizations to compile statistics on aircraft damage, identifying the different types of ground support equipment involved. While the identification of the equipment suggested possible causal factors, the method of data collection did not allow for much quantification, analysis or interpretation. More recently the Aerospace Psychology Research Group (Trinity College, Dublin) has been collaborating with one successor to that group, the IGHC (IATA Ground Handling Council) Ramp Safety Group to develop a more comprehensive and consistent accident reporting scheme. This scheme is based on a new accident report form which includes details of the circumstances of the accident, the personnel involved, the identification of causal and contributory factors (including in particular human factors) and the consequences of the accident (damage, injuries, costs). A preliminary trial of this scheme encompassed aircraft damage accidents for a six month period beginning January 1993. An overview of the main features of a sample of the data to-date (580 accidents), together with some information from individual company documentation, is provided below. A more complete and regular periodic analysis of the IATA/APRG data will be available in due course.

The IATA/APRG database was initially set up only for reports in which damage occurred to aircraft, facilities or equipment; of these reported accidents only about two percent actually resulted in personal injury or death. Other databases (e.g. Kerkloh, 1992) indicate that minor acute injuries are the most common source of lost working days, particularly bruises and strains and particularly involving feet, fingers, legs and hands. These arise from operations such as manual loading and baggage handling. In addition, there are permanently disabling injuries which can affect 3–5% of the ground handling workforce annually (Kerkloh, 1992). Indications from other data sources (unpublished) are that back injuries are a major source of chronic disability. These data suggest that people, equipment and aircraft are damaged or injured in fundamentally different types of accident. Furthermore the IATA/APRG data indicate that there is very little overlap (less than 5% of the total) between aircraft damage accidents and equipment damage accidents.

Additional understanding of the circumstances of accidents in the IATA/APRG database comes from an analysis of the specific operation being performed when the accident occurred. For aircraft damage accidents the most commonly reported sources of damage arise from the positioning of equipment such as container loaders and conveyor-belt loaders adjacent to the aircraft, the process of loading and unloading the aircraft and aircraft towing. The majority of equipment damage accidents occur while the vehicle is *en route*.

The reporting system groups contributory factors into five broad categories:

(i)     regulations or procedures not followed;
(ii)    equipment use or defects;
(iii)   organizational;
(iv)    behaviour; and
(v)     physical circumstances.

By far the most common contributory factors in aircraft damage accidents were in the first category, specifically failure to follow standard operating procedures or safety regulations. Problems relating to regulations and procedures (either not following them or lack of appropriate procedures) were identified in over half of aircraft damage accidents. An equally common group were in the behavioural category: spatial misjudgement, poor judgement and failure to see were involved in over half of aircraft damage accidents. Poor discipline and inadequate supervision, taken together, were involved in over a quarter of the accidents. Incorrect use of equipment and poor weather conditions were each reported in 10% or more of the accidents.

Factors contributing to equipment damage accidents followed a similar pattern: approximately two thirds involved some visuo-spatial or judgement problem (failure to see, spatial misjudgement, poor judgement or distraction); poor discipline and excess speed were also fairly common. Failure to follow regulations and procedures were represented in over half the equipment damage only accidents.

It should be noted that in most accidents several contributory factors were represented. Thus, for aircraft damage accidents many of the visuo-spatial contributory factors were instanced in accidents where standard operating procedures or safety regulations were not followed (over one third) or poor discipline was a factor.

Two broad points should be made in relation to interpretation of these preliminary results. First, the quality of the data is entirely dependent on the quality of the reporting process. These accident reports were made by ramp managers and supervisors from organizations from around the world. The constraints under which accidents were assigned to particular event- or condition-categories undoubtedly varied. While there could be no direct control over the consistency of the reporting process, consider-able efforts were made to ensure that the category labels were clear and well understood by the participants in the exercise. Secondly, identi-fication of contributory factors should not be seen as a definitive allocation of causes. The main relevance of this procedure is to make it easier to begin to design countermeasures which can help to eliminate or ameliorate a problem. It should always be born in mind that the assignment of contributory factors necessarily involves an element of interpretation and susceptibility to reporter biases. A more refined analysis, including analysis of narrative sections of the accident reports, will enable a more complete picture to be built up as the research effort continues.

## Implications for the management of ramp safety

*Responsibility for safety*

From the IATA/APRG accident reports three broad factors stand out: failure to adhere to regulations or standard procedures (or lack of suitable procedures), visual-spatial difficulties and problems of supervision and discipline. The ramp is an unstable environment in a number of different ways: from the point of view of the human factor it does not, to say the least, comprise a well-designed work environment and the problems range from complex activity and congestion in a poorly delineated environment to the problems of controlling the movement of large objects in space with inadequate information and cues. The preferred method in the industry for controlling this instability is to standardize the way the task should be done through the development of standard operating procedures. Clearly, in many instances this strategy is failing: regulations and standard procedures are not being followed and there frequently appears to be a problem of discipline or supervision. There may be a number of underlying reasons for this: for example, the standard procedures and regulations drawn up may not be appropriate or adequate, or the task may be inherently intractable to standardization. However, it is difficult not to suspect that underlying at least part of the problem there may be several unresolved and contradictory goals.

Perhaps the most common unresolved contradiction that is expressed in one form or another by ramp managers concerns the requirement for safety on the one hand and, on the other, the requirement to maintain a punctual turn-around of the aircraft, despite any delays or disruptions caused by unforeseen events. The way this contradiction is played out is frequently along the following lines: in principle, standard operating procedures allow the operative no discretion in relation to the correct method of performing the operation. But in practice, in the exigencies of the situation with a job to be done which is important to the effectiveness of the organization, the operative is *de facto* given discretion as to how closely to adhere to the standard procedure. In this implicit risk management scenario, when there is no mishap the payoff is good and this is presumably the most frequent occurrence. When there is an accident however, the payoff is less clear and may depend as much on the disciplinary climate and procedures in the organization as it does on the direct physical consequences of the accident. These consequences (costs, delay, passenger inconvenience, etc.) may well be diffused throughout the organization and exert little or no contingent control (direct or indirect) over the behaviour of the operative.

This analysis is quite congruent with the increasingly common interpretation that accidents frequently happen when the operative takes on additional discretionary responsibility, demanded by the situation to get the job done (Robinson, 1982; Taylor, 1991). Robinson (1982) develops this argument from a socio-technical systems theory perspective, attacking the 'one best way' notion of 'scientific management' and suggesting a possible way out of the double-bind confronting the worker of 'being punished for a failure to "keep things going" and punished if the methods he adopts leads to an accident'. He starts with the assumption that 'people are required in the system largely for their discretionary, decision making abilities and that they will take the necessary responsibility if given the correct organization and support' (p. 129). The merit of this argument is that it seeks to bring out into the open the contradiction between safety and production goals which in many organizations is latent and unresolved. By formulating the problem in this way, Robinson has identified a possible solution.

Robinson's perspective poses the question about the proper basis for exercising discretion in terms of the competence, experience and the explicit responsibility of the operative. It is difficult to address this question with any precision with the evidence available, but it is clear that training standards for ramp operatives throughout the world vary considerably from a nominal two or three days to a structured programme of skill acquisition which may take three years. In many airports with seasonal traffic, a large part of the ground handling workforce are casual, temporary workers, and in many organizations the explicit selection criteria relate solely to strength, physical capacity and literacy. A lot of ramp work, especially in baggage handling and loading, is regarded simply as unskilled and manual in nature. Where a minimalist approach is taken to recruitment criteria and training standards, the operative is not given the required support to exercise his discretion with responsibility. Within such a system the effective managerial control is wielded by the supervisory level. With these broad considerations in mind it is interesting to note the relative frequency of lack of discipline or poor supervision which are instanced in the accident reports. Geller *et al.* (1989) take the issue even further by arguing that effective safety management arises from the front-line worker taking over 'ownership' of safety, displacing the concept of safety as a goal that is imposed by management from above.

One of the barriers to developing this approach is the relative lack of understanding, on the part of management of the role of human and organizational factors in the generation of accidents and of how they should optimally be managed. The great majority of accidents on the ramp have a human factor involvement, and while this might be acknowledged in principle, the 'theory in action' of many ramp managers provides no satisfactory way forward. One of the most common formulations of the problem takes the form: 'If only the ramp workers would treat the ramp equipment and vehicles as they do their own car'.

## The role of accident reporting systems

Because accidents on the ramp are relatively frequent (per km driven, approximately 50 times the road accident rate) one way of developing an understanding of the role of human and organizational factors is to enable the organization to learn from its own accidents by a system for reporting, investigation and analysis and for feeding this information back into the organization. The development of the IATA/APRG accident report form may contribute to this process. For example, the form places new requirements on the participating organizations to collect data, some of which virtually none of them had collected before.

However, the move towards the development of effective data collection, detailed accident investigation and the comprehensive analysis of the consequences of accidents will pose problems for many organizations.

While having a dedicated accident investigator capable of investigating the causal pattern leading to an accident, including the human and organizational parameters, might be the best solution in principle, for most ground handling organizations this is not a possibility. The frequency and severity of such accidents in one organization would not necessarily warrant it. Thus management and supervisory grades will inevitably have an important role to play in recording and investigating accidents, particularly where there is no technical safety specialist available. One challenge for human factors specialists is therefore to develop training and support materials providing a basic accident investigation competence for those who have this responsibility.

## The use of accident information

The flow of information is the lifeblood of a safety system. One of the main sources of information is the operative performing the core production or service function. He or she is the first line of reporting of hazardous situations, incidents, accidents, damage and so on. For example, a focused strategy to elicit all safety relevant information from railway operatives has been one of the key responses of British Rail to a series of major rail accidents (Taylor, 1991). Spagnoli (1993) reports a participative system using working groups to identify causes and solutions for problems in Aeroporti di Roma; Kitagawa (1989) describes the extension of Japanese quality circles to reports on 'near-misses' and experiences of danger.

What is critical to this flow of information is the use to which this information is put. Accident reporting and recording is too often regarded as an end in itself. Any reporting system has to be designed with clear administrative ends in view. Frequently these ends are outside the organization, as when an accident report on a particular form has to be provided to some state agency. A preventive safety management strategy requires that the right range of information is collected which will enable appropriate countermeasures to be formulated and this information needs to be fed back in the appropriate format to all appropriate levels of the organization. Relevant information will include the personnel involved in an accident and the range of outcomes of the accident (damage, injury, direct and indirect costs). However the necessary information flow can conflict with the organization's established 'boundary locations' (Cherns, 1987), especially those characterized by 'segmentalism' (relatively impervious organizational boundaries). Kanter (1984) has argued that segmentalism can act as a barrier to innovation in organizations. The same argument can be applied to an organization's capacity to learn to be safer. Where ground handling is a subsidiary part

81

of a larger organization, as in an airline, it may be difficult to satisfy the principle of optimal boundary location.

Cherns (1987) observes that information in organizations has three uses: for control, for record and for action. While the first is pernicious and the second is necessary (but open to abuse), the third ensures that information is provided to those who require it, when they require it. His analysis is clearly applicable to accident and incident information: use of this information to control, in the sense of holding people punitively responsible for reported errors and mishaps, provides a massive deterrent to report freely and openly. As Johnston (in press) argues, the issue is not whether or not the transgressor is held responsible, but whether this responsibility is dealt with constructively or punitively. A further problem with the use to which accident and incident information is put is that different departments or sections may have varying goals in relation to the overall safety management system. Thus a personnel department may see information on reported accident or incident involvement as a legitimate input into a routine employee appraisal process, without considering the consequences for the operations department. Cherns states that information systems should be designed with and for the primary users – those who have the main responsibility to act and to provide information that is not automatically registered. Placing clear restrictions on the use of such information solely for purposes related to safety is often necessary. One way of doing this is through confidential reporting schemes which may be anonymous. These may be external to the organization (see Chapter 8) or within the organization, and in both cases their independence, confidentiality and imperviousness to leaks or influence need to be assured beyond question.

## The development of countermeasures

The main purpose of accident information is prevention, which involves designing effective countermeasures. Basic risk management principles involve prioritizing countermeasures, not only in terms of the severity and probability of the accidents they are designed to prevent, but also in terms of the costs of those measures and the probability of their effectiveness. While Pidgeon (1991) has argued that for large-scale technological hazards the uncertainties involved in predicting both technological and behavioural outcomes are such as to make it very difficult to model the risks involved, the data that are beginning to be produced should make it possible to develop viable risk management models. These, in turn, could significantly enhance management decision making. However in relation to the impact of unreported damage on flight safety and the risk of a major aircraft disaster, such models may be less applicable.

Human factors on the ramp is about matching, matching equipment, procedures and environmental influences, through design, to the strengths and limitations of the human operator; and matching individuals to the demands of the task through appropriate selection and training methods. It is also about managing social and communication processes so as to reinforce appropriate individual and group behaviour in achieving the goals of safe, effective and efficient ramp services.

There are three general aspects of ramp operations where design should be considered: the infrastructure (buildings, plant, facilities), the equipment and the organization. Many ground handling operations have to cope with an airport infrastructure which was designed for a volume of traffic and a size and complexity of equipment which no longer exists. Few airports have the luxury of starting from scratch in a greenfield site with adequate space and resources. Many airports are developing their infrastructure to expand their capacity within the constraints of an existing site and facilities. In this context it becomes particularly important for designers and those who commission designs to understand the consequences for safety of decisions which may be based on a sub-optimal compromise between alternatives. The designer's terms of reference, implicit or explicit, may not have as a priority ground handling operations (as opposed to passenger facilities or aircraft movements). Even without development work there are important design issues to do with the optimal use of existing facilities, for example the assignment of ramp space to various functions such as apron, roadway and equipment parking. Ashford and his colleagues (Ashford and Wright, 1979; Ashford, Stanton and Moore, 1984) have contributed greatly to the systematization of knowledge concerning airport engineering and airport operations. They provide an excellent basis for further exploration of the relationship of design to safety.

With regard to the design of equipment, the accident statistics emphasize the fact that it is at the interface between ramp equipment and aircraft (in positioning, loading, unloading, servicing, etc.) that there is the greatest potential for expensive aircraft damage and that much of this problem is to do with visual ergonomics. In the light of this evidence it is remarkable that there appear to be no industry equipment standards for feelers or sensors or aids to vision. Clearly this is an area where there is some room for the application of more advanced technology.

Organizational aspects of design such as work systems and scheduling should not be neglected. As an example of the latter, Knauth *et al.* (1983) describe the redesign of a shift system for ramp operatives to reduce the amount of night working, with the goal of improving the overall health and well-being of the workforce.

Matching the operator to the task is a further way of attempting to meet task demands, and traditionally involves the development of selection and training methods. Criteria and assessment methods for selection which have good predictive validity (in accepting suitable personnel and rejecting unsuitable ones) are not universally evident in ramp personnel management practice, where often the main consideration is to prefer a short, strong, stocky man with satisfactory vision who can drive a motor vehicle. Evidence from transportation studies suggests that there is a close association between measures of socially deviant behaviour and crash involvement on the roadway. From a recent review of the evidence (Evans, 1991) it might be concluded that poor learning of socially acceptable behaviour and insensitivity to the effects of social controls may account in part for both criminality and high-risk driver behaviour. It is also evident that young, extraverted males are more likely to be involved in traffic violations and accidents. However, before this evidence triggers a worldwide rush to recruit burly, introverted, middle-aged females for ramp work, it should be noted that the data describe associations between variables, not necessarily causal connections, and that the unreliability of measures of social deviance, extraversion–introversion and so on would inevitably mean that many potentially safe operatives would be excluded, and that a number of potential accident repeaters would be taken on. What is demonstrably needed to resolve these difficulties is a vigorous research programme.

Concerning the development of training methods in the ramp service industry, it is worth noting here that the salience of human factors considerations in matching the individual to task demands highlights the need for major developments in the human factors component of basic training, not only for ramp workers themselves but also for supervisors and managers. And it should be stressed that selection and training procedures are not necessarily simply to be administered at employee entry and then dispensed with. Selection occurs on a moment to moment basis when a particular individual is assigned to perform a specific function, thus the individual's competence at the time is obviously something that should be taken into account. A tug driver who is normally perfectly competent at push-back is not necessarily matched to the task at the end of a long and fatiguing day of baggage handling. With regard to training, over a long period of time a ramp worker might not operate a particular piece of equipment and then get into difficulties when suddenly called on to do so. Clearly some form of recurrent training and practice is necessary to minimize such problems as this.

The organizational implications of designing effective preventive countermeasures should not be underestimated. Kerkloh (1992) describes the establishment, in an airport authority, of multidisciplinary work teams crossing internal organizational boundaries (work safety, medical, technical development, engineering, personnel). The task of these teams is to develop and implement plans to redesign the workplace for ramp operatives to reduce the problem of injury and disability. Because part of the workplace is outside the organizations's boundaries (e.g. within the aircraft) the issues addressed even include developing economic incentives for carriers to install automated loading equipment. Such a radical development as this can occur within a management framework which values worker co-determination and participation in workplace, equipment and organizational design. It has a lifetime career perspective on human resource development, for example by being concerned to find productive work for operatives who are no longer capable of heavy manual work, and a commitment to develop management's focus on long-term efficiency in the context of an overall cost-benefit analysis.

Given the variety of types of ground handling organizations and the environments in which they exist, the approach reported by Kerkloh is clearly not applicable 'off-the-shelf' to any ground handling organization. However, successful tackling of the problems of safety on the ramp has to involve all levels of the organization from senior management down through supervisory levels to the operatives themselves. At all these levels there is an educational and training function to create a better awareness of human and organizational factors in ramp safety and the competence to manage these factors effectively. This should encompass consideration of a broad range of countermeasures. It is also important to develop a preventive strategy over a medium-term time frame as opposed to a short-term reactive approach. One initiative which is addressing these objectives is SCARF (Safety Courses for Airport Ramp Functions), which is a five nation joint training action under the EC COMETT programme involving four universities and eight aviation organizations who have a role in ground handling (McDonald *et al.*, 1993). This three year project is developing courses for operatives, their trainers, supervisors and management on safety in ground handling with a particular focus on the human factor.

## References

Ashford, N.J., Stanton, H.P.M. and Moore, C.A. (1984), *Airport Operations*, New York, Wiley.

Ashford, N.J. and Wright, P.H. (1979), *Airport Engineering*, New York, Wiley-Interscience.

Busk, I. (1992), 'Economic consequences of ramp accidents involving aircraft', *International Air Transport Association Seminar 'Creating the Safety Culture on the Ramp'*, Miami, FL, 26–27 October.

De Boeck, F. (1992), 'Practical guidelines for activating ramp safety', *International Air Transport Association Seminar 'Creating the Safety Culture on the Ramp'*, Miami, FL, 26–27 October.

Cherns, A. (1987), 'Principles of sociotechnical design revisited', *Human Relations*, **40** (3), 153–162.

Evans, L. (1991), *Traffic Safety and the Driver*, New York, Van Nostrand Reinhold.

Geller, E. S., Lehman, G. R. and Kalsher, M.J. (1989), *Behavior Analysis Training for Occupational Safety*, Virginia, Make-A-Difference Inc.

Hanus, R. (1992), 'TQM and Safety', *International Air Transport Association Seminar 'Creating the Safety Culture on the Ramp'*, Miami, FL, 26–27 October.

Hawkins, F. H. (1987), *Human Factors in Flight*, Aldershot, Gower.

HSE (1991), *Successful Health and Safety Management*, London, HMSO.

ICAO (1989), *Human Factors Digest No.1: Fundamental Human Factors Concepts*, ICAO Circular 216-AN/131, Montreal, International Civil Aviation Organization.

Johnston, A. N. (in press), 'Blame, punishment and risk management', in C. Hood, D. Jones, N. Pidgeon and B. Turner (eds), *Accident and Design*, London, University College Press.

Kanter, R. M. (1984), *The Change Masters*, London, Allan and Unwin.

Kelman, S. (1981), *Regulating America, Regulating Sweden: A Comparative Study of Occupational Safety and Health Policy*, Cambridge, MA, MIT Press.

Kerkloh, M. (1992), 'The socio-psychological consequences of personal injuries on the ramp – the Frankfurt experience', *International Air Transport Association seminar 'Creating the Safety Culture on the Ramp'*, Miami, FL, 26–27 October.

Kitagawa, H. (1989), 'The role of safety managers and safety officers – an overview', *OECD Workshop on Prevention of Accidents Involving Hazardous Substances: Good Management Practice*, Berlin, 22–25 May. (Cited in B. A. Turner, N. Pidgeon and D. Blockley (1989), 'Safety culture: its importance in future risk management', *Second World Bank Workshop on Safety Control and Risk Management*, Karlstad, Sweden, 6–9 November.)

Knauth, P., Eichorn, B., Löwenthal, I., Gärtner, K. H. and Rutenfranz, J. (1983), 'Reduction of nightwork by re-designing of shift rotas', *International Archives of Occupational and Environmental Health*, **51**, 371–379.

McDonald, N., White, G., Fuller, R., Walsh, W. and Ryan, F. (1993), 'Safety in airport ground handling', *Seventh Aviation Psychology Symposium*, Columbus, OH, 26–29 April.

National Safety Council (1975), *Defensive Driving Course Instructor's Manual, Airport Ramp Service Vehicles Supplement*, Air Transport Services, National Safety Council Driver Improvement Program, Washington, National Safety Council.

Pidgeon, N. (1991), 'Safety culture and risk management in organizations', *Journal of Cross-Cultural Psychology*, **22** (1) 129–140.

Robinson, G. H. (1982), 'Accidents and sociotechnical systems: principles for design', *Accident Analysis and Prevention* **14** (2), 121–130.

Ruckert, A., Romhert, W. and Pressel, G. (1992), 'Ergonomic research study on aircraft luggage handling', *Ergonomics*, **35** (9), 997–1012.

Spagnoli, O. (1993), 'New training methods', *Airports Council International Seminar on Ramp Safety: The Ongoing Challenge*, Rome, Italy, January.

Taylor, R. K. (1991), *Safety on British Rail – Research into the Human Factor*, London, British Rail Research.

Wynne, R. (1989), 'Workplace action for health', working paper no. EF/WP/89/30/EN, Dublin, European Foundation for the Improvement of Living and Working Conditions.

# 5 Human error in aircraft maintenance

*David A. Marx and R. Curtis Graeber*

## Introduction

Charley Taylor and Orville Wright: most of us know Orville Wright as the inventor and pilot of the first 'powered' aircraft, the Wright Flyer. Early in the development of the Wright Flyer, Orville Wright took his airplane to Fort Meyer, Virginia, for flight testing. On that particular trip, Orville brought along an assistant to help maintain the aircraft. It was Charley Taylor, the world's first dedicated Aviation Maintenance Technician (Harris, 1991). Since then, the roles of maintenance and the maintenance technician have grown to an enterprise consuming approximately 1300 USD per departure and 11% of US airline operating expenses (McKenna, 1992).

While the Aviation Maintenance Technician (AMT) has been around as long as flight itself, the AMT has not been the subject of much human factors scrutiny. Instead, most progress in the area of human performance has been accomplished through the qualitative experience of aircraft designers and maintenance specialists, rather than through the analytic techniques of aviation psychologists (Marx, 1992). From Ross McFarland's early books on aviation human factors (1946, 1954), to Wiener and Nagel's *Human Factors in Aviation* (1988), the role of AMT is conspicuously absent. There are numerous possible reasons for the lack of human factors focus in this area, not the least of which is the often-quoted statistic showing that the majority of aircraft accidents can be attributed to pilot error (Boeing, 1993).

Pilots, however, are not the only humans within the aviation system who err. Undoubtedly, Charley Taylor and Orville Wright were faced with the issue of maintenance error and its effects upon Orville's own well being. This chapter will focus on the challenges presented by human error in aircraft maintenance. While there are a variety of human factors issues in aircraft maintenance, it is 'maintenance error' where maintenance, systems safety, and the role of the aviation psychologist all converge.

## Maintenance error

Maintenance error can be defined as an unexpected aircraft discrepancy (physical degradation or failure) attributable to the actions of an AMT. We use the word 'attributable' because maintenance error can take two basic forms. In the first case, the human error results in a specific aircraft discrepancy that was not there before the maintenance task was initiated. Any maintenance task performed on an aircraft becomes the opportunity for human error which can result in some unwanted aircraft discrepancy. Consequently, it is not rare for the maintenance technician who works on an aircraft to actually do more damage than good. Examples include incorrect installation of a line-replaceable unit, failure to remove a protective cap from a hydraulic line before re-assembly, or damaging an air duct used as a foot hold while gaining access to perform a task. The second type of maintenance error results in damage being undetected while performing a scheduled or unscheduled maintenance task designed to detect aircraft degradation. Examples of this second type of maintenance error might include a structural crack which went unnoticed during a visual inspection task or a discrepant avionics box that remains on the aircraft because incorrect diagnosis of the symptoms led to removal of the wrong box.

Unfortunately, within these definitions of error is an underlying theme of who is to blame. This is often the case because of our natural tendency, in searching for some root cause, to look for the last person who touched the broken object. In the case of aircraft maintenance, many will look to the technician even when his instructions were wrong or his training or environment inadequate. It must be understood that maintenance error, while discussed within this chapter in terms of the technician working on the aircraft, may actually find its root cause in the errors of the aircraft designer, maintenance manager, work card developer or fellow technician.

Catastrophic events such as the Aloha Airlines 737–200 structural failure and the United DC-10 centre engine disc failure, both of which implicated maintenance as the probable cause of the accidents, have focused increased attention on maintenance error (NTSB, 1989, 1990). Accident statistics reveal that maintenance and inspection are factors in 12% of major aeroplane accidents (see Table 5.1).

**Table 5.1   Significant accident causes and percentage present in 93 major accidents**

| Cause of accident | Presence (%) |
|---|---|
| 1. Pilot deviated from basic operational procedures | 33 |
| 2. Inadequate cross-check by second crew member | 26 |
| 3. Design faults | 13 |
| 4. Maintenance and inspection deficiencies | 12 |
| 5. Absence of approach guidance | 10 |
| 6. Captain ignored crew inputs | 10 |
| 7. Air traffic control failures or errors | 9 |
| 8. Improper crew response during abnormal conditions | 9 |
| 9. Insufficient or incorrect weather information | 8 |
| 10. Runway hazards | 7 |
| 11. Air traffic control/crew communication deficiencies | 6 |
| 12. Improper decision to land | 6 |

(Source: Sears, 1986)

Many, if not most, of these accidents involved human error. In some cases the error itself was the primary cause of the accident, whereas in other cases the resulting maintenance discrepancy was just one link in a chain of events that led to the accident.

The British Civil Aviation Authority (1992) has published a listing of the most frequently occurring maintenance discrepancies (see Table 5.2).

**Table 5.2   Top eight maintenance problems listed in order of occurrence**

1. Incorrect installation of components
2. The fitting of wrong parts
3. Electrical wiring discrepancies (including cross-connections)
4. Loose objects (tools, etc.) left in aircraft
5. Inadequate lubrication
6. Cowlings, access panels and fairings not secured
7. Fuel/oil caps and refuel panels not secured
8. Landing gear ground lock pins not removed before departure

(Source: UK CAA, 1992)

Not surprising, most of these discrepancies involve human error of the first type described above (maintenance-induced discrepancy). The second type of error discussed above (failure to detect degradation) does not prominently show up in the statistics described in Table 5.2. To better understand the relative importance of this second error type, studies have shown that as many as 60% of the avionics boxes removed from aircraft are subsequently confirmed as not having caused the aircraft problem

(Ruffner, 1990). Most disturbing about this statistic is the corollary: in many cases of fault diagnosis, the faulty part remained on the aircraft for at least one more flight.

While the focus of this chapter is aircraft safety, it must be recognized that maintenance error has significant economic implications as well. In one study, one-third of all equipment malfunctions were attributed directly to poor prior maintenance or improper application of a maintenance procedure (Ruffner, 1990). In another recent study of commercial airline operation, it was found that 50% of engine-related flight delays and cancellations were caused by improper maintenance. Each flight line delay costs the customer approximately 10000 USD per hour, and each flight cancellation costs the customer approximately 50000 USD (Gregory, 1993).

## The aircraft maintenance technician

To gain a better understanding of maintenance error, it is essential to

**Table 5.3  Comparison of system maintenance and system operation characteristics**

| Characteristic | Maintenance | Operation |
| --- | --- | --- |
| Task initiator | Systems failure | Mission events |
| Types of tasks | Mostly discrete | Discrete, continuous |
| Temporal attributes of tasks | Often sequential, Seldom concurrent | Sequential, concurrent |
| Problem solving requirements | Moderate–High | Low–Moderate |
| Environment conditions | Highly varied | Relatively constant |
| Required workspace | Large | Small |
| Required postures | Highly varied | Relatively constant |
| Physical accessibility requirements | Moderate–High | Low–Moderate |
| Visual accessibility requirements | Moderate–High | Moderate–High |
| Physical strength requirements | Moderate–High | Low–Moderate |
| Mobility requirements | Moderate–High | Low–Moderate |
| Tool manipulation requirements | Moderate–High | Low |
| Workload components | Visual, kinaesthetic, cognitive, physical | Visual, auditory, kinaesthetic, cognitive, psychomotor |
| Contributors to workload | Time pressure | Concurrent tasks |

understand the tasks and environment of the aircraft maintenance technician. While investigating human factors methodologies for improving the maintainability of military aviation systems, John Ruffner (1990) compiled a comparison of job characteristics relating to system operation and system maintenance personnel (see Table 5.3).

Probably the most significant difference between system maintenance and system operation is what Ruffner refers to as the 'task initiator'. In a line maintenance environment, a sizable portion of the maintenance is driven by the failure of very complex systems, making the maintenance system itself stochastic, i.e. driven by random events (Inaba and Dey, 1991). For a given line technician, the average time between maintenance actions on the same equipment can be months or even years. It is not unusual for a maintenance technician who receives and dispatches aircraft at a line station to have work assignments on ten different aeroplane types.

These conditions, plus the seemingly random nature of system failures (at least through the eyes of a single technician), make the maintenance technician's job one that requires a high degree of problem solving capability across a wide range of maintenance situations usually under significant time pressure. This fact becomes significant to aviation psychologists who rely heavily on task analysis and scenario development. Unlike the normative flight crew environment, where the majority of tasks are repeated each flight, the line maintenance technician's job is often hard to model and, therefore, difficult to study in a systematic manner. However, AMT human factors issues in the shop or hangar maintenance environment, where tasks are more repetitive and structured, may benefit from the methodologies utilized for the analysis of human factors issues for pilots.

## Maintenance error versus pilot error

Pilot error and maintenance error, while both 'human errors', have some unique characteristics. Push the wrong button or pull the wrong knob, and the pilot will nearly always experience the effects of the error before the aircraft completes its flight. If a pilot is blamed for an accident or incident, that individual is nearly always one of the pilots on the flight deck at the time of the accident or incident. While this important characteristic may seem obvious for flight crew error, it does not always apply to aircraft maintenance error.

In contrast to the 'real-time' nature of pilot error, maintenance errors are often not identified or caught at the time the error is made. In some cases, the maintenance technician who makes an error may never know what he or she has done. Detection of the error could occur days, months or years after the error was made. In the case of the 1989 Sioux City DC-10 engine disc failure, the suspected inspection failure occurred 17 months before the aircraft accident.

When maintenance error is detected, usually through some system malfunction, we often only know the resulting aeroplane discrepancy. What is rarely known is *why* the error occurred. Unlike the flight operations environment, in aircraft maintenance there are no equivalents to the cockpit voice recorder or the flight data recorder to preserve the details of the event. Additionally, maintenance self-report programmes have not progressed to the sophistication of those within the flight environment. (Because of the differences between pilot error and maintenance error discussed above, maintenance self-report programmes may never be able to reach the sophistication of those within the flight environment.) Thus, in most cases we simply do not have the proper data to discuss maintenance error in terms of a specific type of human error. Instead, we resort to discussing such errors in terms of the aircraft discrepancy. For example, consider a New York-based line maintenance technician who has forgotten to install an anti-vibration clamp on an engine-mounted hydraulic tube. Three months later, the tube deteriorates from fatigue in flight, resulting in the loss of one hydraulic system. Upon landing in London, another maintenance technician inspects the engine and finds that the anti-vibration clamp is missing. Does he know why? Most likely not; because the error had taken place three months ago in New York. Consequently, the human error gets recorded as 'clamp missing'.

This irretrievability of 'scene of the error' causal data represents a significant problem for airlines, manufacturers and human performance investigators who are accustomed to searching for a specific cause. Looking at Table 5.1, it can be seen that pilot error has been broken down into specific performance failures such as poor crew coordination and miscommunication with air traffic control. On this same chart, however, maintenance and inspection receives only one line. Notwithstanding all of the errors possible in the maintenance of a complex aircraft, every maintenance-related accident falls within that single line. Except for major accidents that are exhaustively re-created, the identification of maintenance-related causal factors beyond this highest level is rarely seen.

## An example: The installation error

As shown in the UK CAA study, installation error represents one of the most frequently occurring maintenance errors. We will discuss one installation error, in particular, which aptly illustrates the role that human factors awareness can have in helping to prevent maintenance-induced aircraft discrepancies and thereby improve overall system reliability and safety.

On May 5, 1983, Eastern Airlines Flight 855 departed Miami International Airport *en route* to Nassau, Bahamas. Twenty minutes into this short flight, while descending through 15 000 ft, the low oil pressure light for the No. 2 engine illuminated. As a precaution, the No. 2 engine was shut down, and the captain decided to return to Miami and land. Approximately three minutes later the pilot radioed the following message to Miami Center:

> and Miami Center Eastern ah eight fifty-five ah we have ah some rather serious indications ah we have ah indications of all three oil pressures on all three engines ah ah down to zero we believe it to be faulty indications ah since the chances of all three engines having zero oil pressure and zero quantity is almost nil however that is our indications in the cockpit at the present time.

While backtracking to Miami, the flight engineer contacted airline maintenance personnel to determine if there was a common electrical source which could affect the engine instruments. The airline's Miami Technical Center called back to Flight 855 to notify them that the No. 2 AC bus was the common power source for the oil quantity instruments. The flight engineer checked the appropriate circuit breaker and found no discrepancies. Eighty miles from Miami, the No. 3 engine failed. At this point the crew realized that the indications of zero oil pressure were correct. Fifty miles from Miami, the third and final engine failed. Frantic attempts were made to restart all three engines. Twenty-two miles from Miami, while descending through 4000 ft, the crew was fortunately able to restart the No. 2 engine. Eastern Flight 855 made a one-engine landing at Miami International Airport with the No. 2 engine producing a great deal of smoke.

Those studying human factors applications to aircraft maintenance are quite familiar with this classic case of maintenance error. Upon investigation, the US National Transportation Safety Board (NTSB) found that all three master chip detector assemblies had been installed without the required O-ring seals (NTSB, 1984). The chip detectors are magnetic elements within the oil system which can be removed and inspected for metal chips with the expectation of catching engine deterioration before an engine failure (Figure 5.1). The night preceding the accident, a foreman assigned two mechanics a work card requiring removal of all three engines' master magnetic chip detectors and installation of new chip detectors. At 1.30 am, one mechanic went to the cabinet in the foreman's office to pick up three master chip detectors to replace the three he planned to remove from the aeroplane. When the mechanic did not find any chip detectors in the cabinet, he went to the stock room and drew three chip detectors from supply. The mechanic gave one chip detector to the other mechanic to install that chip detector on the centre engine which was more difficult to access. The first mechanic installed the other two chip detectors on engines 1 and 3. After installation of these two chip detectors, the mechanic went to the cockpit and motored the three engines for about ten seconds. The engines were then inspected for observable leaks at the chip detectors.

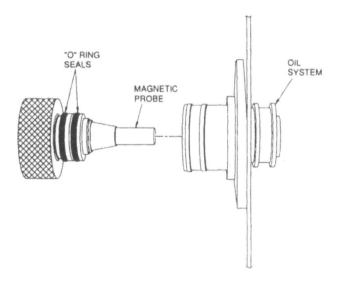

**Figure 5.1   Engine chip detector**

Unfortunately for the passengers, crew and mechanics involved, the chip detectors taken from the store room did not have the O-rings installed. The work card given to the mechanics, however, did include a step to install the O-ring seals. For many people who have analysed this accident, the cause is obvious. The NTSB concluded that the master chip detectors were installed without O-ring seals 'because the mechanics failed to follow the required work card procedures, and because they failed to perform their duties with professional care expected of an A&P mechanic.' Notwithstanding the conclusions of the NTSB, one can take a different look at this accident and arrive at different conclusions.

Chip detector installation was not a new task for these mechanics. It was estimated by the airline that each of the two mechanics had successfully performed over 100 chip detector changes. Additionally, the two mechanics involved in the incident were given a work card that specifically required that they install the O-ring seals on the chip detectors. However, at the Miami base, there were informal procedures existing which were not written on the work card. While the NTSB report is not entirely clear, it appears that the well-meaning mechanics of the first shift were inserting the O-rings on the replacement chip detectors beforehand as a convenience for the night crew. Thus, when the night crew changed the chip detectors on the aircraft, they never expected that they had to perform the O-ring installation step of the procedure. After the accident, the general foreman was asked if he knew of any mechanics who routinely inserted O-ring seals on the replacement master chip detectors. He said that he did not know of anyone who did. The night general foreman was apparently aware of previous master chip detector installation problems and knew that mechanics were not routinely inserting O-ring seals on replacement master chip detectors, yet he failed to take positive action to require compliance with the procedure as prescribed. The Director of Line Maintenance was unable to explain why the mechanics had never replaced the O-ring seals. One finding of the NTSB was that the mechanics 'had the responsibility to install O-ring seals'. Quite ironically, the subsequent finding in the NTSB report states that 'the mechanics had always received master chip detectors with "installed" O-ring seals and had never actually performed that portion of the requirements of work card 7204.' From a human factors perspective, these two findings raise a simple question. Can we expect reliable performance from a technician who, for the first time, must perform a task for which he has always been responsible but has never been 'allowed' to perform?

Master chip detector installation problems were not new for this airline, or the industry as a whole. In the two years prior to this accident, the airline recorded eleven cases of oil loss or O-ring seal problems associated with the completion of the chip detector installation work card. In the nine months preceding the incident, five events occurred at Miami, two of which led to engine shutdowns. Investigation by the airline's engineering department confirmed that essentially all airlines were using similar procedures. The airline concluded that the procedure was still appropriate and that the installation problems were the result of 'personnel errors'. In December 1981, after three L-1011 aeroplanes experienced in-flight engine shutdowns due to loss of oil, the airline issued a special training procedure and a revision to the procedure as attempts to both reduce the frequency of mis-installation and capture the error after it occurs. The special training procedure, intending to reduce the error rate, was never seen by the two mechanics involved in this accident.

Additionally, the engine motoring check, intended to capture any mis-installed chip detectors, was inadequately specified on the work card, thus allowing the mechanics to perform the check and not discover the missing O-rings.

For the maintenance manager, the engine manufacturer, and the federal regulator, the issue at hand was how to manage the error of chip detector mis-installation. A few different intervention strategies appeared to be both applicable and effective, although to different extents. The special training procedure was implemented as an error reduction strategy, apparently with limited effectiveness since many chip detectors were mis-installed even after the special training procedure was issued. The task of engine motoring, to be performed after chip detector installation, was put in place specifically for detecting mis-installation of the chip detectors (error capturing). Unfortunately, in this case the instructions for the engine motoring were vague and did not specifically state the required time the engines must be motored to discover any mis-installation.

As a result of the Miami incident, there has been a regulatory change in the US intended to prevent such an accident from occurring again. Today, for Extended Twin Engine Operations (ETOPS), the US Federal Aviation Administration (FAA) prohibits maintenance tasks from being performed on multiple elements of any ETOPS-critical system. In simpler terms, the FAA prohibits applying the same maintenance task to both engines prior to a given flight. This intervention strategy (error tolerance) is based on the assumption that the error will occur, and that the safety of the flight is best maintained by ensuring that the error can affect only one of the engines. Ironically, this rule today only applies to twin-engine extended range operations, and not three-engine (such as the L-1011 in the 1983 Miami event) or four-engine aeroplanes.

## Challenges for the future

Relative to its potential for harm, human error in aircraft maintenance has been remarkably well-managed. Lessons learned over the past 90 years of aviation have rapidly made their way into the methods of aircraft and maintenance systems design. Based upon the safety and economic data provided earlier, however, there appears to be significant potential for improvement.

As previously described, the situational complexity of maintenance error can range from errors as simple as a mechanic not re-installing the oil filler cap to error scenarios as complex as the 'one-in-a-million' loss of all three engines in the Miami L-1011 accident. On one end of this spectrum are the significant breakdowns within the maintenance system. These are the errors that cause aeroplanes to crash, typified by some strange and unforeseen sequence of events. In most cases, not only was the primary maintenance task mis-performed, but many levels of defence had to be penetrated in order for the error-tolerant maintenance system to break down so significantly.

Throughout the middle of the spectrum are those systematic errors that can be more readily tied back to some deficiency in the design of the aircraft or the management of the maintenance process. The maintenance community has become increasingly adept at dealing with these errors through re-design and process change. For example, line-replaceable units (LRUs) are most often designed today with different size or shape electrical and fluid connectors to eliminate cross-connection errors upon re-assembly. On the operational side, maintenance departments have established sophisticated systems to ensure that work started on one shift is properly turned over to the next shift.

At the other end of the spectrum are the everyday errors (e.g. B-nut not torqued, lockwire not installed, access panel not secured). Such errors continue to frustrate designers and maintenance managers because they are associated with such elemental pieces of equipment that re-design of the equipment or changes to the maintenance system seems impractical, if not impossible. These errors are seldom life threatening; however, their operational and economic impact can still be significant. For example, suppose a maintenance technician forgets to torque a screw or nut that he has installed only finger-tight. What is the change that can reduce the error rate? Would it be to remove all nuts and screws from the aeroplane? Would it be to require duplicate torquing for all nuts and screws on the aircraft? Regardless of the economic environment faced by manufacturers or commercial airlines, neither of these changes would have much chance of implementation. These errors are not so much the result of system deficiencies, but more so a reflection of inherent limitations in the technology of both aircraft design and maintenance systems. Theoretically, to reduce removal and installation errors we would need to design aeroplanes with just a few components, rather than the three or four million parts within a typical large commercial jet transport. However, today's technology requires that we must use B-nuts and lockwire on the aircraft. As a result, we can expect that sooner or later, due to improper execution of a maintenance task, each of these parts will inadvertently be left off a departing aeroplane.

To take the next significant step in maintenance error reduction, three issues should be worked. First, maintenance data need to be put into a form that will allow us to study the human performance aspects of maintenance. Secondly, the cultural gap between the maintenance community and the aviation psychologist needs to be narrowed. Thirdly, methods and tools should be developed to help aircraft designers and maintenance managers address the issue of human error in a more analytical manner.

## Maintenance data and error classification schemes

Much of the work in the theory of human error revolves around the classification of error. For the cognitive psychologist, there are many classification schemes: slip/lapse/mistake; errors of commission and omission; skill-based, rule-based and knowledge-based errors; and systematic versus random errors. Each of these classification schemes is applicable to errors in any context, including aircraft maintenance. While these classifications are intended to impart order to what appear as random errors, they have, for the most part, been absent from use within the aircraft maintenance community. The problem for those in the 'real world' of maintenance is that establishing the type of error provides little practical help in determining the underlying cause of an error (Barnett, 1987). Without a link between theoretical error classifications and real-world management of maintenance error, it is of little help for the maintenance community to know the distinction between slips, lapses and mistakes.

Another approach for error classification which has been embraced by the aviation industry is to focus on cause or contributing factors. This approach, after all, is how the industry arrived at the startling statistics which attributed such a high percentage of accidents to pilot error. While this may be appropriate for equipment failure, this approach has significant limitations when applied to the issue of human error. In 1991, Boeing conducted a study of maintenance-related accidents occurring over the previous ten years. By reviewing available data, analysts assigned contributing factors to each accident within each of the seven broad categories of performance-shaping factors listed below.

1. Tasks and procedures
2. Training and qualifications
3. Environment/workplace
4. Communication
5. Tools and test equipment
6. Aircraft design
7. Organization and management.

In an attempt to guard against the temptation to place blame, the maintenance technician was deliberately excluded as one of the issues. The overall result, however, was a subjective list of causes placed within one or more of the seven performance-shaping categories. Consequently, placing 'blame' emerged as one of the undesirable, and unavoidable, aspects of this type of analysis, partially because of the sketchy data regarding the human performance aspects of each accident. Two significant issues emerged from this analysis:

1  Can particular biases which analysts are likely to bring to an investigation due to experience, training, or expertise be controlled? For example, would a maintenance instructor be more likely to identify training as a deficiency in a particular accident or incident?
2  What insights can be gained from a study that relies heavily on subjective assessment?

Both of these questions point to the need for improved human performance data collection and investigation techniques that decrease reliance on subjective assessments and are understood and embraced by aircraft designers and maintenance managers.

Historically, it seems that investigations into human performance have simply traced error back to the careless and unprofessional work habits of the individual involved. According to Rasmussen (1990), during investigation of accidents, backtracking occurs until all conditions pertinent to the accident are explained by *abnormal, but familiar events or acts*. If an aircraft component fails, a component fault will be accepted as the prime cause if the failure mechanism appears 'as usual'. After all, human error is something familiar to the investigator: to err is human. Therefore, the investigation quite often stops at the identification of the human who erred. (Thus the often quoted statistic showing the majority of aircraft accidents being caused by pilot error.)

How can we improve our human performance investigations to eliminate these premature judgements against the human? While we are not attempting to discount individual responsibility within a mishap, system safety is best served if focus is put on those elements within the system that are manageable. As Reason points out, what is going on inside the heads of the maintenance workforce is often the hardest of factors to manage (personal communication, June, 1993). Thus, to conduct analyses that will provide value towards system improvement, we must look for attributes of maintenance error that do not simply point to the maintenance technician involved and do not require subjective assessments of deficiency. We must look for factual threads among accidents, incidents and events that will allow members of the maintenance community to work together to improve the overall margin of safety.

This brings us to the third, and most appealing, approach to studying error, which relates to the maintenance process or behavioural task, rather than to the actual human error or causal factors. The UK CAA study, discussed earlier, represents this approach. At the highest level of maintenance processes, for example, we may identify errors associated with:

- Equipment removal
- Equipment installation
- Inspection
- Fault isolation/troubleshooting
- Repair
- Servicing.

Classifications of maintenance error based upon the process or task involved can provide tangible near term benefits. For example, the Aloha 737 structural failure in 1987 led to heightened awareness of the human factors associated with visual structural inspection (NTSB/AAR-89/03). As a result, the US FAA is spending a significant portion of its maintenance human factors funding on non-destructive inspection. However, would an analysis of the safety or economic cost of human error in maintenance point to improper installation as the most significant error? If we look back at the maintenance-related accidents over the last forty years, would improper fault isolation be a larger cause of accidents than improper inspection?

A more in-depth example of this approach for analysing and classifying human error in aircraft engine troubleshooting has proved beneficial to the design of maintenance training systems (Johnson and Rouse, 1982). In this case, the errors were classified according to information processing steps within a particular task of troubleshooting. The basic categories were: observation of system state, choice of hypotheses, choice of procedures and execution of procedures.

This process or behaviourally-oriented classification avoids the pitfalls associated with the cause or contributing factors approach discussed earlier. There is less 'blame' placed within this classification scheme as compared to the previous approaches discussed. Rather than reacting defensively, most people will view this type of analysis as a means of generating simple facts, while pointing the way for improvements within the process.

## Error prevention strategies

In addition to error classification, prevention strategies can also be classified. Classification of error prevention strategies in maintenance is important because it helps to widen the visibility of tools that can be utilized by designers and maintenance managers in the management of maintenance error. We would like to propose three classes of strategies to manage human error in the maintenance of aircraft. Each of these classes is defined in terms of its method for controlling error.

1  *Error reduction.*   Error reduction strategies are intended to intervene directly at the source of the error itself. Examples of error reduction strategies include improving access to a part, improving the lighting in which a task is performed, and providing better training to the maintenance technician. Most error management strategies used in aircraft maintenance fall into this category.

2  *Error capturing.*   Error capturing assumes the error is made. It is a method used to 'capture' the error before the aircraft has departed. Examples of error capturing strategies include post-task inspection, verification steps within a task, and post-task functional and operational tests.

3  *Error tolerance.*   Error tolerance refers to the ability of a system to accept an error gracefully. In the case of aircraft maintenance, error tolerance can refer to both the design of the aircraft itself as well as the design of the maintenance system. Examples of error tolerance include the incorporation of multiple hydraulic or electrical systems on the aircraft (so that a single human error can only take out one system), and a structural inspection programme that allows for multiple opportunities to detect a fatigue crack before it reaches critical length.

Of the three classes of prevention strategies listed above, only error reduction addresses the error directly. Error capturing and error tolerance are more directly associated with system integrity. From a system safety perspective, human error in maintenance does not directly or immediately cause an aeroplane to be unsafe. Until maintenance technicians are working on aircraft in-flight, this will always be the case. It is the aeroplane being *dispatched* with a maintenance-induced discrepancy that is the real cause for concern.

## Bridging the gap: The maintenance community and the aviation psychologist

Over the past 15 years, the aviation psychologist and the pilot community have grown to speak in a common language. Today, much of the human factors work in the flight deck arena is accomplished through the

101

interdisciplinary teaming of pilots, engineers and aviation psychologists. Concepts such as mode error and Crew Resource Management (CRM) have become common ground on which the aviation psychologist and the operational community could work together to improve system safety.

With few exceptions, however, the aircraft designer, the maintenance technician, and the aviation psychologist are still worlds apart. If we look at the Eastern Airlines chip detector example, we can ask if the aviation psychologist would have been able to identify better intervention strategies than those undertaken by the airline. Reason, who has been addressing the issue of maintenance error, points out that much of the human factors effort to date, especially in aviation, has been directed at improving the immediate human-system interface (Reason, 1990). Error reduction has been the basis for effective human factors activities. The chip detector mishap, though, was just one of the everyday errors that involve relatively elemental components of the aircraft that have little chance of being changed. Reason contends that the most productive strategy for dealing with active errors is to focus upon controlling their consequences rather than striving for their elimination. This is the approach taken by the US FAA in requiring the staggering of engine maintenance tasks when flying under ETOPS rules.

With the ultimate goal of reducing maintenance-caused accidents, the aviation psychologist must move beyond man–machine interface to more of a systems analysis approach. For example, there are two major steps within error analysis. The first step, which we will call 'contributing factors analysis', is concerned with understanding why the error occurred. This step is well-suited to the skills of the aviation psychologist. For example, identifying why the AMTs in the Miami incident did not install the O-rings can be studied through a human factors perspective. The second major step in error analysis, which we will call 'intervention strategies analysis', is concerned with identifying the aircraft or maintenance system changes that should occur to effectively address the maintenance error.

Developing the strategies for addressing future occurrences of the maintenance error requires skills that often extend beyond those of the human factors engineer or psychologist. To develop specific intervention strategies requires an understanding of system constraints, criticality of the error and its resulting discrepancy, as well as error management practices unique to the domain of aircraft maintenance. For example, strategies such as the staggering of engine tasks to reduce exposure to multiple engine loss are more likely to result from interdisciplinary teaming than from any independent consultation by an aviation psychologist. If there is a lesson here, both Reason's comments and the Miami accident point to the need for aviation psychologists and the maintenance community to join together to work the issue of maintenance error. Through interdisciplinary teaming, we can begin looking at how maintenance error can be managed within the larger context of system safety.

## Methods and tools development

The preceding section discussed the attributes of teaming the maintenance community with the aviation psychologist. Since the start of aviation, those within the maintenance community have been continuously improving the safety and economy of flight. In the arena of aircraft maintenance, this has been largely accomplished without the help of aviation psychologists, through what Norman (1990) calls 'naive psychology', or that human performance analysis practised, in our case, by aircraft designers and maintenance managers representing the human by virtue of their own experience as human beings. Clearly, however, we are approaching, or have passed the point where 'naive psychology' will suffice. The design of the human interface for a sophisticated on-board maintenance system is a task that requires greater analytical skills and knowledge about human cognitive performance than those acquired through years of experience as a human being. However, as human factors practitioners increase their involvement in maintenance error analysis, we must not lose sight of the fact that the bulk of error analysis and management today, as it will be in the future, is performed by aircraft designers, manual designers, maintenance trainers and maintenance managers. Thus, the maintenance community must look to the aviation psychologist as a resource to help understand the inherent capabilities and limitations of the AMT. As a resource, the aviation psychologist should focus on the development of sound methods and tools that can be transferred to the design and operational environments. Through better methods and tools, the goal of improved error management will be achieved in a more rapid and systematic manner.

## References

Barnett, M. L. (1987), *Factors in the Investigation of Human Error in Accident Causation*, College of Maritime Studies, Warsash, Southampton, UK.

Boeing (1993), *Statistical Summary of Commerical Jet Aircraft Accidents, Worldwide Operations, 1959–1992*, Seattle, WA, Boeing Commercial Airplane Group.

Gregory, W. (1993), 'Maintainability by design', *Proceedings of the Fifth Annual Society of Automotive Engineers: Reliability, Maintainability, and Supportability Workshop*, Dallas, TX.

Harris, B. L. (1991), *Challenges to United States Tactical Air Force Aircraft Maintenance*

*Personnel: Past, Present, and Future*, Air Force Institute of Technology, Wright Patterson AFB, OH.

Inaba, K. and Dey, M. (1991), *Programmatic Root Cause Analysis of Maintenance Personnel Performance Problems*, US Nuclear Regulatory Commission, Washington, DC.

Johnson, W. B. and Rouse, W. B. (1982), 'Analysis and classification of human errors in troubleshooting live aircraft power plants', *IEEE Transactions on Systems, Man, and Cybernetics*, **12**(3), 389–393.

Marx, D. (1992), 'Looking toward 2000: The evolution of human factors in maintenance', *Proceedings of the Sixth Federal Aviation Administration Meeting on Human Factors Issues in Aircraft Maintenance and Inspection*, Alexandria, VA.

McFarland, R. A. (1946), *Human Factors in Air Transport Design*, New York, McGraw-Hill.

McFarland, R. A. (1953), *Human Factors in Air Transportation, Occupational Health and Safety*, New York, McGraw-Hill.

McKenna, J. (1992), 'Airlines study human factors role in boosting efficiency of mechanics', *Aviation Space & Space Technology*, January 27, 54–55.

NTSB (1984), *Aircraft Accident Report, Eastern Air Lines, Inc., L-1011, Miami, Florida, May 5, 1983*, (NTSB/AAR-84-04), Washington, DC, National Transportation Safety Board.

NTSB (1989), *Aircraft Accident Report, Aloha Airlines, Flight 243, Boeing 737-200, N73711, Near Maui, Hawaii, April 28, 1988*, (NTSB/AAR-89/03). Washington, DC, National Transportation Safety Board.

NTSB (1990), *Aircraft Accident Report, United Airlines Flight 232, McDonnell Douglas DC-10-10, Sioux Gateway Airport, Sioux City, Iowa, July 19, 1989*, (NTSB/AAR-90/06), Washington, DC, National Transportation Safety Board.

Norman, D. A. (1990), *The Design of Everyday Things*, New York, Doubleday/Currency Edition.

Rasmussen, J. (1990), *Human Error and the Problem of Causality in Analysis of Accidents*. Riso National Laboratory, Roskilde, Denmark.

Reason, J. (1990), *Human Error*, Cambridge, UK, Cambridge University Press.

Ruffner, J. W. (1990), *A Survey of Human Factors Methodologies and Models for Improving the Maintainability of Emerging Army Aviation Systems*, US Army Research Institute for the Behavioral and Social Sciences, Alexandria, VA.

United Kingdom Civil Aviation Administration (1992), 'Maintenance error', *Asia-Pacific Air Safety*, September.

Wiener, E. L. and Nagel, D. C. (1988), *Human Factors in Aviation*, San Diego, CA, Academic Press.

# 6 Passenger safety

*Helen C. Muir*

## Introduction

The public demand for air transportation has increased steadily over the last two decades. The civil aviation industry has adapted and grown to meet this demand. The predicted further increase in demand into the next century has led airframe manufacturers to start to give consideration to designs for airframes carrying as many as 800 or 1000 passengers. The issue of passenger safety in air travel has always been a high priority within the industry, but with the introduction of increasingly large numbers of passengers in one aircraft any accident will have a major impact. For this reason the issues of primary safety, that is the prevention of occurrence of accidents, and secondary safety, that is improving the survival rate, will have major importance in the years to come.

## Aircraft accidents

The worldwide accident statistics indicate that the number of accidents has decreased over the last two decades (Boeing Commercial Airplanes, 1989). Unfortunately, the dramatic reduction in the overall accident rate has not been accompanied by an equivalent reduction in the fatality rate of those on board an aircraft which is involved in an accident (Taylor, 1989). One of the objectives of new or modified safety regulations, requirements or procedures must be to increase the probability of survival in aircraft accidents.

Aircraft accidents may be classified according to a number of criteria, the most critical of which being whether the accident was *survivable*.

Utilizing this classificatory system, it is possible to assign accidents to one of three groups:

- Those which are *fatal* or *non-survivable*. Accidents in which none of the passengers or crew survive (for example, the Air India 747 in 1985 (Indian Civil Aviation Department, 1985) and the Pan Am 747 in 1988 (Air Accidents Investigation Branch, 1990), in which the crash forces were of such severity that all on board were killed instantly).
- The *non-fatal* or *survivable*, in which all the passengers and crew survive (for example, the TriStar which overran the runway in 1985 at Leeds-Bradford Airport (Air Accidents Investigation Branch, 1985)).
- The *technically survivable*, a grouping which includes the accident at Manchester Airport in 1985 (Air Accidents Investigation Branch, 1989), and the accident at Los Angeles Airport in February 1991 (World Airline Accident Summary, 1991). Accidents in which some of the passengers or crew survive.

Since approximately 90% of aircraft accidents are categorized as survivable or technically survivable, steps have recently been taken by the UK's CAA, some European Airworthiness Authorities and the FAA in the United States in an attempt to reduce the number of fatalities. These improvements have included the introduction of floor proximity lighting and fire-blocking materials. The behaviour of passengers and their impact on emergency evacuations has also come under scrutiny. It is anticipated that with a comprehensive understanding of behaviour in highly stressful and disorientating conditions, steps can be taken to improve the probability of a successful evacuation of all passengers from the aircraft.

As yet, little research effort has centred on the impact of passenger behaviour on aircraft emergencies. However, it has been possible to extrapolate information from other disaster situations, such as building fires and earthquakes. This information, along with reports from survivors of recent aircraft accidents, has been used to build up a representation of the types of responses which passengers adopt and the impact of such behaviours within the cabin, particularly in those emergencies which involve smoke and fire.

# Passenger behaviour in aircraft emergencies

*Motivation to escape*

Whilst no two accidents can ever be the same, it is possible to learn from comparisons of the causes of accidents, their location and the environmental conditions present, the types of passengers on board and their responses to the emergency. For instance, there were many similarities between the accident at Manchester in 1985 (Air Accidents Investigation Branch, 1989) and the one which occurred at Calgary in 1984 (ICAO, 1984), in that they were both caused by an engine fire at take-off. However, they differed in one important respect, namely that at Manchester there were 55 fatalities whereas in Calgary everyone survived. We know that in some aircraft accidents everyone files out of the plane in a rapid although orderly manner – for example, in the evacuation of a British Airways 747 at Los Angeles in 1987 as a result of a bomb scare. In other accidents, however, the orderly process breaks down and confusion in the cabin can lead to blockages in the aisles and at exits, with a consequent loss of life.

From the reports of a number of accidents it is possible to build up a picture of the exits typically used by passengers who survive an emergency where there is smoke and fire (Figure 6.1).

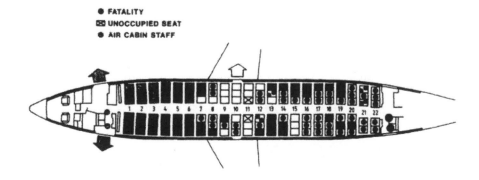

**Figure 6.1  Simulation of survivor escape door and seating fatalities**

The information in Figure 6.1 indicates:

- That some passengers exit by their nearest door, as would be expected.

- That other passengers do not exit by their nearest available door but travel for considerable distances along the cabin, e.g. extreme cases of back to front.
- That other passengers apparently near exits, do not survive.
- We also know that blockage can occur in the aisles and at exits in some accidents, when this does not occur in evacuation demonstrations for certifications.

Why and in what circumstances do these behaviours occur? Do people panic and freeze, give up, get crushed by other people from behind or around, do they have their seat backs pushed onto them? There are in fact a great many questions which as yet we are not able to answer about the behaviour of people in emergencies, including the important question of why in some accidents the passengers evacuate in an orderly manner, and in other accidents their behaviour is disorderly.

It is suggested that one of the primary reasons for the differences in behaviour between the orderly and disorderly situations must rest with the individual motivation of the passengers. In some accidents, as in the aircraft certification evacuations, all of the passengers assume that the objective is to get everyone out of the aircraft as quickly as possible, and they therefore all work collaboratively. In other emergencies, however, the motivation of individual passengers may be very different, especially in the presence of smoke and fire. In a situation where an immediate threat to life is perceived, rather than all passengers being motivated to help each other, the main objective which will govern their behaviour will be survival for themselves, and in some instances, members of their family. In this situation when the primary survival instinct takes over, people do not work collaboratively. The evacuation can become very disorganized, with some individuals competing to get through the exits. The behaviour observed in the accident which occurred at Manchester (Air Accidents Investigation Branch, 1989), and other accidents, including the fire at Bradford City, UK (Taylor, 1990), supports this contention.

### Factors influencing behaviour

Information obtained from accident experience suggests that fire and smoke are the most serious environmental factors to affect an aircraft accident. The presence of either is one of the primary reasons for initiating an evacuation, for example the Air Canada DC-9 descent into Cincinnati Airport in 1983 (NTSB, 1984) following the discovery of an in-flight fire in the aft lavatory. Equally, post-impact fire has a dramatic effect on the efficiency of the evacuation and the types of behaviours displayed by passengers.

Smoke and fire have the potential to limit the number of exits available for egress and produce toxic fumes, factors which will consequently induce certain behavioural responses.

*Limited number of exits*   If smoke and fire are outside the aircraft when an evacuation is initiated the number of exits which can be used is often limited. A limited number of exits obviously increases the demand on available escape routes. Accounts from survivors of the British Airtours Boeing 737 at Manchester in August 1985 (Air Accidents Investigation Branch, 1989) and the 737 at Los Angeles in 1991 (World Airline Accident Summary, 1991) indicated that passengers egressed over seat backs and forcibly pushed themselves towards available exits. Consequently, human blockages occurred adjacent to the overwing exit and at the vestibule area of the galley. These blockages dramatically decreased the efficiency of the evacuation, as passengers were overcome by smoke and were trampled by others in the anxiety ridden rush for the exits.

*The presence of toxic fumes*   In addition to the impairment of breathing and vision which can occur when smoke is present in the cabin, the toxic fumes which emanate from cabin fires also have the potential to influence psychological functioning, which may affect the behavioural responses of individuals in an emergency evacuation.

## Behavioural responses

The behavioural responses which can be experienced with passengers involved in an aircraft accident include fear, anxiety, disorientation, depersonalization, panic, behavioural inaction and affiliative behaviour.

*Fear*   Fear is a primary response when survival is threatened. The two reactions to fear are 'fighting' and 'flight'. We know that either of these reactions can occur in an aircraft accident. The most frequently reported reaction is that of 'flight', for instance by rapidly evacuating the aircraft. However, in response to a situation such as an in-flight cabin fire the 'fight' response may be induced.

*Anxiety*   Anxiety is a response experienced by the majority of passengers in an emergency situation. This is due to the fact that the situation is frequently perceived as potentially life-threatening. Because in this situation passengers are required to make a series of novel and difficult responses, it is hardly surprising that the optimum egress does not occur. In the emergency situation simple tasks such as unfastening the lap belt become more difficult for passengers and in fact they sometimes revert to the actions which would release a car seat belt.

*Disorientation* Disorientation can be experienced as a result of factors such as the reduction in visibility caused by dense black smoke, or the fact that the airframe may have come to rest on its side or at some strange angle. The disorientation will not only increase levels of anxiety among passengers but may also cause them to enter areas of the aircraft from which there is no escape.

*Depersonalization* People who have encountered life threatening events often say that the passage of time slows while mental activity increases. For some people, detaching themselves from the actual situation, and acting as an 'observer', means that they feel better and are able to think and respond effectively. Such depersonalization may account for the reactions of passengers during the period prior to three premeditated evacuations, studied by Robson (1973). Of the 268 individuals, cabin staff classified 35% as calm, 47% as mildly agitated, 2% as very agitated, with less than 1% exhibiting signs of panic.

*Panic* Panic may be defined as uncontrollable and irrational behaviour. In practice it is believed that true instances of panic among passengers involved in aircraft accidents are relatively rare. When the behaviour of passengers in emergencies is studied in detail and their changing perceptions of the situation as the accident develops are followed, the behaviour exhibited can usually be interpreted as a series of rational responses. For instance, it cannot be considered to be inappropriate for people to put enormous energy and drive into escaping from fire. In fact Robson (1973) indicated that less than 1% of passengers displayed responses akin to uncontrollable panic, a result which is also reported from the analysis of other disaster situations.

*Behavioural inaction* The evidence from aircraft accidents seems to indicate that there are many more instances of behavioural inaction than of panic. The analysis of four disasters led Allerton (1964) to conclude that between 10% and 25% of people did little or nothing to escape from danger. A number of fatalities on an Air Canada DC-9 accident in 1983 (NTSB, 1984) were located in seats which had been allocated to them before take-off. Equally, a number of passengers on board the taxiing Boeing 747 at Tenerife in 1977 (World Airline Accident Summary, 1977) were judged by their fellow passengers to make little attempt to escape from the burning aircraft.

*Affiliative behaviour* Affiliative behaviour involves movement towards the familiar. In an aircraft accident movement towards the familiar frequently occurs. Passengers' attachment to their hand luggage has often been observed, with many passengers insisting on taking their personal belongings with them when undertaking an emergency evacuation. It seems the perceived value of the contents obviously out-weighs the risk they believe they will encounter if they take it with them.

## Factors influencing passenger safety and survival

Snow *et al.* (1970) suggested that the factors which influence survival in aircraft accidents can be broadly classified into four groups:

1 *Configurational*   The standard features of the aircraft cabin which may influence access to exits and hence evacuation flow rates, e.g. seating, number and location of exits.
2 *Procedural*   This includes the experience and training of the crew and other rescue personnel, e.g. fire crew, which can influence the evacuation procedures.
3 *Environmental*   These are the features of the cabin and external conditions which influence the survivability and evacuation time, e.g. heat and toxic smoke in the cabin, light and weather conditions externally.
4 *Behavioural*   These include the psychological, biological and cultural attributes of individual passengers which influence their behaviour as individuals and as members of a group, e.g. sex, age, prior knowledge and experience, fitness, physical and mental health.

A large research programme has been conducted in the UK into emergency escape behaviour which has included consideration of: configuration factors, by the evaluation of changes to the space available adjacent to the exits on passenger evacuation rates; procedural factors, by a comparison of the presence of cabin staff or alternatively an acoustic tone indicating the location of the exit on evacuation rates; environmental factors, by an evaluation of the effects of non-toxic smoke in the cabin on evacuation rates; and behaviour factors, by the development of a technique which simulated the disorganized behaviour known to occur in a serious life threatening emergency such as an aircraft accident.

## Models of human behaviour in emergencies

Various models have been developed of human behaviour in fires. The model developed by Canter *et al.* (1990) appears to have the most validity for the understanding of human behaviour in fire incidents. In this model human behaviour is seen as passing through a number of identifiable stages with the possibility of various routes from one stage to the next. The fire response is seen as having three broad stages:

1   The individual receives initial cues and investigates or misinterprets these initial cues.
2   Once the fire is apparent the individual will try to obtain further information, contacts others or leaves.
3   Thereafter the individual will deal with the fire, interact with others or escape.

The model has been applied to behaviour in multiple occupancy and hospital fires. This model has not so far been applied in an aviation context. To obtain the data required for the model it would be necessary to contact the surviving passengers, crew and rescue personnel following an aircraft accident involving fire and obtain the following information: the point at which the occurrence of the fire was recognized, location of the occupant, ongoing behaviour, sequence of actions, perception of the situation including other people's behaviour, physical circumstances of the fire and time estimates. Other information about the person's related past experiences together with background information such as the role of the person, the configuration of the aircraft and the fire damage would also be required. Using the model, data from a range of types of aircraft fire could be analysed. For instance, data could be obtained from a small in-flight cabin fire which was contained and data from fire which followed a crash landing. A potential difficulty is that in the latter situation passengers typically have little more than two minutes from the onset of the fire before conditions in the cabin becoming life threatening.

A number of computer-based egress models which are designed to simulate the evacuation of large numbers of individuals from an enclosure are being developed and applied in an aviation context. The model developed by Galea *et al.* (1993) appears to be the first which has used behavioural data from simulated aircraft emergencies in order to determine the accuracy of predictions from the model. The model tracks the trajectory of each individual as they make their way out of the aircraft, or are overcome by fire hazards.

The model comprises five core interacting components – the movement, behaviour, passenger, hazard and toxicity submodels. The passenger submodel uses a collection of 22 attributes which define the passenger's physical and psychological state as well as his or her progress through the aircraft. In the present implementation of the model, the defining attributes are: name, sex, age, weight, condition, mobility, agility, travel speed, volume of air breathed, incapacitation dose, response time, drive and patience.

A long-term objective within the aviation industry will be to use egress models to slowly replace the need for human testing for the evaluation of the emergency facilities on new aircraft.

## Research into behavioural and configurational factors

In any research programme in which the objective is the human factors evaluation of safety systems for use in an emergency (from aircraft, motor vehicles, fires in buildings, etc.) the researchers are faced with a dilemma: how to introduce sufficient realism into the experimental programme, whilst at the same time not putting people at serious physical and perhaps mental risk. This trade-off between safety and realism is a challenge at the design stage of any investigation. Information from the literature regarding human behaviour in accidents indicates that where there is a serious threat to life and only a limited opportunity for escape, not only is everyone very frightened, but that it is natural for individuals to compete with each other to survive.

For both ethical and practical reasons it is not possible to put volunteers into a situation of fear and threat for the purpose of research; for instance, it would not be acceptable to take a group of volunteers on a flight and then tell them that an emergency has occurred, and video their behaviour. However, a technique used in laboratory work in behavioural science is to offer an incentive payment to subjects. This is done in an attempt to influence the motivation and performance of individuals either individually or in groups.

In the experimental programme conducted in the UK (Muir *et al.*, 1989), an incentive payment system was developed to introduce an element of competition. A series of evacuation exercises were performed in which an incentive payment was given to the first half of the subjects to leave the aircraft. Volunteers recruited from the public were paid £10 attendance fee to perform four emergency evacuations from an aircraft, with a £5 bonus paid to the first half of the volunteers to exit the aircraft on each evacuation. The experiment was later replicated without the use of bonus payments so that an evaluation could be made of the influence of additionally motivated behaviour on evacuation rates.

The main objective of the evacuation test programme was to investigate the effects of changes in cabin configuration adjacent to the emergency exits, on the rate at which passengers could evacuate an aircraft. The configurations evaluated involved a range of apertures through a bulkhead leading to floor level exits, and a range of seating configurations adjacent to a Type III overwing exit.

To introduce as much realism as possible, evacuations took place from a Trident aircraft parked on an airfield. On boarding the aircraft, volunteers were met by members of the research team trained and dressed as cabin staff. Following a standard pre-flight briefing, volunteers heard taped noise of the engine start up, taxi down the runway and finally the sound of an aborted take-off, followed by the voice of the captain telling them to undo their seat belts and get out.

The exits to be used were opened by the cabin staff, or members of the research team. This was to ensure that the evacuation times were not influenced by the variable time taken for passengers to open doors (Rasmussen and Chittum, 1986).

Ramps were mounted at the doors for passengers to walk onto. This ensured that volunteers did not hesitate before leaving the aircraft. It also reduced the potential risk of injury which could have been sustained if evacuation chutes had been employed, thus eliminating another possible compounding variable. The behaviour and evacuation times of the volunteers were recorded using video cameras (with time bases) mounted inside and around the exits from the cabin.

Regarding safety, only volunteers who claimed to be reasonably fit and were between the ages of 20–50 were recruited. On arrival all volunteers were given a medical examination. They were asked to complete a questionnaire indicating that they had (i) fully understood the purpose of the trials, (ii) that the medical information which they had supplied was correct, and (iii) that they were satisfied with the insurance cover. A doctor and the airfield fire service were present at all times. A system of alarms was introduced to stop any trial should a real emergency occur.

**Table 6.1   Competitive and non-competitive tests. Mean evacuation times (seconds) for the thirtieth volunteer to exit over the six bulkhead conditions**

| Bulkhead aperture | Competitive trials | | Non-competitive trials | |
|:---:|:---:|:---:|:---:|:---:|
| | *Mean* | *SD* | *Mean* | *SD* |
| 20″ | 26.3 | 2.9 | 25.1 | 2.0 |
| 24″ | 24.5 | 5.8 | 21.8 | 1.4 |
| 27″ | 23.2 | 7.1 | 23.7 | 2.7 |
| 30″ | 18.4 | 1.9 | 23.4 | 0.0 |
| 36″ | 17.2 | 3.1 | 21.4 | 3.4 |
| PGR | 14.7 | 1.4 | 17.6 | 0.5 |

PGR = Port galley removed

**Table 6.2  Competitive and non-competitive mean evacuation times (seconds) for the thirtieth volunteer to exit over the six overwing conditions**

| Vertical projection between seat rows | Competitive trials | | Non-competitive trials | |
|:---:|:---:|:---:|:---:|:---:|
| | *Mean* | *SD* | *Mean* | *SD* |
| 3″ | 71.4 | 15.0 | 53.2 | 1.8 |
| 6″ (OBSR) | 53.2 | 10.0 | 39.6 | 2.5 |
| 13″ | 55.9 | 10.3 | 39.9 | 3.3 |
| 18″ | 53.7 | 8.2 | 37.2 | 0.2 |
| 25″ | 54.9 | 11.5 | 40.8 | 2.7 |
| 34″ | 62.3 | 8.1 | 35.3 | 0.6 |

OBSR = Outboard seat removed

Volunteers were recruited from the public, in groups of approximately 60, to perform a series of emergency evacuations. A total of 2262 volunteers took part in the evacuations from a Trident aircraft. The evacuations times for the range of configurations tested are shown in Tables 6.1 and 6.2.

In the competitive evacuations statistical treatment of the data indicated that as the aperture between the bulkheads was increased, the evacuation rate increased, leading to a reduction in the time for the first 30 individuals to evacuate the aircraft ($F_{5,11}$ = 10.5 $p$ < 0.001). The individual comparisons of means indicated that there was a significant difference between the mean times when the aperture in the bulkhead was 27″ or less, and the mean times when this aperture was 30″ or greater.

In the non-competitive evacuations the means suggested that increasing the width of the aperture through the bulkhead leads to a small reduction in the evacuation times. Statistically there was no significant difference between the mean evacuation times for the first thirty to evacuate the aircraft ($F_{5,11}$ = 3.2 ns) through the six configurations, however this result may have been due to the fact that only two evacuations were conducted through each configuration.

The mean times also show that in the tests of the 20″ and 24″ bulkhead apertures the times for 30 people to exit were a little faster on the non-competitive trials. For the remaining widths, the times were faster in the competitive trials. No significant difference was found between the means for the competitive and non-competitive evacuations ($F_{5,1}$ = 0.2 ns) perhaps because the total of non-competitive evacuations was only 12 as opposed to the 56 competitive evacuations. Overall the results suggested that the blockages known to occur in some emergency evacuations can be significantly reduced when the width of the passageway through a bulkhead is greater than 30 inches.

The data from the competitive evacuations indicated that the seating configuration had a significant effect on the mean evacuation times ($F6,1 = 7.0\,p < 0.001$). Individual comparison of means indicated that the time for the first 30 volunteers to egress through the configuration involving a 3″ vertical projection was significantly longer than the evacuation times for all of the other configurations.

In the non-competitive evacuations the data indicated a significant difference between the mean evacuation rates for the various configurations ($F5,11 = 16.84\,p < 0.01$). Individual comparisons of means indicated that the seating configuration involving a 3″ vertical projection gave rise to significantly increased evacuation times when compared to any of the other configurations.

As can be seen from the means, the times to evacuate 30 passengers were slower in the competitive trials for all of the configurations tested ($F5,1 = 37.99\,p < 0.001$). The removal of the outboard seat meant that rather than there being a single aisle with a 6″ vertical projection adjacent to the exit, which would be comparable to the other conditions, there were two aisles with 6″ vertical projections leading to the exit.

The minimum seating configurations specified by the UK Civil Aviation Authority in Airworthiness Notice No. 79 in 1986 (6″ and 13″ vertical projection between the seat rows) were shown to have significantly increased the rate at which passengers can evacuate through a Type III overwing exit in an emergency. Blockages were found to occur in evacuations involving a 3″ vertical projection between the seats (pre AN79). The 6″ vertical projection with the outboard seat removed (an AN79 alternative) led to a rapid evacuation flow rate, but had a tendency to give rise to blockages and the opening and disposing of the exit hatch was found to be more difficult in this configuration. The results suggested that the optimum distance between the seat rows either side of the exit would involve a vertical seat projection of between 13″ and 25″. A CAA report (Muir *et al.*, 1989) provides a full description of the methodology and results obtained from the programme of competitive and non-competitive evacuations.

## Research into environmental factors

In 1989, the CAA (UK) sponsored a preliminary investigation of the influence of the presence of non-toxic smoke in the cabin on evacuation behaviour and rates (Muir *et al.*, 1989). The results from the experimental programme (see Tables 6.3 and 6.4) indicated that in a non-competitive evacuation the presence of non-toxic smoke in the cabin significantly reduced the rate at which volunteers were able to evacuate the aircraft. The initial results also appeared to indicate that in the presence of non-toxic smoke, changing the configuration of the cabin adjacent to the exits may influence the rate at which passengers are able to evacuate an aircraft in an emergency (Muir *et al.*, 1989).

**Table 6.3   Competitive and non-competitive evacuations in smoke. Mean evacuation times (seconds) for the thirtieth volunteer to exit over the four bulkhead conditions**

| Bulkhead aperture | Competitive trials | | Non-competitive trials | |
|---|---|---|---|---|
| | *Mean* | *SD* | *Mean* | *SD* |
| 24″ | 32.0 | 6.1 | 40.1 | 2.9 |
| 30″ | 28.3 | 3.1 | 38.0 | 4.4 |
| 36″ | 30.1 | 5.1 | 32.6 | 7.0 |
| PGR (72″) | 26.8 | 3.1 | 55.9 | 2.0 |

PGR = Port galley removed

**Table 6.4   Competitive and non-competitive evacuations in smoke. Mean evacuation times (seconds) for the thirtieth volunteer to exit over the four overwing conditions**

| Vertical projection | Competitive trials | | Non-competitive trials | |
|---|---|---|---|---|
| | *Mean* | *SD* | *Mean* | *SD* |
| 6″ (OBSR) | 70.7 | 16.2 | 59.7 | 9.9 |
| 13″ | 55.3 | 9.2 | 51.7 | 14.9 |
| 18″ | 57.2 | 6.7 | 49.5 | 7.0 |
| 34″ | 51.5 | 6.2 | 57.9 | 3.2 |

OBSR = Outboard seat removed

In the second phase of the evacuation research with non-toxic smoke present in the cabin, passengers competed with one another for bonus payments in an attempt to simulate an emergency situation (Muir *et al.*, 1992). It was found that as the width of the aperture between the bulkheads was increased the rate of egress tended to increase, although the differences did not reach statistical significance (see Table 6.3).

However, a significant difference in egress rates was found when four alternative vertical projections between seats adjacent to a Type III exit were tested (see Table 6.4).

The results from this programme of evacuations were compared with those from the non-competitive evacuations previously performed in non-toxic smoke. The presence of a competitive element was found to have had a significant effect upon egress rates for the evacuations through the bulkhead. However the competition did not affect the evacuation rate through the Type III exit. In the latter case, the four vertical projections between seats were found to explain the differences in escape times. The overall effect of the smoke was to reduce the speed of the evacuation. This effect was increased when passengers had relatively long distances to travel through the cabin to an operational exit. From information derived from the questionnaires administered to participants after the evacuations it was found that participants mainly relied upon the sense of touch to assist in their evacuation from the aircraft when smoke was present in the cabin.

It was concluded from this experimental programme that the use of incentive payments to produce a competitive evacuation has the potential to provide both the behavioural and statistical data required for the assessment of design options or safety procedures for use in emergency evacuations which maximize the degree of realism. Nevertheless the technique should be used sparingly since it can be potentially hazardous for volunteers.

## Research into procedural factors

*Cabin staff procedures and acoustic signals*

A recommendation of the UK Air Accidents Investigation Branch in their report on the accident at Manchester Airport in 1985 (Air Accidents Investigation Branch, 1989) was that an assessment should be made of the viability of 'audio attraction' signals to assist passengers to locate operational exits. A prototype system for producing acoustic signals at exits was developed and a series of evacuations from a stationary aircraft was conducted. The objective of this research was to determine the influence of acoustic signals at exits on the rate at which 'passengers' evacuate an aircraft in the presence of non-toxic smoke. The cabin staff did not assist with the evacuation. This gave an important opportunity for an assessment of the contribution of the cabin staff to evacuation rates, as these had assisted in the earlier trials.

The mean evacuation times are shown in Tables 6.5 and 6.6. For the

**Table 6.5    Mean evacuation times (seconds) for the thirtieth person to exit through a 30″ bulkhead aperture**

| Condition | Mean | Standard deviation |
|---|---|---|
| Non-toxic smoke<br>No cabin staff<br>No acoustic signal | 75.30 | 14.43 |
| Non-toxic smoke<br>No cabin staff<br>Acoustic signal | 68.64 | 14.23 |
| Non-toxic smoke<br>Non-competitive | 37.9 | 4.4 |
| Non-toxic smoke<br>Competitive | 28.3 | 3.1 |
| Clear air<br>Non-competitive | 23.4 | 0.0 |
| Clear air<br>Competitive | 18.4 | 1.9 |

**Table 6.6    Mean evacuation times (seconds) for the thirtieth person to exit through the overwing exit with a 13″ vertical projection between the seat rows**

| Condition | Mean | Standard deviation |
|---|---|---|
| Non-toxic smoke<br>No cabin staff<br>No acoustic signal | 73.96 | 12.64 |
| Non-toxic smoke<br>No cabin staff<br>Acoustic signal | 65.99 | 8.3 |
| Non-toxic smoke<br>Non-competitive | 51.6 | 14.9 |
| Non-toxic smoke<br>Competitive | 55.3 | 9.1 |
| Clear air<br>Non-competitive | 39.9 | 3.3 |
| Clear air<br>Competitive | 55.9 | 10.3 |

evacuations through the aperture between the bulkheads, the results demonstrated that the presence of an acoustic signal had no significant effect upon egress rates, although from a point at which over ten volunteers had evacuated, the evacuations with the acoustic location signal were faster. In addition, although the egress rates through the Type III exit were consistently faster when the location signal was in operation, the differences were not found to be significant. It was therefore concluded that the use of acoustic exit location signals does not have a significant effect upon egress rates (Muir and Bottomley, 1992).

The mean evacuation times clearly indicate the importance of the role of cabin staff since the mean evacuation time in the non-competitive non-toxic smoke condition was 37.9 s with cabin staff assistance and 75.30 s without cabin staff present. An acoustic signal indicating the operational exits did appear to assist volunteers to locate the exits and increase the speed of the evacuation from 75.30 s to 68.64 s, although this difference was not statistically significant. A similar but less marked effect could be observed from the mean evacuation times through the Type III overwing exit. In the non-competitive non-toxic smoke evacuations with cabin staff assistance it took 51.6 s to evacuate 30 passengers, without cabin staff the same evacuation took 73.96 s and in the presence of an acoustic signal 65.99 s. These differences between the evacuation times without cabin staff and the evacuations with an acoustic tone were not significant.

## The presentation of safety information

Another area of concern has been that of the information passengers successfully obtain from the pre-flight briefing and their knowledge of the safety procedures.

In 1989 an investigation to determine the most effective ways in which passengers could be encouraged to pay more attention to safety procedures was conducted (Fennell and Muir, 1992). A total of 166 passengers responded to a questionnaire survey which investigated the influence of passenger attitudes towards the safety briefing, their perceptions of the role of the cabin attendants and their perceptions of the severity of aircraft emergencies on their motivation to attend to safety procedures. Passengers' opinions of the effectiveness of possible alternative introductions to the safety briefing indicated that an approach in which passengers are informed of the importance of their knowing how to carry out safety procedures would be more likely to encourage attention to the safety briefing and the safety card. The cabin attendants were perceived to be primarily responsible for passenger safety in an emergency, suggesting that the lack of attention to safety information on the part of some passengers may be attributable to a belief that they need not assume responsibility for their own safety.

Almost 80% of passengers involved in the survey thought that the operators should encourage passengers to be more safety conscious. The passengers suggested ways in which this could be achieved and these included tighter control over the stowage and quantity of cabin baggage, the banning of smoking, alcohol and duty free goods, making safety briefings more interesting or varied and the promotion of safety education.

A second programme was conducted to investigate passenger comprehension of airline safety information. Two experimental studies were conducted to investigate:

- The effectiveness of safety cards for conveying safety information to passengers;
- The effect of varying the content of information presented in safety briefings on passenger attention.

In both the experimental studies, volunteers boarded a stationary aircraft and were given a safety briefing. An emergency situation was simulated and the volunteers were instructed to put on their lifejackets, and then to brace for an emergency landing.

Volunteers' knowledge of the less complicated safety briefing card information, such as the location of the oxygen masks and when and how to inflate the lifejacket, was generally high. However, volunteers' knowledge of more complex procedures, such as the correct method of donning the lifejacket, and of operating the overwing and main exits, was more limited. A visual demonstration was shown to significantly increase the likelihood that volunteers would know the correct method of operation of the oxygen mask and the correct method of donning the lifejacket and that they could adopt an effective brace position. A comparison of lifejacket donning times indicated that volunteers who donned their lifejacket four hours after having seen a standard safety briefing were not significantly slower than those who donned the jackets 5–10 minutes after the briefing. Volunteers' opinions indicated that emphasis on the importance of passengers knowing how to operate items of safety equipment in briefings would not discourage the majority of them from flying and would be likely to increase attention to safety briefings.

A number of human factors problems were identified as affecting volunteers' ability to carry out safety procedures quickly and effectively. For example, the lack of specific information (in all of the briefings investigated) led to problems in locating and retrieving the lifejacket from under the seat. Inadequate instructions led to the loss of valuable time as

121

passengers tried to find out how to open the lifejacket container and identify the inside and outside of the jacket. These problems indicate the need for more specific information to be included in the safety briefing and on the card to ensure that the correct method of operating safety equipment and the appropriate procedures to adopt are obvious to passengers.

Although air travel was considered by passengers to be the safest form of transport, aircraft accidents were perceived to be less survivable than accidents involving other forms of transport. Previous findings that passengers tend to underestimate their chances of survival in aircraft accidents were supported by passengers' relatively low perceptions of their survival chances in eight aircraft emergency situations. To improve the accuracy of passengers' perceptions of aircraft accident survivability a more realistic image of aircraft safety is required. The public need to be made aware through the media that the majority of aircraft accidents are survivable and the information contained in safety briefings and on safety cards may save their lives (Fennell and Muir, 1992).

## Future developments

In this chapter, some of the research which has been conducted in the field of cabin safety has been described. In the future, cabin safety research involving collaboration between European nations, and with the FAA, will be required since it is only from collaborative research initiatives that the data which can form the basis for international regulations can be obtained. Future research should include:

1   A revised procedure and new methods for the testing and certification of aircraft safety equipment and evacuation capability.
2   The continued development of behavioural and computer models to predict passenger behaviour and evacuation rates in a range of scenarios.
3   A review of criteria for cabin staff selection in view of their primary safety function.
4   The inclusion in cabin staff training of:

- basic aircraft technical information;
- joint technical/cabin crew courses where instruction and practical assessment are based on the total crew resource management concept;
- involvement in technical crew LOFT exercise programmes to develop insight into flight emergencies including take-off and landing from a flight safety perspective;

5    The presentation of safety information to include:

- the best form and length of presentation;
- the best medium of presentation;
- the vital contents;
- the retention capability of passengers;
- the role of the cabin staff;
- how to encourage passenger appreciation of the value of the briefings.

6    The potential contribution of airport safety centres where passengers can familiarize themselves with and practise the operation of the safety equipment.

In the last decade aviation psychologists have made important contributions to safety by providing information which has been used as the basis for safety regulations and also to assist airframe manufactures and operators.

The predicted increase in demand for air travel together with the proposed development of new larger airframes indicates that in the future the contribution of aviation psychologists to cabin safety will become of increasing importance.

## Acknowledgements

The author wishes to express her thanks to the members of the Cranfield Team who have contributed to this programme, particularly to C. Marrison, D. Bottomley, J. Fennell, J. Hall, A. Evans and D. Harris. She would also like to acknowledge the contributions made by A. Thorning, N. Butcher, L. Virr and other members of the CAA (UK).

## References

Air Accidents Investigation Branch (1985), *Report on the accident to TriStar L-1011 GBBAI at Leeds/Bradford Airport on 27 May 1985*, Air Investigation Branch Bulletin Special S1/85, London, HMSO, Department of Transport.
Air Accidents Investigation Branch (1989), *Report on the accident to Boeing 737–236 series 1 G-BGJL at Manchester International Airport on 22 August 1985*, Aircraft Accident Report 8/88, London, HMSO, Department of Transport.
Air Accidents Investigation Branch (1990), *Report on the accident to Boeing 747 N739PA at Lockerbie, Dumfriesshire, Scotland on 21 December 1988*, Aircraft Accident Report 2/90, London, HMSO, Department of Transport.

Allerton, C. W. (1964), *Mass Casualty Care and Human Behaviour*, Columbia, Medical Annals of the District of Columbia.

Boeing Commercial Airplanes (1989), *World Travel Demand and Airplane Supply Requirements*, Seattle, WA, Boeing Publications.

Canter, D., Breaux, J. and Sime, J. (1990), 'Domestic, multiple occupancy and hospital fires', in D. Canter (ed.), *Fires and Human Behaviour*, London, UK, David Fulton Publishers.

Fennell, P. J. and Muir, H. C. (1992), *Passenger Attitudes Towards Airline Safety Information and Comprehension of Safety Briefings and Cards*, CAA Paper 92015, London, UK, Civil Aviation Authority.

Galea, E. R. P., Galparsoro, J. M. and Pearce, J. (1993), 'A brief description of the exodus evacuation model', *18th International Conference on Fire Safety, January, USA*.

*ICAO (1984), Report on the accident to Boeing 737 C-GQPW at Calgary Airport on 22 March 1984*, ICAO Summary 1984–2, Montreal, Canada; International Civil Aviation Organization.

Indian Civil Aviation Department (1985), *Report on the accident to Boeing 747 VT-EFO which crash landed in the Atlantic Ocean off the SW coast of Ireland on 23 June 1985*, Accident Summary.

Muir, H. C. and Bottomley, D. (1992), *Aircraft Evacuations: A Preliminary Series of Aircraft Evacuations to Investigate the Influence of Acoustic Attraction Signals Located Beside the Exits*, CAA Paper 92002, London, UK, Civil Aviation Authority.

Muir, H. C., Bottomley, D. and Hall, J. (1992), *Aircraft Evacuations: Competitive Evacuations in Conditions of Non-Toxic Smoke*, CAA Paper 92005, London, UK, Civil Aviation Authority.

Muir, H. C., Marrison, C. and Evans, A. (1990), *Aircraft Evacuations: Preliminary Investigation of the Effect of Non-Toxic Smoke and Cabin Configuration Adjacent to the Exit*, CAA Paper 90013, London, UK, Civil Aviation Authority.

Muir, H., Marrison, C. and Evans, A. (1989), *Aircraft Evacuations: The Effect of Passengers Motivation and Cabin Configuration Adjacent to the Exit*, CAA Paper 89019, London, UK, Civil Aviation Authority.

NTSB (1984), *Report on the accident to DC-9 C-FTLU at Cincinnati Airport on 2 June 1983*, (NTSB-AAR-84–9), Washington, DC, National Transportation Safety Board.

Rasmussen, P. G. and Chittum, C. B. (1986), *The Effect of Proximal Seating Configuration on Door Removal Time and Flow Rates Through Type III Emergency Exits*, Memorandum No. AAM-119–86–8, Washington, DC, Federal Aviation Authority.

Robson, B. M. (1973), *Passenger Behaviour in Aircraft Emergencies*, Technical Report No. 73106, London, UK, Royal Aircraft Establishment.

Snow, C., Carrole, M. J. and Allgood, M. A. (1970), *Survival in Emergency Escape from Passenger Aircraft*, Report No. FAA AM 7016, Washington, DC, Civil Aeromedical Institute.

Taylor, A. F. (1989), *Aircraft Accidents*, Cranfield, UK, Cranfield Institute of Technology.

Taylor, P. (1990), *Hillsborough Stadium Disaster: Final Report of Enquiry by Rt Hon Justice Taylor*, London, UK, HMSO, Home Office.

World Airline Accident Summary CAP 479, *Report of accident to Boeing 747 at Tenerife on 27 March 1977*, London, UK, Civil Aviation Authority.

World Airline Accident Summary CAP 479, *Report on the accident to Boeing 737-N388US at Los Angeles Airport on 1 February 1991*, London, UK, Civil Aviation Authority.

# Part 2

# Learning from Accidents and Incidents

# 7 Investigation of human factors: The link to accident prevention

*Peter G. Harle*

## Introduction

For decades accident investigators have recognized the importance of the human being in accident causation. Indeed, the focus of accident investigations has traditionally been on the unsafe acts committed by individuals which ultimately culminated in tragedy. Hence, normal, healthy, qualified and experienced crews operating airworthy aircraft have often been at the focus of accident investigations. While investigators might reveal a number of unfortunate circumstances surrounding the accident, typically blame is centred on those closest to the accident; had they properly assessed the situation, the accident could have been avoided. When no tangible technical evidence can be found to explain the occurrence, investigators have found it difficult to deal with the contextual human issues that might facilitate understanding why these ordinary people were unable to safely cope with the circumstances in which they found themselves.

Accident investigation reports usually depict clearly *what* happened, *when*, and especially *who* caused the accident; but in too many instances the reports stop short of fully explaining *how* and *why* the accidents occurred. For example, by stating that a pilot did not follow the rules implies that the rules are well-founded, safe and appropriate. The result is that investigation reports often limit conclusions to phrases such as 'pilot error', 'failed to see and avoid', 'improper use of controls', or 'failed to observe and adhere to established standard operating procedures (SOPs)'.

Effective accident prevention measures require more than identifying who is culpable. To find safer ways of operating aircraft, a better understanding of the context in which these ordinary people faced accident-conducive circumstances is required. While individuals generally precipitate the triggering events which ultimately cause the accident, accident prevention depends on someone altering the conditions under which these individuals are performing their assigned duties. Thus, humans are not only the source of accidents; they are the key to accident prevention.

Better understanding of the context in which individuals are prone to less than optimal human performance can be significantly enhanced by thorough, competent investigation of these factors in aviation occurrences. Investigation of human factors, then, must be an integral part of any investigation of an accident or incident; as such, the collection and analysis of human factors information must be just as methodical and complete as the collection and analysis of information pertaining to the aircraft, its systems, or any of the other traditional areas of investigation. The human factors investigation requires trained and disciplined investigators who apply their skills in a systematic way. The size and scope of the human factors investigation depend on the circumstances of the occurrence, but it must be coordinated and integrated with the other elements of the investigation. The human factors investigation is not a luxury that is added to the end of an investigation, time and resources permitting.

## A systems approach to accident causation

Traditionally, investigators have examined a chain of events or circumstances which ultimately led to someone committing the unsafe act that triggered the accident. Following this approach, accident prevention efforts are concentrated on finding ways of reducing the risk that the unsafe acts will be committed in the first place. However, the unsafe acts or triggering events leading to an accident seem to occur randomly; these same unsafe acts may have been committed hundreds of times before without ill effect. Hence, safety efforts to reduce or eliminate random events may be ineffective.

Analyses of accidents demonstrate again and again that the circumstances were ripe for the accident. A compelling alternative approach has been proposed by James Reason (1990). Reason emphasizes the importance of the latent situation extant when the unsafe acts were committed. This latent situation potentially contains many weaknesses at all levels of the aviation system, e.g. equipment design problems, poor operating procedures, maintenance failures, communications failures, organizational deficiencies, personnel selection, training and scheduling problems, etc. The unsafe acts or triggering events for the accident occur within this overall situational context – and are often exacerbated by it. While the triggering events ultimately are helpful in defining the accident, the underlying situational context determines whether or not there will be an accident. Hence, safety efforts should concentrate on eliminating the latent weaknesses in the system that facilitate the triggering events or unsafe acts.

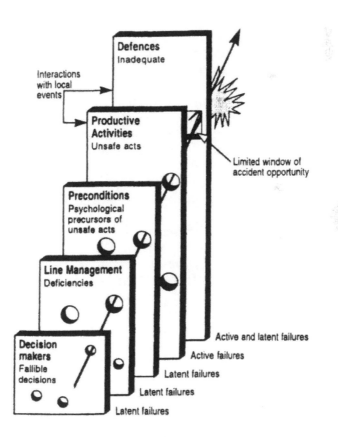

**Figure 7.1 Modified version of James Reason's model of accident causation, showing the various human contributions to the breakdown of a complex system**
(Source: © Reason, 1990; reprinted with the permission of Cambridge University Press)

Reason views industry as a complex productive system (see Figure 7.1). One of the basic elements of the system is the *decision makers* – those who are responsible for the strategic management of available resources to

129

achieve and balance two distinct goals: the goal of safety, and in the case of aviation, the goal of on-time and cost-effective transportation of people and cargo. These decision makers include the manufacturer, the carrier's corporate management and the regulatory authorities. A second key element is *line management* – those who implement the decisions made by upper management. These players adopt and implement the everyday activities and procedures for maintenance and operations, e.g. such company practices as duty days, crew pairing, training and supervision, etc. Upper-management decisions and line-management actions create many of the *preconditions* in which the workforce attempts to safely and effectively carry out their responsibilities. For example, serviceable equipment must be provided, the workforce has to be skilled and knowledgeable, and environmental conditions have to be safe. But, the line personnel too have certain preconditions they take into the workplace, e.g. well rested and motivated, free of stress from outside situations, etc. Reason's model also recognizes the *defences* or safeguards in place to prevent injury, damage, or costly interruptions of service in the event that some unsafe act is committed.

In aviation there are strict rules, high standards and sophisticated monitoring equipment in place to safeguard the aviation operating system. Because of technological progress and this excellent network of defences, accidents seldom originate exclusively from the errors of operational personnel (front-line operators) or as a result of major equipment failures. Instead, they result from the interactions of a series of inherent flaws in the system. Many of these failures are not immediately visible, and they may have delayed consequences. Just as installation of a defective component may not be detected until after it fails, poor management decisions or line-management practices may not be detected until there is a thorough investigation of an accident.

Failures can be of two types, depending on the immediacy of their consequences. An *active* failure is an error or a violation which has an immediate adverse effect. These errors are usually made by the front-line operator. A pilot raising the landing gear lever instead of the flap lever exemplifies this failure type. A *latent* failure is a result of a decision or an action made well before an accident, the consequences of which may lie dormant for a long time. Such failures usually originate at the decision-maker, regulator, or line-management levels, i.e. with people far removed in time and space from the event. Following the merger of two companies, a decision not to provide training to standardize operating procedures might create a latent failure. Failures can also be introduced at any level of the system by the human condition; for example, through poor motivation or fatigue resulting in pilot inattention, forgetfulness, poor information processing, etc.

Latent failures, which originate from questionable decisions or incorrect actions, although not harmful if they occur individually, can interact to create 'a window of opportunity' for a pilot, air traffic controller or mechanic to commit an active failure which breaches all the defences of the system and results in an accident. Thus, the front-line operators may become 'the inheritors of all the system's defects'. They are the ones dealing with the situations in which their actions, technical problems, or adverse conditions will reveal the latent failures present in a system.

The potential for an accident is created when human actions interact with the latent failures present within the system so as to breach all the defences. The task of the investigator is not only to identify what actions were made by operational personnel, but also to identify why these actions led to the breakdown in defences resulting in the accident. This requires determining the related latent failures present at all levels within the aviation system, including those created by management. Fortunately, these latent failures in the system can often be identified by investigating incidents.

## Investigation of incidents

The decision to investigate an occurrence is often guided by the severity of the outcome. If there is significant loss of life or property, there is more likelihood that an investigation will be ordered. However, the underlying causal factors may be just as evident in less serious occurrences. Hence, the determination whether to investigate or not should be guided as much by the potential for meaningful accident prevention measures, as by the magnitude of the loss.

Most accidents originate in actions committed by reasonable, rational individuals who were acting to achieve an assigned task in what they perceived to be a responsible and professional manner. They have probably committed these same unsafe acts before without negative consequences because the existing conditions at the time did not favour an interaction of the flawed decisions or deficiencies present in the system. Under different circumstances, the consequences of any accident might have been less serious and resulted only in an incident.

There are many incidents occurring every day which may or may not require reporting to the investigation authority, but which come very close to being accidents. Because there is no injury or little damage, they might not be investigated. This is unfortunate, because an investigation of an incident can often produce better accident prevention results in terms of revealing the latent failures in the system, than can the investigation of an accident.

In incidents, injury, damage and liability are generally reduced, and there is less publicity associated with these occurrences. In principle, more information should be available (e.g. live witnesses, undamaged flight recorders, etc.), and, without the threat of substantial damage suits, there should be less of an adversarial atmosphere involved in the investigation. Thus, investigators should have a better opportunity to identify the underlying human performance issues. There should be more likelihood of discovering why the incidents occurred and, equally, how the defences in place prevented them from becoming accidents. In an ideal world, preventive measures to ameliorate these latent deficiencies can be initiated before an accident occurs.

Knowledge derived from incidents, whether they are investigated in depth or not, can provide a significant insight into accident prevention. This realization has led several nations to establish confidential reporting programmes whereby individuals can report their perceptions of unsafe conditions or share personal experience; the evidence emerging from these constitutes a rich source of data on human factors in aviation.

## Who should investigate?

From early days, the investigation of aviation accidents has been conducted by technically competent personnel, trained as generalists. They have been investigating the highly technical and complex aspects of occurrences – including the advanced technology considerations of today's 'glass cockpits'.

Recognizing the significant contribution that human failures have in accident causation, investigators must be knowledgeable of and skilled in the application of the basic principles of human performance. They must be able to recognize human error and its underlying context, and they must be capable of the same sound data gathering and rigorous analysis as they apply for technical problems. Just as they need not be qualified as metallurgists or computer design engineers to conduct the technical investigation, they need not be physicians, psychologists, sociologists, ergonomists, etc., to conduct the human performance portion of the investigation.

Since aviation accidents generally involve normal, competent, certified crews, the generalist investigator is often well-suited to gather the facts and analyse the performance of their peers. In a small percentage of occurrences, the human performance may be sufficiently atypical to require the expertise of specialists to provide the investigators with guidance and assistance in interpreting the facts of the occurrence.

To fulfil their role in the investigation of human factors, generalist investigators require relevant training recognizing the interdisciplinary nature of human factors, the fundamental areas of examination, data that should be collected, data sources and collection methods including interview techniques, and analytical techniques, etc. This training should also include general guidance on the type of specialists that are available to assist in the investigation of human factors, where they can be found and when it would be appropriate to employ them. Given the requisite training, the experienced accident investigator should be able to conduct all but the most specialized aspects of the human performance investigation.

## What information should be collected?

Investigators must gather information which will enable the construction of a detailed chronology of each significant event known to have occurred prior to the occurrence and, if appropriate, afterwards. In the context of the human factors investigation, this chronology must place particular emphasis on the behavioural events, and the effects these behaviours may have had on the occurrence sequence. In addition, investigators must gather that information which will permit making reasonable inferences as to factors which may have influenced or motivated a particular behaviour. This information should include all the relevant 'preconditions' under which the operations were being conducted. In short, investigators must collect sufficient information to enable them to determine *what* happened and *why*, from a human factors perspective. The data collected must encompass the decisions, actions and behaviour of all the people concerned with the particular occurrence, not just the actions and active failures of the operational personnel. By systematically analysing all the latent failures in the system, a full understanding can be gained of how the 'window of opportunity' for the occurrence was created.

### The SHEL model

To facilitate the data collection task, the conceptual SHEL model is a helpful tool (see Figure 7.2). First developed by Edwards (1972), and later modified by Hawkins (1984, 1987), the SHEL model places emphasis on the human being and its interfaces with the other components of the traditional man–machine–environment system.

Each component of the SHEL model (Software, Hardware, Environment, Liveware) represents the foundations or the building blocks of human endeavour in a modern technological context. The human element (Liveware) is the centrepiece of the model, interacting directly

133

S = Software (procedures, symbology, etc.)
H = Hardware (machine, equipment, etc.)
E = Environment (internal and external)
L = Liveware (humans)

Note: In this model, the match or mismatch of the interface of the blocks is just as important as the characteristics of the blocks themselves. A mismatch can be a source of human error.

**Figure 7.2   The SHEL model**
(Source: Adapted from Hawkins, 1987)

with each of the others. The edges of this human block are not simple and straight, so other blocks must be carefully matched to them if stress and eventual breakdowns (as manifested by accidents or incidents) are to be avoided.

The human factors investigation must identify where mismatches between components existed and contributed to the occurrence. Therefore, the data collected during the investigation must be sufficient to enable a thorough examination and analysis of each of the SHEL components to determine if and where deficiencies in the interfaces between components existed.

Following is a brief description of the line of thought that the investigator must pursue in examining the various components of the system, their relationships and the potential latent failures extant at the time of the occurrence.

*Liveware – The individual*   The central liveware component, the individual, is the hub of the SHEL model. The data that should be collected to address this central component can be broken down into four factor categories: physical, physiological, psychological and psychosocial.

- *Physical factors*   Was the individual physically capable of performing the required task? Were there any physical impediments or limitations to successful performance of the task? If so, how did these physical or sensory limitations create difficulties or illusions that may have affected performance?

- *Physiological factors*   To what extent did the individual's general health, nutrition, disease, tobacco, alcohol or drug use, stress or fatigue level, or other lifestyle considerations influence the crew's performance or judgement?

- *Psychological factors* To what extent did the psychological background that individuals brought with them to work affect their ability or capacity to cope with the circumstances at the time of the occurrence? Here such things as the sufficiency, relevancy and currency of their training, knowledge and experience must be weighed against their ability and confidence to perform the required tasks at that level of workload. There may have been misperceptions created by the circumstances, or even serious visual or vestibular illusions which could impair or bias judgement and decision making. Workload may have exceeded the individual's information processing or attention capacities. The individual's psychological fitness may have been affected by a recent attitudinal change or mood swing, possibly affecting motivation and judgement, or even capability to cope with the stress of an emergency situation.

- *Psychosocial factors*   To what extent were pressures brought on the individuals by their social system, both in their work and their non-work environment? Did an argument with a supervisor, a death in family, personal financial problems etc. influence the individual's approach to the situation, his ability to handle stress or unforeseen events?

*Liveware – Liveware interface*   The liveware–liveware interface is the relationship between the individual and other persons in the workplace. In aviation, the interface between members of flight crews is currently receiving extensive attention under the rubric of Crew Resource Management (CRM)(Wiener *et al.*, 1993). Crew interactions and the interactions between individuals and their management regarding issues of personnel, supervision and labour relations could significantly affect overall performance. Thus, the investigator should assess crew compatibility in terms of personality, experience level and working habits. What kind of working environment did management foster, how did their policies affect working conditions, and was there effective supervision in implementing these policies? Similarly, union activities may have influenced worker performance or managerial attitudes towards the workers. A long strike or a corporate merger can create organizational tensions which can adversely affect the interface between crew members or with management for years after the event itself.

*Liveware–Hardware interface*  The liveware–hardware interface is the relationship between the human and the machine. The investigator must examine the cockpit/work station configuration, display and control design, seat design and configuration, man–machine dynamics, etc.
For example, how did the physical work environment impact on crew information processing: did similarities, differences or peculiarities in design or layout influence response time, action sequencing, habit patterns, workload or orientation?

*Liveware–Software interface*  The liveware–software interface is the relationship between the individual and all the supporting systems found in the workplace, e.g. the regulations, manuals, checklists, publications, standard operating procedures and computer software design. Here, the investigator must consider the currency and accuracy of the relevant publications, their 'user friendliness' in terms of format, vocabulary, clarity, etc. Similar considerations apply when assessing the effects cockpit automation may have had on crew performance. In addition, cockpit automation may have altered the nature of cockpit duties as well as the relation between crew members; workload may have increased during some phases of flight, possibly affecting crew attitudes towards their work and each other.

*Liveware–Environment interface*  The liveware–environment interface is the relationship between the individual and the internal and external environments. The internal environment includes such considerations as temperature, ambient light, noise, vibration, air quality, etc. The external environment includes such things as visibility, turbulence, terrain, etc. Further, the aviation system operates within the context of broad political and economic constraints, and those aspects of environment are included in this interface. Thus, the investigator must give consideration to the adequacy of the physical facilities and supporting infrastructure, the economic situation, regulatory effectiveness, etc. While the cockpit work environment may be creating pressures to take short cuts, inadequate infrastructure support or economic pressures may be compromising the efficacy of the crew's decision making.

## How much information is enough?

How many peers, relatives, supervisors, etc. of the pilot should be interviewed? How far back in time should personal activities be investigated? To what extent should interpersonal relationships be examined? At what point does past behaviour cease to influence current behaviour? How high in management should the investigation progress?

There are no clear criteria as to how far back the investigator should probe through the many layers. In constructing a detailed accident tree, the questioning process of 'why?' and 'so what?' can be carried to extremes. In practice, it is reasonable for the investigator to stop where corporate and regulatory authorities no longer have any power to change the underlying situation in the interests of accident prevention.

Investigators must focus on the factors most likely to have influenced actions. However, in practice, the dividing line between relevance and irrelevance is often blurred in dealing with issues about human factors. Data that initially may seem to be unrelated to the occurrence could prove to be extremely relevant after relationships between particular events or elements of the occurrence are established as a result of further analysis.

The field phase of an investigation requires a continuous, selective, analytical process. Even in the absence of a formal methodology, investigators soon develop some form of ongoing reasoning process. Field investigators learn to continuously formulate and test hypotheses to explain the known or suspected facts. This ongoing analytical process determines the direction of the investigation. As dead ends are reached, new avenues are explored. The more compelling is the evidence, the more rigorous is the follow-up examination for confirmation.

Since the investigator is seldom a human factors specialist, checklists are used to ensure thoroughness in the investigation of the relevant issues; they help the investigator to organize and prioritize the gathering of evidence of a perishable nature (ICAO, 1993, appendices.) However, since most occurrences are by their nature unique and diverse, the checklists cannot be used as strict protocols for the rigid step-by-step conduct of the human factors investigation.

Pursuit of the human factors aspects of the investigation to the degree necessary in controlled research is not warranted; indeed, academic curiosity may distract the investigator from his purpose of finding causal and contributory factors in order to prevent accidents. Nor is it necessary in most national investigation jurisdictions for the established facts, analysis and conclusions of investigators to stand the test of a court of law. The task is to explain how the causal event sequence was initiated and why it was not interrupted before the mishap. The critical question is *why* – not *who* was to blame. If the information does not help explain *why*, then it is not relevant to the purpose of the investigation.

Although the number of causes and contributing factors revealed in an investigation will be a function of the resources allocated to the investigation, an open-ended approach to the investigation may only be justified following a major air disaster. For smaller investigations, the available resources may mean that efforts must be concentrated on the principal individuals, and that less information is collected on the more peripheral players in an occurrence. Since the purpose is accident prevention, the resources dedicated to an investigation should be guided by the scope for safety action. Deeper investigation should be confined to those safety issues with potential for further preventive action.

## Information sources

Information relevant to the investigation and analysis of the human factors aspects of an aviation occurrence can be acquired from a variety of sources. Primary sources that relate specifically to the occurrence under investigation include the remaining or similarly configured physical equipment, documentation, audio/flight recorder tapes, interviews, medical/post-mortem examinations, direct observation of aviation personnel activities, and simulations. Secondary sources include aviation occurrence databases, technical literature and human factors professionals/specialists.

### Primary sources

Hardware evidence is most often associated with the aircraft but may also involve other work stations and equipment used by aviation personnel (e.g. air traffic controllers, aircraft maintenance and servicing personnel). Specific sources include aircraft wreckage, similarly configured aircraft, manufacturer's data, company records and logs, maintenance and servicing equipment, air traffic control consoles, etc.

There is often documentation spanning the complete spectrum of SHEL interfaces: e.g. personal records/logbooks; certificates, licences; company personnel and training records; aircraft flight manuals; company manuals and standard operating procedures; training manuals and syllabi; company training and operational schedules; regulatory authority records; weather forecasts, records and briefing material; flight planning documents; medical records; reports of post-mortem examinations.

In addition to preserving a record of communications, audio recordings (air traffic control and CVR tapes) can provide evidence on the state of mind of individuals, and possible indications of stress or fatigue. Within the company's flight recorder monitoring programmes, there may be a wealth of normative information about crews' operating procedures.

Flight data recordings (and sometimes ATC radar tapes) may provide useful information for determining the sequence of events and examining the liveware–hardware interface. In addition to traditional flight data recordings, in new generation aircraft, maintenance recorders and some electronic components with non-volatile memories are also potential sources of information pertinent to the examination and analysis of the liveware–hardware interface.

In the absence of measurable data, interviews may be the single source of information, and, in that light, investigators need to be skilled in interview techniques. Interviews conducted with individuals either directly or indirectly involved in the occurrence can be used to confirm, clarify, or supplement data from other sources.

Direct observation of actions performed by aviation personnel in the real environment can reveal important information about the potential human factors involved. Observations can be made of flight operations activities, flight training activities, maintenance activities and air traffic control activities.

Simulations permit reconstruction of the occurrence and can facilitate a better understanding of the sequence of events which led up to the occurrence, and the context from which involved personnel perceived the events. Computer simulations can be used to reconstruct events using data from flight recorders, air traffic control tapes and other physical evidence. Often a session in an aircraft flight simulator or reconstruction of a flight in a similar aircraft can offer valuable insights.

*Secondary sources*

During or after the field phase of the investigation, additional empirical data can be collected which will eventually facilitate the analysis of the factual information collected in the field. These supporting data can also come from several sources.

Basic psychological and sociological references provide good sources of information about general human performance, but they seldom address human behaviour in conditions comparable to the operational environment of commercial aviation. However, in recent years, professionals in the human factors community have provided some valuable source materials which do address aviation operational issues.

Investigators cannot be experts in every field of human factors related to the aviation operational environment. They must, therefore, be willing to consult when necessary with professionals outside their area of expertise at any time during an investigation. These professionals include, but are not restricted to:

- *medical officers:* to analyse the impact of medical conditions the investigator has discovered;
- *psychologists:* to help analyse the impact of environmental, operational and situational factors on the motivation and behaviour of the pilots, controllers, maintainers, etc.;
- *sociologists:* to help evaluate the social factors that have an impact on the interactions and performance of pilots, controllers, passengers, etc.;
- *sleep researchers:* to evaluate the quality of rest available to the individual, and the impact of a particular work–rest duty cycle and circadian factors on the expected performance of the operator;
- *ergonomists:* to assess the impact of design and layout characteristics on the user's actions and information processing capabilities, etc.

Some of the most useful secondary sources of supporting factual information come from databases directly related to the aviation operational environment, because these can often be generalized to the factual data pertaining to a specific aviation accident. Aviation safety databases based on accident/incident data are maintained by many nations and ICAO. Some nations maintain confidential reporting systems, and some aircraft manufacturers maintain occurrence data on their products – all information directly related to the aviation operational environment.

These aviation occurrence databases warrant further examination as a key source for the human factors investigator. But, before using any data, the prudent investigator will assess the limitations inherent in the database; consideration should be given to the source of the data, its intended purpose, the definitions used in recording the data, its currency, etc. Unfortunately, specific information about the sample characteristics of the database(s) being examined and the exposure rates of aircraft or pilots in similar situations are often not available. Hence, it may be difficult to make any conclusions about the probability of similar accidents or incidents occurring again (see Chapter 8).

## Analysis of data

Having completed the task of collecting all relevant human factors information surrounding an occurrence, the investigator must make judgements in analysing the data to arrive at meaningful and supportable conclusions. While the courts may use data uncovered by the investigator to attach blame and liability for the accident, the investigator should direct the analysis towards measures to prevent recurrence. Thus, the framework adopted for analysing the occurrence data should lead to safety action as the principal output. For example, data reduction to simple statistics in itself serves little useful purpose without further evaluation of the practical significance of the statistics in defining a problem which can be resolved.

Investigators have been quite successful in analysing much of the measurable data collected during an investigation – even as they pertain to human factors, e.g. strength requirements to move a control column, lighting requirements to read a display, ambient temperature and pressure requirements, etc. Unfortunately, many critical human factors do not lend themselves to simple measurement and are thus not entirely predictable. As a result, much human factors information does not allow an investigator to draw indisputable conclusions.

A manager from the Boeing Commercial Airplane Company has identified several other limiting factors which have typically reduced the effectiveness of human performance analysis (Fadden, 1985). These include:

- The lack of normative human performance data to use as a reference against which to judge observed individual behaviour.
- The lack of a practical methodology for generalizing from the experiences of an individual crew member to an understanding of the probable effects on a large population.
- The lack of a common basis for interpreting human performance data among the many engineering, scientific, and management disciplines which make up the aviation community.
- The ease with which humans can adapt to different situations, further complicating the determination of what constitutes a breakdown in human performance.

The logic required to analyse some of the less tangible human performance phenomena is sometimes different from that required for other aspects of the investigation. Schleede (1981) argues that traditionally investigators are comfortable using deductive argument which produces 'conclusive evidence of the truth . . .'. Investigators feel secure because the validity of their conclusions is self-evident and cannot be challenged by peers or superiors.

Deductive methods are relatively easy to present and lead to convincing conclusions. For example, a measured windshear produced a calculated aircraft performance loss, and a conclusion was reached that the windshear exceeded the aircraft's measurable performance capability. Some measurable human factors as heart attack, drug/alcohol impairment, hearing, eyesight, etc. lend themselves to deductive argument.

However, when the validity of their conclusions cannot be tested conclusively, and the investigators must deal with analysis based on probabilities and likelihoods, investigators tend to become cautious and reluctant. Caution is commendable, but investigators must adopt strategies for overcoming this problem.

Some human factors issues such as complacency, fatigue, distraction or judgement, i.e. the intangible human performance factors, are less readily measured. Hence, inductive reasoning using probabilities and likelihoods is used. Inferences are drawn, and some degree of speculation is involved. Inductive conclusions can be challenged depending on the reasoning process used by the investigator and the weight of evidence available.

The Bureau of Aviation Safety in Australia applies the following step-by-step reasoning process to ensure that all reasonable possibilities are considered, while at the same time reducing the investigator's task to manageable proportions (ICAO, 1993). First the investigator attempts to establish the probability of the *existence* of some human factors condition. By comparing the circumstances of the occurrence against what is empirically known about these issues and their underlying causes, the investigator determines the probability that one or more of these human factors conditions existed. Secondly, the investigator must establish the probability that this particular human factors condition *influenced* the sequence of events leading to the occurrence. Having examined what is empirically known about the effects of the human factors conditions determined to exist in the occurrence, the investigator must assess the extent to which these conditions may have contributed to the sequence of events leading to the occurrence. Just because a factor such as fatigue, distraction or complacency can be demonstrated to be present does not mean that it had any real effect on the outcome.

This analytical approach relies on an accumulation of evidence which may not allow indisputable conclusions to be drawn, but which will often allow conclusions of probability to be drawn. Both steps of the approach will require the development and testing of hypotheses. In some ways the use of conclusions of probability is similar to the legal profession's use of 'circumstantial evidence'. In other words, in the absence of indisputable evidence, the circumstances suggest the presence of a given human factor.

The strength of this approach is that it forces the human factors investigator to draw conclusions only on the basis of empirical knowledge and verifiable evidence. It ensures that the investigator considers all of the likely factors. In the absence of evidence on which to base indisputable conclusions, it also allows the investigator to draw *some* conclusion rather than no conclusion. In using the word probable, it also honestly acknowledges that the conclusions are not indisputable.

Since occurrences are seldom the result of a single cause, the human factors analysis must take into account the fact that seemingly insignificant, apparently unrelated factors can be viewed in combination to reveal a sequence of related events that culminate in an accident. The SHEL model provides a systematic approach to examining the constituent elements of the aviation system as well as the interfaces between them. The interactive system suggested by Reason also provides an excellent framework by which investigators can ensure that the analysis of human factors addresses all levels of the production system. The human factors analysis must not focus only on the active failures of front-line operators but must include an analysis of the 'fallible decisions' at all levels which interacted to create the occurrence 'window of opportunity'.

## Writing the human factors report

Having completed the gathering and analysis of the relevant facts, the investigator must prepare the report of the investigation. Irrespective of the quality of the investigation, its impact will only be as good as the final public report. Indeed, a poor report may even undermine the quality of the investigation if its analysis and logic are incomplete or flawed. The agencies responsible for taking corrective action to remedy safety deficiencies confirmed through the investigative process will focus on the report's weakness, and no preventive action will be initiated. Nowhere is this more true than in the report's credible treatment of human factors. Since few readers of aviation accident reports have a good grasp of the role of human factors in accident causation, the report must often educate and convince them. (ICAO's Annex 13 prescribes a suitable format for an accident report; this guidance allows considerable flexibility to accommodate a factual presentation of most any human factors considerations –including any necessary elaborations to explain significant deviations from expected human performance (ICAO, 1988).)

Since the fundamental purpose of the investigation is the prevention of accidents and incidents, the report should identify the hazards uncovered and offer safety recommendations to eliminate those hazards. The report should be written so that the reader, be it pilot, mechanic, manager or regulator, can relate to the hazards reported. Thus, the various 'target' audiences of the aviation community should be able to see themselves as potential creators of similar unsafe conditions and adopt preventive strategies of their own. It is important that investigators should also understand that the most influential readers are those responsible for the implementation of the safety recommendations in the report. If they are not convinced by the report, then preventive actions may not be taken.

143

Having established the relevant facts, the investigator develops the reasons why the circumstances created an opportunity for the accident. Any assumptions used in the course of the investigation must be identified and justified to explain clearly the reasoning process. That which has not been determined must be evaluated as to potential impact, as well as any controversial and contradictory evidence. For the less tangible human factors issues, not only must the existence of the factor be established, but its potential impact on the occurrence outcome must be assessed. All reasonable hypotheses as to cause and contributing factors should be stated and evaluated.

While the focus of the investigation is on causal and contributory factors, other findings with a significant potential for safety action should also be reported. The use of probability language may be called for when stating findings relating to human performance. When the weight of the evidence is such that a definitive statement cannot be made, then investigators must state findings as positively as possible, with the appropriate degree of confidence and probability. Even still, sometimes the circumstances of the accident may be such that no firm conclusion can be drawn about causes. In such a case, some of the more likely hypotheses may be discussed, but the cause may have to be reported as *undetermined*.

Determination of cause remains one of the trickier tasks for the investigative authority. Historically, investigators have concentrated on the proximate cause, such as the active errors of the pilot or controller. Laying bare the latent failures in the system that explain *why* the human error occurred has thus far eluded many investigative bodies. However, accident prevention depends on how well the real, systemic underlying causes are convincingly reported.

## Accident prevention

To prevent accidents, follow-up action must be taken in response to the hazards identified in the course of investigations. According to ICAO, accident prevention must aim at all hazards in the system, regardless of their origin (ICAO, 1984). While the emphasis is on formulating recommendations, the more difficult task is clearly identifying the safety deficiencies warranting follow-up safety action. The focus of the investigator at this point must be on problem definition. Only after the problem has been clearly identified and validated can reasonable consideration be given to corrective action (Wood, 1979).

The Reason model provides guidance in the formulation of preventive measures just as it provides guidance for accident investigation. Since many of the psychological precursors and unsafe acts are the consequences of decisions made further up the line, it makes sense to concentrate preventive measures on hazards created or ignored by the higher levels of management. If an accident or incident report focuses on the specific error of an individual pilot, air traffic controller or engineer, to the exclusion of a consideration of higher level decisions, it will do nothing to address the underlying responsibilities for identifying, eliminating or mitigating the effects of hazards.

Validating safety deficiencies in human performance can be demanding. When dealing with clear-cut factual findings such as errors in publications, material deficiencies through design errors, etc., the validation phase may be relatively short. However, for potential safety hazards involving the less tangible human factors issues (e.g. the effects of fatigue on crew performance, the consequences of a company's putting pressures on pilot decision making, etc.), the factual evidence may be more difficult to acquire, and the effects of their inter-relationships are more difficult to assess.

For many human performance phenomena, the evidence from a single occurrence may be insufficient to validate a safety hazard. Hence, the investigator must evaluate the data available from similar occurrences (perhaps on a worldwide basis) to demonstrate the probable impact of a particular phenomenon on human performance in the investigation in question. For example, it may be determined that crew coordination between a pilot and co-pilot was poor, in part because the two pilots were inexperienced both on aircraft type and with the operation. Disciplining or dismissing the pilots involved in the occurrence would do nothing to eliminate the larger problems of crew pairing, not only in the company but in the aviation system at large. To establish the existence of a systemic safety problem, the investigator should reveal other occurrences where a link had been established between crew coordination problems and company crew pairing practices. Thus, comprehensive review of the professional literature is usually warranted.

Unfortunately, when dealing with hazards deriving from human factors, completely eliminating the safety deficiency is often not practical. For example, human limitations in short-term memory are not likely to be expanded. Thus, the tendency has often been to prescribe coping strategies. For example, if the investigator's analysis examines the reasons for the situation which led to short-term memory limitations becoming a causal factor, then safety action to modify the situation may be feasible, e.g. by incorporating procedural system defences or incorporating warning systems (see Chapter 3).

Since safety hazards with respect to many human factors may be extremely difficult to validate, further study of the perceived hazard by more competent authorities may be recommended. In this way, the investigator can proceed with the confidence that the investigation report is not the final word on particularly difficult safety issues. Industry's recognition of the importance of Crew Resource Management (CRM) illustrates this point. In a number of accident investigation reports, the National Transportation Safety Board in the US had identified hazards resulting from the lack of effective flight deck management (see Kayten in Wiener *et al.*, 1993). The problem was thus validated through the investigation and analysis of many accidents. Eventually, some of the larger airlines not only recognized there were potential problems in the cockpit, but also designed and implemented CRM courses to eliminate these problems. CRM training is now widely accepted and available throughout the industry.

## Database requirements

Since the events of a single accident or incident seldom convincingly demonstrate the presence of a fundamental safety deficiency with respect to human factors, the analysis of similar occurrences of similar occurrences is necessary. For such a validation process to be effective, all relevant information from previous similar occurrences would have to be adequately recorded for future reference. Indeed, one of the many reasons why progress has been slow in initiating appropriate preventive actions for many human factors issues is inadequate recording of this type of information.

Whether or not the human factors data gathered in an investigation are clearly linked to the causes of the specific occurrence, these data should be recorded in a human factors database to facilitate the analysis of future occurrences. For ICAO contracting states, the principal database for recording such information is ADREP, a computerized framework for coding and recording pertinent information. The ADREP system records a series of factors which describe *what* happened, as well as a series of explanatory factors which explain *why* things happened.

Since human error or shortcomings in performance are usually factors in accidents, ADREP provides a sound framework for recording human factors data. Unfortunately, to date ADREP has not been consistently used to record pertinent data from incidents or even from some states' accidents. Nevertheless, the architecture of ADREP is continually being updated to better meet the needs of the global aviation community; its value will be predicated upon the quality of human performance investigations completed and diligent data recording.

# Summary

Accident prevention is critically linked to the adequacy of the investigation of human factors issues. The aviation system contains a complex of interrelated latent deficiencies at all levels of the system. Almost daily there are benign incidents which demonstrate the existence of these potential system failures. From time to time, healthy, experienced, competent personnel commit an unsafe act that breaches all the system's built-in defences, and an accident occurs. Unfortunately, often it is not until a competent investigation of such a major accident is conducted that the latent safety deficiencies in the system are identified, validated and communicated to responsible authorities for corrective action.

# Acknowledgements

This chapter is an abridged version of a submission prepared by the staff of the Transportation Safety Board (TSB) of Canada for consideration by the International Civil Aviation Organization's (ICAO) Flight Safety and Human Factors Study Group. The final text was published by ICAO in March 1993, as *Safety Digest No. 7: Investigation of Human Factors in Accidents and Incidents*. While this chapter incorporates the input of several participants in the ICAO study group and several TSB staff members, Peter Harle accepts responsibility for any errors in fact or logic.

# References

Edwards, E. (1972), 'Man and machine: Systems for Safety', *Proceedings of British Airline Pilots Association Technical Symposium*, London, British Airline Pilots Association.

Fadden, D. M. (1985), 'Key factors in achieving effective human performance assessment', *ISASI Forum*, October.

Hawkins, F. H. (1984), 'Human factors in education in European air transport operations', in *Breakdown in Human Adaptation to Stress. Towards a multidisciplinary approach, Vol. 1*, Commission of the European Communities, The Hague, Martinus Nijhoff.

Hawkins, F. H. (1993), *Human Factors in Flight*, 2nd edn, (ed.) Orlady, H. W., Aldershot, Ashgate.

Hudson, P. T. W. (1991), 'Prevention of accidents involving hazardous substances: The role of human factors in plant operation,' *OECD Environment Monographs*, 44.

ICAO (1984), *Accident Prevention Manual*, 1st edn, Montreal, International Civil Aviation Organization.

ICAO (1986), *Manual of Aircraft Accident Investigation* 4th edn, Montreal, International Civil Aviation Organization.

ICAO (1988), *International Standards and Recommended Practices on Aircraft Accident Investigation*, Annex 13 to the Convention on International Civil Aviation, 7th edn, Montreal, International Civil Aviation Organization.

147

ICAO (1993), *Flight Safety and Human Factors Digest No. 7: Investigation of Human Factors in Accidents and Incidents*, Montreal, International Civil Aviation Organization.

Reason, J. (1990), *Human Error*, New York, Cambridge University Press.

Schleede, R. L. (1981), 'Human performance investigation – The investigator-in-charge and report writer roles', *ISASI Forum*, Winter.

Wiener, E. L., Kanki, B. G. and Helmreich, R. L. (1993), *Cockpit Resource Management*, San Diego, CA, Academic Press.

Wood, R. H. (1979), 'How does the investigator develop recommendations?' *ISASI Forum*, Winter.

# 8 Using voluntary incident reports for human factors evaluations

*Sheryl L. Chappell*

## Introduction

The last few decades have brought vast improvements in hardware and facilities to the aviation industry, from the increased reliability and performance of new generation aircraft to innovations in air traffic control equipment and procedures. In the meantime, the incidence of human error has remained stubbornly persistent: approximately 75% of all aviation accidents and incidents are attributable to human failures in monitoring, managing and operating systems.

At the forefront of the quest to understand factors affecting human performance is the aeronautical human factors practitioner. The mission of the human factors practitioner is to improve aviation safety through optimization of both systems and procedures. A tool of increasing value to this human factors analysis is the use of aviation incident data.

It is true that incidents may be detrimental to the system in terms of both safety and efficiency. However, the premier value of incident knowledge is its potential role in preventing accidents. Diehl (1991) has shown that the rate of occurrence of some types of incidents is proportional to their accident rate. Those causal factors that frequently result in unsafe situations may also be assumed likely to culminate in disastrous accidents.

The human factors experts who created the Federal Aviation Administration's National Plan for Human Factors in the United States recognized the power of incident data:

The relevance of human factors research and development efforts to aviation

149

problems depends on the ability to identify trends, both positive and negative, in the operational environment (Federal Aviation Administration, 1991, p. 9).

*Answering 'what' with 'why'*

Not only do incident data provide a metric of aviation system safety, but they also offer insights from incident participants as to the underlying factors, sequences of events and conditions associated with safety anomalies. These data can, and should, play an important role in creating policies and procedures for the operation of aircraft and air traffic control.

In tracing the causal chains of aviation accidents, safety investigators and researchers have generally been effective in determining *what* happened. They have been less effective in determining the *why* of events – why people acted as they did, why a system failed, why a human erred. Incident data are a unique means of obtaining first-hand evidence on the factors associated with mishaps from event participants themselves.

Who are the participants in safety incidents? Aeronautical incident reports come from many sources, including pilots, air traffic controllers, flight attendants, mechanics and ground personnel. Reporters are willing to admit errors and quite often are able to identify the sequence of events that resulted in the incident. Reporters provide insights into their perceptions and behaviours. They describe the causal relationships between stimuli and their actions. They also provide their interpretations of the effects of various factors affecting human performance, including fatigue, interpersonal interactions and distractions. Many reporters are able to offer valuable suggestions for remedial action.

## Voluntary and confidential systems

In general, an incident may be defined as an event that involves an unsafe, or potentially unsafe, occurrence or condition, for example, two aircraft passing 100 ft apart. An incident does not involve personal injury or significant property damage. When an incident occurs, the individual(s) involved may or may not be required to submit a report. The requirement to report varies with the laws of the nation in which the incident occurred.

A *voluntary* incident report is submitted by the reporter without any legal, administrative, or financial requirement to do so. In a voluntary reporting system, regulatory agencies may offer an incentive to report. Fines and penalties may be waived for unintentional violations that are reported. The reported information may not generally be used against the reporters in enforcement actions.

A *confidential* reporting system protects the identity of the reporter. Reports may or may not be submitted anonymously to a confidential system. If the identity of the reporter is known at the time of submission, it is either removed or protected from distribution. Voluntary confidential incident reporting programmes promote the disclosure of human errors, provide the benefit of situations described with candid detail, and enable others to learn from mistakes made.

Incident reports are collected by agencies in many countries and are held in databases to be used by practitioners and scientists to improve aviation safety. These databases are a treasury of information which is unavailable from any other source. Voluntary, confidential reporting programmes are probably the best available source of data on human error. Lauber (1984) discusses the research uses of aviation incident data, stating that an incident database 'is a veritable gold mine of information waiting to be tapped. It is, without a doubt, the most comprehensive source of information about human operator error in existence.'

This chapter addresses incident reports that are submitted voluntarily and confidentially. Its aim is to facilitate the use of incident data as a rich source of information in aeronautical human factors, and to promote the use by policy makers and human factors practitioners, resulting in an increase in aviation safety.

## The use of incident data

### Characteristics of incident data

Incident data are ideally suited for proving the existence of a safety issue, understanding its possible causes, defining potential intervention strategies, and tracking the safety consequences once intervention has begun. Many voluntary incident reporting systems hold contributions from a variety of sources, including pilots, air traffic controllers, flight attendants, mechanics and passengers.

Incident records generally contain key words under categories which can be searched, for example: the role of the reporter (e.g. pilot or air traffic controller), what type of incident occurred, where and when it happened, the type of weather conditions, the role of the individuals involved, their qualifications, and the type of aircraft and flight operation. In addition, the records often include a description of the incident in the reporter's own words and the circumstances which contributed to the loss of safety. If multiple reports are received of the same incident, some systems make all descriptions available. Thus, for example, the perspect-

151

ives of different crew members and air traffic controllers may be utilized to better understand the incident.

An incident report may paint a complete picture of the safety issue and suggest remedial action or identify a hazard which may not have been known through other sources. For example, the signs or markings along a taxiway may not clearly identify the hold short position for an active runway. More often, however, the safety issues are more insidious. This chapter will focus on those instances where a study is required to clarify the underpinnings of the problems and investigate the optimal remedy for the situation. When a safety hazard is suspected, the incident data can often provide facts that will confirm the suspicion and suggest solutions. These data can be invaluable by providing insights lacking from other sources.

## Examples of incident report usage

Incident data have been used by human factors professionals to create cockpit and air traffic operating procedures, display and control design, and to better understand basic human performance associated with the operation of aircraft and air traffic control. Proper use of the incident data can provide unique insights into safety issues for which follow-up laboratory research can be conducted. Topics investigated using incident data include spatial disorientation, visual illusions during approaches, controller perception, pilot judgement, pilot response to in-flight emergencies, the impact of automation on cockpit operations, crew resource management, fatigue, crew distraction and cockpit warning systems. Many scientists have found incident data to be essential in scoping their research programmes. Issues that were judged to be important safety concerns have gained momentum and focus from the body of related incidents. Other studies have changed direction towards more valid issues after investigating an incident database.

Incidents have also been used extensively by operational human factors professionals. These data have suggested procedures which were more resistant to interruptions, and have provided insight into the level of proficiency of pilots using automated systems. The data have also been a key to uncovering specific communications problems between pilots and air traffic controllers, including ambiguous and misunderstood clearances. Operational human factors professionals also find incident data valuable in communicating a concept. For example, if a safety issue arises within an airline, the flight operations department may find it useful to provide an example of an actual incident in their safety newsletter to establish credibility for the expressed concern. Simulation training scenarios that are based on actual occurrences tend to be taken more seriously by controllers and pilots.

Incident data have been used in very directed studies, for example, to evaluate the efficacy of a new warning system. Ground proximity warning systems were mandated for airline aircraft in the United States. Loomis (1981) evaluated incidents in which these warnings prevented aircraft from striking the ground. Despite reported false alarms from the ground proximity warning system, the warnings were effective in preventing this type of accident. Another study (Mellone and Frank, 1993) revealed that the collision avoidance system mandated for air carrier aircraft in the United States (TCAS II) affected the air traffic control system to a much greater extent than was anticipated. Air traffic arrival and departure procedures at some airports had to be modified to accommodate aspects of the system to reduce the number of false alarms.

More insidious safety issues have also been effectively addressed with incident data. Information transfer problems are persistent in the aviation system. These problems have been addressed from multiple aspects using incident data, including information transfer between air traffic controllers at separate locations (Grayson, 1981), between pilots and air traffic controllers (Grayson and Billings, 1981), between pilots within the cockpit (Foushee and Manos, 1981), and information transfer in emergency air-ground communications (Porter, 1981).

## The strengths of using incident data for human factors evaluations

*Information from participants* Incident data provide detailed information from the participants in the events which often is not available following an accident. Self-reported incidents have many other advantages. The reporter can provide valuable insight into the factors that contributed to an unsafe situation. The information is in the form of narratives of the sequence of events, from the perspective of the flight crew or air traffic controller. Telephone conversations between the reporter and the safety analyst augment and clarify the information submitted, when necessary. A comparison of the descriptions of an event by the flight crew and the participating air traffic controller illustrate the pertinence of these perspectives. Reporters are subject to biases and perceptions that result from their perspective of and contribution to a situation. This is both a strength and weakness of self-reporting.

*Large numbers of observations* Aviation accidents are rare events which are investigated more thoroughly than incidents; many valuable lessons have been learned from these tragedies. However, the limitations of small samples exist when using accident data: the wrong conclusions may be

153

drawn or features of the real picture may be obscured. Incident data offer a large quantity of detailed information provided by the participants in the events. For example, at the time of this writing, the voluntary incident database in the United States, the Aviation Safety Reporting System (described in Reynard *et al.*, 1986) held 148 987 reports of which 55 762 described altitude deviations occurring in the previous six years. The availability of such a large sample of reports on many topics allows the investigator to carefully select those that will be studied in depth. Reports that do not contain an adequate level of detail may be disregarded.

*Ecological validity*   Another positive advantage of using incident data in human factors evaluations is the fact that these incidents really happened, and that they occurred in the normal operating environment. In other words, the incident is ecologically valid and not merely a laboratory artefact. A severe shortcoming of laboratory research is its lack of context. Many times a strong finding in a laboratory study dissolves in the real world, where the stimuli are very rich and the tasks are subject to interruption and time sharing. The importance and influence of contextual effects continues to be more appreciated by behavioural scientists (see, for example, Vicente, 1990).

An additional application of incident data is in the construction of realistic scenarios for training and research simulations. If the flight crew is made aware that the incident really happened, they are likely to take the simulation more seriously. For example, a study was conducted to measure pilot performance in avoiding a collision with another aircraft, using a collision avoidance system (Chappell *et al.*, 1989). Actual traffic conflicts taken from an incident database were simulated in the experimental scenarios. The validity of the traffic conflicts was appreciated by the subjects who participated in the study and the industry evaluation team who utilized the study's findings in the design of the collision avoidance system.

## The limitations of using incident data for research

*Information not validated*   In some countries, voluntary, confidential reports can be fully investigated and information from other sources brought to bear on the incident. However, the confidentiality of other systems precludes any additional investigation and reports are thus unverified. But even when the information cannot be substantiated, the reports may be reviewed by experts in the various domains of aviation (e.g. flight crew performance). If large numbers of reports on a topic are available, it is reasonable to assume that consistently reported aspects are likely to be true. It is doubtful that a large number of reporters would exaggerate or report erroneous data in the same way.

Reporters may have a tendency to understate their errors and blame the occurrence on other parties. Incidents may be embellished to benefit the reporters. For example, the controllers at an airport tower facility may inflate the number of traffic conflicts to support the addition of radar, which would result in an increase in their salary. When these reports are analysed, these factors are often very apparent to the experienced report analysts and their suspicions can be reflected in the analyses of those reports.

*Reporter biases*   There are two factors which bias voluntary incident data: *who* reports and *what* gets reported. The demographics of who submits reports will result in a faction of the aviation community being over-represented in the incident database: reporters must be familiar with the programme, they must have access to reporting forms or phone numbers, and they must be motivated to report. A pilots' or controllers' organi-zation may support the reporting system and make reporting forms available to their members. Other pilots and controllers who are not members of the organizations may find it more difficult to contribute safety reports.

This reporting bias can be seen in the United States where air carrier flights account for one third of all flying hours and 6% of the accidents, yet air carrier pilots contribute 75% of the reports to that country's voluntary reporting system. In addition to knowledge of the incident programme and the availability of reporting forms, several motivational factors may contribute to this skewed representation. The immunity offered to reporters in the United States may motivate air carrier pilots to submit reports with more regularity than general aviation pilots who do not fly professionally. The incident may be reported by more than one crew member in multi-crew aircraft. Air carriers also participate in the air traffic control system to a greater degree than general aviation pilots; thus their flights receive greater monitoring by humans and electronics. All these factors probably contribute to potential bias in the incident data.

What events get reported can also be subject to biases. These biases can affect the type and the number of reports received. When an individual makes a procedural error, generally that individual takes responsibility for ensuring the error is not repeated. If there were no significant consequences of the error and regulatory immunity is not needed, an individual may anticipate little benefit in reporting this type of event. When another individual makes an error, the only method for affecting a solution may be to report their error. For example, a pilot may receive an unsafe clearance from an air traffic controller and feel the only recourse is

155

to notify the incident reporting programme. Other factors affect what gets reported to a voluntary system. Events in the media, such as an accident, raise awareness and sensitivity to specific safety issues. These events have been seen to create temporal spikes in the numbers of reports on a type of occurrence. The media and other means may be used to stimulate reports of a certain type. If a new procedure or technology is implemented, the pilots and controllers may be encouraged to report on their experiences with that system.

Enforcement policies also affect reporting, especially where immunity is offered to reporters. When the Error Detection Program was initiated in the *en route* airspace of the continental United States, loss of separation was recorded automatically by the computer. In airspace where the required vertical separation was 1000 ft, the system would alert and record separation that was less than 700 ft. The flight numbers and the actual altitudes were preserved in the record. With this system, pilots received violations for altitude deviations with more regularity. To receive immunity for these deviations, pilots began to report these occurrences in greater numbers.

Biases also stem from what is not reported. Reporters must be cognizant of a factor to submit a report. Errors that go undetected are not reported. For example, it is unlikely that an illusion would be reported, unless it were identified through some means. There is also a tendency to report only operational problems, not safety improvements. When a new procedure goes into effect that cures a hazard, only the decline in incident reports marks that improvement. If the new procedure benefits one faction of the industry (e.g. the pilots) and creates problems for another (e.g. the controllers), only the controllers are likely to submit safety reports. The old adage that no news is good news generally applies to safety data.

Despite these known reporter biases, studies have revealed similarities between voluntarily submitted reports and other sources of safety data. Wilhelmsen (H. Wilhelmsen, personal communication, 31 January 1992) compared voluntary incident data, mandatory reports of air traffic operational errors and accidents. He found similar scenarios in all the databases and he found the relative frequencies of the scenarios to be very similar. In another study, Lyman (1981) found the geographical distribution of voluntarily submitted incidents to be significantly correlated ($r = 0.91$) with the distribution of mandatory incident reports.

## Using incident reports – the process

*Scoping the issues*

Aeronautical human factors studies begin by setting the scope of the project. The dimensions of the topic must be explored to identify the pertinent underpinnings. Incident data are ideally suited to this process, in many studies. An initial phase is devoted to understanding the issues to be investigated, formulating hypotheses, and selecting the best methodology to evaluate these issues. Incident data can also help substantiate the need for further study on a topic. The original concept of cockpit resource management was defined through the study of incident reports describing management, communication and leadership problems. Once the concept evolved, the incident reports suggested an effective orientation for the programme. The incident reports were also used as examples in communicating this important concept to the operational community.

Few studies have the good fortune of being well specified at the onset, and those that are well specified are sometimes found to be off target once the study commences. For example, a recent study began as an investigation of flight crew perceptions/misconceptions of the mechanical systems on advanced technology aircraft. This study was to culminate in a full-mission simulation. A survey of incident data revealed few misconceptions by flight crew members regarding mechanical systems; however, there was evidence of much confusion about the autoflight systems. The project was redirected to address this important safety issue. Other studies have begun by attacking a specific issue where safety is believed to be less than desired. Upon delving into the topic using incident data, other facets of the safety issue are discovered. For example a study of aircraft de-icing problems (Sumwalt, 1993) discovered that in addition to the problems of getting aircraft de-iced within a few minutes of take-off, there was a substantial number of problems in detecting the ice. One impediment to detection was the de-icing fluid covering the cabin windows. Another factor emerging from examination of the icing incidents was that pilots erroneously believed that the snow would blow off on take-off.

Studies of all types can benefit from the use of incident data to scope the issues. One airline, in an attempt to reduce the number of altitude deviations evaluated altitude deviation reports and discovered that a common element was the incorrect setting of the altitude in the altitude window. By requiring both pilots to verify the setting of that value, the number of altitude deviations was reduced. Aircraft simulation studies on the use of new information displays can use incident data to understand how tasks of interest are currently being conducted and identify the shortcomings of the existing technology. Additionally, dependent measures may be suggested by the incident reports. A study of pilot performance in response to a traffic collision avoidance system (Chappell *et al.*, 1989) began with a survey of the evasive actions reported by pilots. These actions were used as a baseline to evaluate the efficacy of the collision avoidance system.

## Retrieving the data

Retrieval of incident data can generally be accomplished sending a formal request to the agency managing the database describing the topic of interest in the greatest detail. The database experts translate the request into classifications used by the database and retrieve the pertinent reports. Databases containing voluntary incident reports commonly have both data fields and textual descriptions (e.g. time of day of the occurrence and narrative information about the sequence of events in the reporter's own words). Both the data and text can usually be searched to retrieve relevant safety reports. If a topic is not directly addressed by the data or text fields, an iterative approach can be employed. The reports resulting from the first search can be screened to determine which search characteristics yielded the most applicable reports, i.e. what terms did applicable reports have in common that could be used to retrieve additional relevant reports. These terms can then be tuned for the second search and the process repeated for further refinement, until the desired reports are retrieved.

Reports retrieved that are not relevant (false positives) present a nuisance, but not a research problem. These can be screened in the compilation process and discarded. However, it is quite reasonable to assume that almost any search strategy, unless very straightforward, will result in false negatives. These present a significant problem for research, in that the nature and significance of the reports *not* retrieved is not known. As a simple example, an individual may wish to evaluate how close aircraft come to each other in traffic conflicts occurring in the airport traffic area. If a search is performed and only those incidents in that airspace labelled 'airborne conflicts' are retrieved, the yield may be limited and the investigator will conclude that aircraft do not pass very close to each other. In reality, safety analysts commonly use the term 'near mid-air collision' to describe those incidents with extremely close miss distances. These reports would not have been retrieved with the above search strategy.

Caution should always be used when employing incident data to determine the prevalence of a safety problem. The failure to retrieve a substantial number of reports on a particular topic does not necessarily mean there are few reports on that topic in the database. The number and nature of false negatives or misses cannot be known.

*Compiling the data*

When performing an evaluation using incident data, the investigator commonly must subject the incident reports to further analysis than was performed in the initial coding. The extent of effort involved in this process will vary greatly from study to study. If the number of reports on a particular topic is of primary interest, that number may be tallied with little or no further analysis. If, on the other hand, the underlying causes of a particular human error are of interest, the reports will likely require further scrutiny. An additional coding process will be necessary. This is generally very time-consuming and may possibly threaten the resources available for other phases of the project.

An example of recoding of incident reports is found in Degani *et al.* (1991). These researchers were interested in what information was used to detect an altitude deviation and by whom it was detected. Possible differences between traditional aircraft and advanced technology 'glass' aircraft were to be identified. All the reports retrieved from the Aviation Safety Reporting System (Reynard *et al.*, 1986) involved altitude deviations. These reports contained fields that differentiated between traditional and glass aircraft. The information about how the altitude deviation was detected, and the agent that detected it, was available only from reading the text narrative of the reporter. This information was then coded and analysed. In addition, reports of turbulence causing excursions from assigned altitudes were excluded, since this type of altitude deviation was outside the scope of the study.

The standardized narrative is a technique of recoding incident reports that was developed by Thomas (Thomas and Rosenthal, 1982) in another study of altitude deviations. This type of coding enables the researcher to retain much of the structure and content of the reporter's narrative, while constraining the information sufficiently to use quantitative techniques. To construct a standardized narrative, the reporter's description of the events is rearranged and rephrased to fit one of several standard sentence types. As the narratives are interpreted, the number of standard types is adjusted to preserve the level of detail that is appropriate for the study. The power of this technique is that elements can emerge that were not anticipated. This is not necessarily true in studies where *less* rigorous coding is performed. An interesting finding by Thomas was that pilots listening to and interpreting a clearance were often wrongly influenced by what clearance they were expecting. This expectation was the result of printed navigational procedures or prior experience in that airspace and was an inherent weakness in the communication process. This finding was not anticipated when the study began and would not have been included in the coding process.

Another technique has proven to be useful in providing more information than the reporter originally submitted in the incident report. This technique involves a structured telephone interview with the reporter. For example, Orlady and Wheeler (1989) were interested in learning about pilots' evaluations of their training for advanced technology, glass cockpits. A number of flight crews who had submitted reports of various types of incidents, but all involving glass cockpit aircraft, were contacted and asked a series of questions related to the training for their aircraft. This technique is time consuming but enables the researcher to acquire much more information about an issue, including topics beyond the incidents being reported.

## Analysing incident data

Once the data have been retrieved and compiled, the analysis can begin. Analyses of incident data may take many forms. There are quantitative and qualitative techniques that are useful in evaluating safety issues using incident reports (see later sections).

As with any analysis, the most important step is to formulate meaningful null and alternative hypotheses. This is often a precarious task for incident data evaluations. The hypotheses must be couched as the probability of an event being reported, not the probability of its occurrence. Many times it is inappropriate to apply inferential analytical techniques; only descriptive statistics are valid. (For a discussion of the applicable statistical tests, please see below.)

## Drawing conclusions

When drawing conclusions from the analysis of voluntary incident data, one must be very cautious. It is not appropriate to infer the prevalence of incidents. There is no knowledge of the total number of incidents occurring, only the number being reported. There is also no knowledge of the nature of those incidents that are not reported. For example, the database may hold more traffic conflicts that occurred at airports with operating air traffic control towers than at airports with no control tower. Therefore, one could erroneously conclude that a traffic control tower results in more traffic conflicts. There are several faults in this logic:

1 *Correlation versus causality*   To prevent incidents we must determine their causes. There is a great temptation to attribute a coexistence of

two factors to cause and effect. The fact that more reports describe traffic conflicts at airports with towers, does not imply that the towers are the cause of the conflicts.

2  *Normalization by number of opportunities*  As a general rule, airports with towers have more traffic. Therefore there are more opportunities for aircraft to come too close to each other. Assuming a linear relationship between the number of flights and the number of traffic conflicts, the frequency of traffic conflicts should be normalized by the number of flights at the respective airports, i.e. the percentage of flights involved in a traffic conflict at tower and non-tower airports should be compared.

3  *Reporting biases*  There are at least two reporting biases that come into play in this example (see also the previous section on reporter biases). Pilots expect air traffic controllers to prevent traffic conflicts at tower airports. When a conflict occurs, pilots may view it as a problem with the air traffic system and therefore feel it is useful to notify the responsible safety organization. At non-tower airports, pilots are responsible for seeing and avoiding the other aircraft. If they fail to do so, the pilots may feel culpable and perceive little benefit in reporting the incident; particularly, if no one else is in a position or likely to follow up on the error for reporting or enforcement purposes. As discussed earlier, a second reporting bias pertains to who reports the incidents. Due to exposure to reporting systems and other factors, airline pilots are more likely than general aviation pilots to report incidents. Since a larger proportion of the airports served by airlines have traffic control towers, one would expect to have a larger number of reports of conflicts at airports with towers. This example demonstrates the need for caution when formulating conclusions from incident data.

The most important factor in drawing conclusions from analyses of voluntary incident data is that the relationship between incidents that are reported and those that occur is not known. The reports are not necessarily representative of the population of events, therefore the only valid conclusion relates to the incidents reported, not the population of events. It is necessarily true that if an attribute is reported it will also be present in the population of all events, in some number. This presence in itself may be of sufficient consequence to draw important conclusions regarding safety. For example, an analysis of incidents revealed that distractions arising from non-essential activities contributed to safety problems (Monan, 1978). It was not possible to determine what percentage of the problems resulted from non-essential activities; however, the safety consequences were sufficient enough to warrant corrective action. The 'sterile cockpit' rule was implemented, which prohibited non-essential activities at altitudes below 10 000 ft.

For many topics the sample of reports submitted can be assumed to be stable. It is probably quite reasonable to assume, for example, that the distribution of altitudes at which altitude deviations are occurring in air carrier operations is represented by the altitude deviations reported to a voluntary reporting system. It is difficult to construct an argument that a report would be more likely to be filed at one altitude than another in the same type of flight operation. However, if an incident database contains more reports from air carrier than general aviation pilots, one cannot consider these data as representative of all flying activities. Many general aviation aircraft operate at lower altitudes than those flown by air carriers. Therefore the distribution of altitudes at which altitude deviations are reported by air carrier pilots should not be used to draw conclusions regarding all altitude deviations.

If a stable sample of reports can be identified, time series analyses and pre-post analyses may be legitimate. These can be quite powerful as a metric of safety. When a new procedure was put into effect for arrival into Denver, Colorado, many reports were received describing altitude deviations. The Aviation Safety Reporting System notified the Federal Aviation Administration of these problems. A revision of the navigational chart was made, which resulted in a decrease in the number of reports received pertaining to that particular arrival. It is reasonable to conclude that the revision ameliorated the problem with altitude deviations on that arrival. This conclusion is supported by the fact that reports continued from another location where no chart revision was made. Note that when looking at the number of reports over time, caution must be taken regarding the influence of such factors as media and enforcement policies. (See the previous section on reporter biases.)

## Analysis methods for incident reports

### Qualitative uses of incident data

Single incidents can be very informative as to the hazards of the aviation system. This fact was brought under the spotlight in 1974 when a TWA flight crew misinterpreted an approach chart and had a misconception of certain air traffic control procedures. The crew descended below the minimum safe altitude and struck the ground. The investigation uncovered an identical incident (with no fatalities) by a crew of another airline six weeks earlier. This incident was reported to that airline, the issue was addressed and disseminated to that airline's flight crews;

however, there was no easy way to circulate the information to other flight crews. These events stimulated the creation of a national incident reporting programme in the United States.

A case study of a single incident may be all that is required, for example, to change a cockpit procedure or an air traffic procedure. Often, however, many reports are required for an analysis to determine the scope of a type of event, or the extent to which those events represent a safety hazard. A suspected hazard can be confirmed by incident reports. Note that the failure of a database search to produce incidents of a particular type must not be interpreted as conclusive evidence that the problem does not exist. (See the previous discussion on the retrieval of incident data.)

The most extensive qualitative analysis known to the author is the work of Kraft and Buntine (1993), who analysed 5200 altitude deviation reports and classified those reports based upon the characteristics of the incident and the reporter's type of operation. The number of qualities forming the category of reports and the strength of the cohesiveness of that category were determined in their analysis process. This technique represents a unique approach to analysing incident data that revealed some unexpected and enlightening classes of altitude deviations.

*Quantitative uses of incident data*

The most common and often the only valid quantitative analyses of incident data are descriptive, rather than inferential. (See the next section on statistical tests.) Probably the most salient analysis involves a categorization of incident types and a report of the relative frequencies of those types. For example, Chamberlin (1991) studied rejected take-off incidents reported by flight crew members. The 168 reports of this type became the population of occurrences, about which conclusions were drawn. The study identified 94 reports of rejected take-offs that were caused by flight crew errors. These errors fell into five categories. The relative frequencies of the error types were informative to those constructing pilot training programmes.

Several caveats apply when conducting a study of the relative frequencies of incident types. First, an event of each type must be equally likely to be included in the reports retrieved on the topic of interest. As described earlier, the search strategy used to retrieve the reports should not bias the likelihood of one type of report being selected over another type. Secondly, care must be taken in determining the level of aggregation of categories so that the frequencies are large enough to be meaningful while preserving essential details. The appropriate level of aggregation of the data under study should be determined by the data and the target audience. Thirdly, if the categories are not mutually exclusive, for example a report may have several types of errors described, this must be considered in the evaluation. This does not present a problem; however, it may change the unit of measure. In this example the unit of measure becomes errors and not reports.

When looking at the relative frequencies of incidents, it is often necessary to normalize the number of incidents by the number of operations of a particular type. As a result, hazards may be uncovered that were previously masked. For example, Lyman (1981) normalized the number of controller reports from a facility describing an air traffic control error by the number of aircraft serviced by that facility. The two were significantly correlated. Had a facility shown a disproportionate frequency of errors, action directed at that facility could be taken to rectify the anomaly.

A powerful technique when exploring a database for safety issues is to try multiple methods of looking at the data. Hazards are sometimes invisible unless viewed in a particular way. For example, it may not be possible to uncover a safety problem when looking at all the events, but a distribution by location may reveal that there is an unacceptable number of occurrences at a particular location, while other locations have a good safety record. It is valuable to take many cross-sections through the data to uncover factors that may be critical.

## Statistical tests for voluntary incident data

Given the nature of voluntarily submitted incident data, the number and type of valid statistical tests are extremely limited. Often it is inappropriate to test hypotheses; only descriptive statistics are valid. Many statistical techniques customarily employed in the human factors field, such as analysis of variance, are not appropriate for use with voluntarily submitted incident data. The data cannot, in general, be assumed to be normally distributed. The data categories are often not mutually exclusive. Many parameters are not continuous or even ordinal. The population is the incidents that are reported, not the incidents that occur. These factors pose a challenge for the researcher.

In describing a set of data from a voluntary incident database, the frequency and percentage of a particular type of report are useful statistics. For example, perhaps 40% of all reports received from pilots and controllers describe a pilot deviation from an assigned altitude. Of those altitude deviations, maybe 87% involve an overshoot/undershoot when climbing/descending to a newly assigned altitude. Perhaps 10% involve excursions from level flight at an assigned altitude, and 3% fall into the 'other' category. These percentages may be useful to the developer of an algorithm for alerting pilots to altitude deviations. Such statistics may first provide support for a need to improve existing systems, since the relative frequency of this type of incident is high (40%), and the developer may optimize the alerting algorithm for climbing/descending, rather than level flight, due to the prevalence of those incidents.

*Analysis of proportional frequencies*  An extremely useful inferential statistic for incident data is the $\chi^2$ (chi-square). This statistic can be used to determine if two or more incident types differ in the proportions of reports falling into various classifications. Both the incident types and their classifications must be mutually exclusive. In addition, the expected values of the cells must meet the minimums for the $\chi^2$ distribution. Suppose an airline were interested in whether the distribution of *reported* altitude deviations across phases of flight is the same for aircraft having advanced flight management systems as for those without flight management systems. The sample of incidents would consist of all altitude deviations. This sample would be broken into two types: aircraft with and aircraft without flight management systems. The incidents in these two groups would be placed into cells as to the phase of flight (see Table 8.1). Each of these cells is mutually exclusive. The $\chi^2$ statistic can be used to determine if the distribution of altitude deviations across phases of flight is the same for aircraft with and without flight management systems. However, if one were interested in the type of human error that led to the altitude deviation, the $\chi^2$ statistic would not be useful, since an altitude deviation may include more than one error type and therefore the categories would not be mutually exclusive.

**Table 8.1  Altitude deviations for aircraft with and without Flight Management Systems (FMS) by phase of flight**

|        | Climb | Cruise | Descent |
|--------|-------|--------|---------|
| FMS    | 35    | 7      | 47      |
| No FMS | 58    | 12     | 79      |

It may be possible to construct research questions in such a way as to create mutually exclusive categories, even though these may not all be useful. For example, if one were interested in who detected the altitude deviation in aircraft with and without flight management systems, one may have three categories: detected by the pilot flying, the pilot(s) not flying and detected by both the flying and not flying pilots at the same time. If very few incidents occurred where both pilots detected the altitude deviation, one may wish to discard this category in the analysis, but the frequencies should be reported. Note that this sample of incidents includes only those where it is known that at least one pilot detected the altitude deviation.

It may be informative to test the number of incidents of a particular type against an *a priori* expected value. For example, you may have reason to hypothesize that if an altitude deviation was reported it is *equally likely* that the deviation was detected by the pilot flying as the pilot not flying. In this example, there would be an expected value of half the reports in each cell. A $\chi^2$ test can be performed to see if the actual frequencies were the same as the expected frequencies.

If there is a low observed value in a cell, Fisher's Exact Test can be performed for 2 by 2 conditions or the extension of this test for $R$ rows by $C$ columns, where $R = C$. This test looks at the actual probability that the observed distribution of frequencies occurred, in addition to the probability of more extreme distributions.

*Analysis of trends* To monitor safety, it is often desirable to detect if a particular type of incident is on the increase. Techniques used in quality control are applicable for detecting trends in incident data. The rate of reports describing a particular incident type can be plotted over time. These data will follow the Poisson distribution. The data should vary around the mean of the database to date within ±2 standard deviations (see Figure 8.1). If a datum exceeds this limit, e.g. it is higher than expected, it should be investigated. Naturally a high number of reports for a particular period may be due to chance.

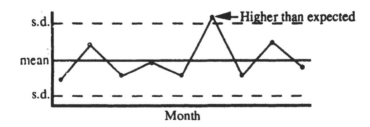

**Figure 8.1   The number of incidents each month**

Another type of test for trend analysis is a Runs Test. When the number of incidents of a particular type are measured over time, a sequential increase or decrease is a run. The number of successive points that increase/decrease is the length of the run (Figure 8.2). There are many formulas for evaluating runs; one such rule suggests that a run of seven or more sequentially increasing or sequentially decreasing points should be

**Figure 8.2  A run of eight sequential increasing points**

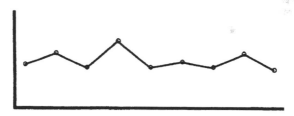

**Figure 8.3  Alternating runs of length 1**

investigated (E. L. Wiener, personal communication, 23 November 1992). The test for runs can be two-tailed; too few sequential increasing or decreasing points is also informative and may suggest a compensation is taking place (Figure 8.3).

In general, one should use caution when applying statistical tests to voluntary incident data, especially those tests where assumptions are made about the underlying distribution of the population. Statistical tests may be useful tools for exploring the data, but often are not appropriate for reporting definitive results.

## Sources of aeronautical incident data

Organizations throughout the world collect incident reports that are available to those concerned with human factors. These organizations include governments, airlines and universities. It is advisable to contact the civil aviation authority in your country to inquire as to the availability of this information. Often these data are readily available to individuals and are provided at little or no cost.

The largest source of voluntary aeronautical incident data in the world comes from the Aviation Safety Reporting System (ASRS) operated by the National Aeronautics and Space Administration in the United States (see Reynard *et al.*, 1986, for a description). Since 1975, the ASRS has collected over 250 000 incident reports from pilots, air traffic controllers, flight attendants, mechanics, ground personnel and others. Each report contains information identifying the type of flight, the air traffic control facility, the nature of the airspace, the experience level of the reporter, the type of incident, and often a detailed narrative description of the circumstances which contributed to the loss of safety. The confidential, non-punitive nature of the reporting system stimulates reports that are revealing of human error and systemic weaknesses. This organization will provide information on a wide range of human factors issues. Requests for ASRS data should be directed to:

Aviation Safety Reporting System
Post Office Box 189
Moffett Field, California 94035–0189, USA

At the time of publication, systems similar to, but smaller than, the ASRS have been instituted in Great Britain, Canada, Australia, New Zealand and Germany. The International Civil Aviation Organization also provides copies of its reports on accidents, although these must be requested by government agencies.

# References

Aviation Safety Reporting System (1977), *Human Factors Associated with Profile Descents*, (NASA Quarterly Report Number 5), Moffett Field, CA, NASA.

Chamberlin, R. W. (1991), 'Rejected takeoffs: Causes, problems, and consequences', *Proceedings of the Sixth International Symposium on Aviation Psychology*, Columbus, OH, Ohio State University.

Chappell, S. L., Billings, C. E., Scott, B. C., Tuttell, R. J., Olsen, M. C. and Kozon, T. E. (1989), *Pilots' Use of a Traffic Alert and Collision Avoidance System (TCAS II) in Simulated Air Carrier Operations*, (NASA Technical Memorandum 100094, Vol. I and II), Moffett Field, CA, NASA.

Degani, A., Chappell, S. L. and Hayes, M. S. (1991), 'Who or what saved the day? A comparison of traditional and glass cockpits', *Proceedings of the Sixth International Symposium on Aviation Psychology*, Columbus, OH, Ohio State University.

Diehl, A. E. (1991), 'Human performance and systems safety considerations in aviation mishaps', *The International Journal of Aviation Psychology*, 1 (2), 97–106.

FAA (1991), *The Federal Aviation Administration's National Plan for Human Factors*, Washington, DC, Federal Aviation Administration.

Foushee, H. C. and Manos, K. L. (1981), 'Information transfer within the cockpit: Problems in intracockpit communications', in C. E. Billings and E. S. Cheaney (eds), *Information Transfer Problems in the Aviation System*, (NASA Technical Paper 1875), Moffett Field, CA, NASA.

Grayson, R. L. (1981), 'Information transfer in the surface component of the system: Coordination problems in air traffic control', in C. E. Billings and E. S. Cheaney (eds),

*Information Transfer Problems in the Aviation System*, (NASA Technical Paper 1875), Moffett Field, CA, NASA.

Grayson, R. L. and Billings, C. E. (1981), 'Information transfer between air traffic control and aircraft: Communications problems in flight operations', in C. E. Billings and E. S. Cheaney (eds), *Information Transfer Problems in the Aviation System*, (NASA Technical Paper 1875), Moffett Field, CA, NASA.

Kraft, R. B. and Buntine, W. (1993), 'Classification and beyond: Exploring the ASRS database', *Proceedings of the Seventh International Symposium on Aviation Psychology*, Columbus, OH, Ohio State University.

Lauber, J. K. (1984), 'Research uses of aviation incident data'. Oral presentation.

Loomis, J. P. (1981), 'The performance of warning systems in avoiding controlled flight into terrain (CFIT) accidents', *Proceedings of the 1981 International Symposium on Aviation Psychology*, Columbus, OH, Ohio State University.

Lyman, E. G. (1981), *ATC contingency operations in the enroute flight regime*, (NASA Contractor Report 166231), Moffett Field, CA, NASA.

Mellone, V. J. and Frank, S. M. (1993), 'Behavioral impact of the TCAS II on the national air traffic control system', *Proceedings of the Seventh International Symposium on Aviation Psychology*, Columbus, OH, Ohio State University.

Monan, W. P. (1978), *Distraction – A human factor in air carrier hazard events*, (ASRS Quarterly Report 9), Moffett Field, CA, NASA.

Orlady, H. W. and Wheeler, W. A. (1989), 'Training for advanced cockpit technology aircraft', *Proceedings of the Fifth International Symposium on Aviation Psychology*, Columbus, OH, Ohio State University.

Porter, R. F. (1981), 'Information transfer during contingency operations: Emergency air-ground communications', in C. E. Billings and E. S. Cheaney (eds), *Information Transfer Problems in the Aviation System*, (NASA Technical Paper 1875), Moffett Field, CA, NASA.

Reynard, W. D., Billings, C. E., Cheaney, E. S. and Hardy, R. (1986), *The development of the NASA Aviation Safety Reporting System*, (Reference Publication 1114), Moffett Field, CA, NASA.

Sumwalt, R. L., III (1993), 'Air carrier ground deicing/anti-icing problems', *Proceedings of the Seventh International Symposium on Aviation Psychology*, Columbus, OH, Ohio State University.

Thomas, R. and Rosenthal, L. (1982), *Probability Distributions of Altitude Deviations*, (NASA Contractor Report 166339), Moffett Field, CA, NASA.

Vicente, K. J. (1990), 'A few implications of the ecological approach to human factors', *Human Factors Society Bulletin*, **33** (11), 1–4.

# Part 3

# New Theoretical Models

# 9 Behaviour analysis and aviation safety

*Ray Fuller*

Historians of psychology, characterizing the end of the twentieth century, will no doubt find remarkable the massive thrust of research effort focused on the mind and its operations, with the once dominant emphasis on behaviour relegated at best to a back seat, a seat which some would describe as rear-facing. If I have the tenor of the *Zeitgeist* right, behaviourism has come to be regarded as an impoverished paradigm, useful mainly for predicting the behaviour of less 'sophisticated' animals, and those animals would not include von Frisch's bees or Kohler's apes. This state of affairs is remarkable precisely because 50 years earlier almost the very opposite was the case. Researchers eschewed mentalism for the objective analysis of observable behaviour and the systematic study of conditions under which it occurred.

It is, of course, difficult not to become excited by contemporary theoretical developments. The proposition that the human brain can form concepts simply as an emergent property of its functional anatomy, without intention or active learning or heuristics or algorithms or any conscious strategies or whatever, and the demonstration of this possibility in a suitably contrived computer simulation, surely puts to the sword once and for all that sinister homunculus who heretofore was inevitably to be found lurking on the bottom line. The concept of mental models, internal representations of domains of experience, which can run forwards and backwards, which enable the pilot to be ahead of his aircraft or the engineer to understand the failure of a component, provides a rich, plausible and economical metaphor for so much that was explicitly ignored by the behavioural tradition. Furthermore, mental represent-ations are beginning to provide a powerful framework for describing a

173

major part of the competence of the expert and they thus also provide a new challenge for the training of expertise.

As cognitive psychology lurches forward it becomes progressively easier to be dismissive of its behavioural predecessor as being largely irrelevant; an essential part of the evolution of psychological thought perhaps, but not at all adapted to the demands of an inner mental life or to the metaphors provided by the era of the microchip. Nevertheless, it is perhaps timely to review where the fundamental concepts of behaviour analysis might relate to some of the broader concepts of the new cognitive approach. It is the contention of this chapter that the methods and theory of behaviour analysis, far from being displaced by contemporary trends, actually complement them. In particular it is in its contribution to the analysis of motivation and learning that a behavioural approach has enduring potential and this chapter will attempt to demonstrate this contribution through an examination of aspects of aviation safety, on both the flight deck and the ground.

## The neglect of behaviour analysis

Behaviour analysis, the application of behaviour theory, provides a functional account of behaviour, focusing on the functions served by acts and the conditions under which those functions are served. With this perspective behaviour analysis attempts to throw some light on why people behave the way they do, under the conditions in which they do it. It is based on the premise that behaviour is ultimately under the functional control of its consequences. Put simply, people behave in such-and-such a way to obtain pleasant or rewarding consequences and to avoid unpleasant or aversive (i.e. punishing) ones.

Despite the obvious relevance of this simple behavioural formulation to the understanding of factors such as motivation, the effects of incentives and deterrents and the effects of past experience on future behaviour, virtually no systematic attention has been paid to it by aviation psychologists. One can conjecture a number of reasons for this. Within psychology, the behavioural framework developed largely on the basis of observations in the animal laboratory, most specifically the operant chamber, and its relevance to human behaviour has, with perhaps the exception of behaviour modification and certain educational techniques, been made more apparent through extrapolation than demonstration. Behaviour theory has also been criticized for its exclusion of consideration of the role of an inner mental life in the determination of behaviour (including an early neglect of the role of language), a criticism which, though no longer true, has contributed to a decline in its popularity, hastened by the current, predominant, cognitive *Zeitgeist*. Furthermore the predictive power of the theory has been undermined by problems of definition and circularity.

One might hazard a further suggestion that aviation psychologists have emerged predominantly through a human factors and ergonomics tradition, a tradition which engenders the concept of the human agent as a component in a man–machine interaction. This view by and large emphasizes human performance and factors which affect it, paying scant attention to ongoing processes of learning and motivation. It is also possible, of course, that in the arena of aviation safety, motivation might never have even been at issue, simply because the consequences of unsafe behaviour for aircrew, passengers and people on whom an aircraft might crash are so unquestionably horrific. Perhaps we don't ask about motivation for air safety for the same reasons we don't ask about motivation for breathing.

Concepts of learning and motivation are not even indexed in such standard university ergonomics texts as Oborne (1987) and Grandjean (1988), even though Oborne does give learning theory about a page-and-a-half when introducing models of accident causation. Similarly, the search for a discussion of a functional analysis of behaviour in aviation psychology texts, such as Hawkins (1987), Wiener and Nagel (1988) and Jensen (1989), would be in vain. Perhaps we should not be surprised therefore that in the significant and influential ATA National Plan to enhance aviation safety through human factors improvements, prepared by the Human Factors Task Force in cooperation with industry and government (1989), a functional behavioural approach is conspicuous by its absence.

## Behaviour analysis

Behaviour analysis is concerned with what people do, what happens as a consequence and the conditions under which act–consequence relationships occur. This model is sometimes described as representing a three-term contingency with antecedent events (A), behaviour (B) and consequences (C) as the three terms, together yielding the 'A-B-C' of the behavioural model. The behaviours people learn and are motivated to continue to enact are precisely those which lead to rewarding consequences and which avoid unpleasant, aversive or 'punishing' ones. For example Chappell (Chapter 8) reports how the behaviour of safety incident reporting changed when consequences changed with the initiation of the Error Detection Program for the *en route* airspace of the continental US. This programme detected and automatically recorded

175

the flight identity and altitude of aircraft involved in loss of separation beyond a defined criterion. Pilots began to receive violations for altitude deviations more regularly. However, since the voluntary reporting of such incidents (B) conferred immunity from censure (C), pilots began reporting them with greater frequency. The behaviour of incident reporting increased because it now served to avoid an aversive consequence. Similarly Prince *et al.* (Chapter 13) note that to escape from the aversive experience of stress (C), decision makers may be motivated to engage in 'premature closure' (B), short-circuiting a more comprehensive process and thereby possibly seriously limiting the quality of the decision made. In both of these examples, behaviour is motivated by its consequences and change in those consequences changes behaviour.

In ramp operations, major consequences for the ramp driver are twofold: to reach a particular destination (often as quickly as possible) and, while doing this, to avoid aversive consequences such as collision, loss of vehicle control, damage to freight or unintentional deviation off the roadway. However, despite the aversive nature of such undesirable consequences, drivers all too frequently encounter them, yielding the enormous estimated average accident cost to the civil aviation industry of $20 000 per ramp worker per annum. In exploring why this should occur, a behavioural analysis offers both an explanatory framework and points to those variables which need to be manipulated if driver behaviour is to be modified and the ramp accident toll reduced.

Antecedent events in the three-term contingency are formally described as 'discriminative stimuli' because they prescribe the conditions under which particular behaviour–consequence relationships obtain. Recognition of these conditions involves discriminating them from other conditions which may signal quite different behaviour–consequence relationships. Thus, for example, flashing anti-collision/navigation lights on an aircraft on the ramp act as a discriminative stimulus (A) which signals that approach (B) by a ramp vehicle or worker may lead to a hazardous consequence (C) such as engine ingestion, hearing damage or collision with the aircraft should it edge further forward on the stand.

Sometimes unsafe responses are made in the presence of stimuli which are different from those which should actually set the conditions for the action but which share some deceptively common properties. Thus airways clearance may be heard as take-off clearance, probably the last significant event in a chain which led to the world's worst ever civil aviation disaster (Hawkins, 1987). Another example would be where a secondary instrument is responded to as if it were a primary, because that is what it would have been in the display configuration of a related aircraft familiar to the pilot (see for example 'Looking behind the error', 1990). A more recent example, in this case a ground handling accident (IGHC/

APRG, 1992), involved an operative mistakenly hitting an air-bridge floor levelling button instead of the identical (colour, size and shape) and adjacent button for canopy elevation, ultimately causing the bridge head to collide with and damage the aircraft door. Perhaps the most memorable, if apocryphal, example of such stimulus confusion is the case of the captain suggesting to his glum-looking co-pilot whilst approaching the point of take-off to 'cheer-up', whereupon the co-pilot retracted the landing gear. These phenomena come under the general rubric of stimulus generalization.

## Implications of behaviour analysis for learning

For any procedure to be safely completed, the operative, be he or she pilot or ramp worker, needs to have learned relevant antecedent-behaviour-consequence relationships: what to do under particular conditions to achieve a safe and efficient outcome. A large proportion of these A-B-C relationships are relatively simple and immediately obvious, for example for the ramp driver when to make steering adjustments to drive around obstructions such as an aircraft wing and when to stop to avoid collision with the aircraft. Others, however, are more complex and subtle and far less predictable, such as the selection of an appropriate speed under various ramp surface conditions to avoid losing control of the vehicle. The learning problem in this example, and in general, is that for each set of conditions, the driver has to learn the response–consequence relationship (or contingency).

These relationships do not all have to be learned through direct experience, however; they can be represented symbolically in the form of rules (Skinner, 1988) and transmitted verbally from one person to another, thereby saving an enormous amount of individual learning through 'discovery'. Such rule learning is typically a fundamental component of most training programmes. Nevertheless, rules cannot deal with every contingency that might be experienced and would in any event overwhelm even the most accomplished student. Of necessity, therefore, verbal rules are frequently expressed as generalizations. For example, to cope with the impossible task of specifying the maximum safe speed for every possible condition of ramp surface, one general rule might be that wet, icy, greasy or snow-covered surfaces require slower speeds. Fine tuning of such a rule, and its application on particular surfaces (such as where there are wide painted lines) arises only out of more direct experience of the contingencies (that is, the relationship between what the driver does and its consequences) in a trial and error fashion. Sometimes the error takes the form of an accident. Thus it is not surprising that it is the novice ramp worker, who has had least experience of direct contingencies, who is often over-represented in ramp accident statistics (Fuller, 1991a).

The learning of contingencies is, of course, not helped by the fact that many of them are unreliable. It may lead to aversive consequences to drive recklessly into the baggage hall, if for example the bags have not been securely stacked. Some of the load may fall and require reloading (perhaps after suffering damage). All of this will involve more work and more time and may result in censure from an observing co-worker, supervisor or manager. But then again, the driver may get away with the reckless driving episode. From the perspective of behaviour analysis, the driver has been exposed to a mixture of reward and punishment for the same behaviour. The fact that a specified way of operating should be unstable or unreliable should therefore come as no surprise.

Thus, in relation to the learning of safe performance, behaviour is under the control of its consequences; novice operatives need to learn the conditions (discriminative stimuli) under which particular response–consequence relationships occur; this learning, although supported by verbalized rules, usually requires direct experience and, finally, it is not helped by the fact that contingencies are often uncertain, being either unreliable or inconsistent.

## Behaviour analysis and the motivation of unsafe behaviour

Lack of reliability in the relationships between antecedents, responses and consequences not only makes learning considerably more difficult, but also provides a circumstance for gambling on unsafe behaviour if potentially aversive consequences (to particular responses) do not occur. It is sometimes the case that a required safe behaviour carries with it a 'cost' to the operative (such as working in inclement weather or exerting more effort) and, what's more, incompatible behaviour may be strongly rewarding (such as short cuts enabling escape from inclement weather or the making of less effort: see, for example, Januzzo, 1993).

A primary example of this kind of conflict on the ramp is the potential hazard-avoidance response of reduced speed which, however, carries with it the corresponding cost of a loss of time. Where time has an economic value, as in slot-sensitive turn-arounds, personnel are motivated to save it. Under such conditions they are essentially rewarded on a piece-work system, earning more time the faster they go. Not surprisingly this system is typically associated with greater risk-taking (McKelvey *et al.*, 1973; Fuller, 1989) and with more frequent and more severe accidents (Hale and Glendon, 1987). Because safety has a price (slower, more effort, etc.), if contingencies are uncertain, the operative may gamble on not paying that price. This behaviour is sometimes observed in the form of 'get-home-itis', where short cuts are taken to complete a task more quickly.

It should be noted that from a financial perspective risk-taking arising from the adoption of faster ways of doing things may be commercially adaptive over the longer term in an enterprise, if the net gains from saved time are considered to outweigh the net costs of accidents and other costs, such as decreased job satisfaction, increased stress and poor health (see Fuller, 1991b). Accidents in the last analysis therefore may be a regrettable, but nevertheless, unavoidable consequence of more productive commercial activity. The accident toll may ultimately represent what a particular organization is prepared to tolerate in achieving its social and economic goals (see Wilde, 1988).

Apart from a loss of valued time, there are other potentially aversive consequences of safe behaviour which may be social in nature. The approval and admiration of others constitute powerful rewards and reinforcers of particular behaviours, especially where an individual seeks acceptance by a group or by an individual with high status. Thus safe behaviour on the ramp may be compromised by the novice ramp worker showing-off his newly acquired competence in a reckless display of bravado, such as reversing at speed to the hold of an aircraft. The corollary of this need for social acceptance is that the withdrawal of social approval can be very punishing. Thus a relatively inexperienced co-pilot, unhappy with a decision or procedure taken by the captain, may be reluctant to challenge his/her authority for fear of censure, a social process which CRM training of course seeks to avoid.

## Behaviour analysis and Standard Operating Procedures (SOPs)

As discussed above, various events experienced by aviation personnel can have the effect of inhibiting the establishment and maintenance of safe behaviour. These events may be characterized as traps lurking to ensnare the unwary (see Fuller, 1991c). A telling example is provided in Chapter 5 by Marx and Graeber, in which they discuss the omission by mechanics to install O-ring seals on aircraft engine master chip-detectors, which led to the in-flight failure of all three engines. The unsafe procedure had a history of being successful because the mechanics had routinely been provided with chip detectors already fitted with their seals. Thus the safe procedure (fitting seals) had not been a behaviour they enacted, nor therefore had it been rewarded or reinforced. The general problem is that the 'natural' contingencies of the environment are often not adequate to establish and maintain safe practices. This may be because contingencies are not learned, antecedents of hazards are not reliable, or behaviour incompatible with safety is rewarded. As Harle points out (Chapter 7), an unsafe act may be committed over a prolonged period without aversive consequences until one day it interacts with other deficiencies present in the system and an accident ensues.

One solution to this type of problem is to prescribe required behaviour through a set of rules and then to reinforce rule-following by rewarding compliance or punishing non-compliance (Skinner, 1988). Rules are specifications of how to behave or not behave under particular conditions and as indicated earlier are equivalent to the three-term contingency of behaviour analysis (antecedent–behaviour–consequence), even though frequently the consequences of failure to follow a particular rule are not made explicit (e.g. unsafe outcomes). Thus, for example, the ramp driving rule requiring the testing of brakes on approach to an aircraft on stand does not specify (nor need to specify) the consequences of not carrying out that procedure. The aviation industry is very familiar with this 'rules' solution, typically implemented in the form of standard operating procedures (SOPs). With this perspective, SOPs may be viewed as rules which prescribe required behaviour independently of the existence of natural and reliable environmental controls.

Such rules are precisely what Orasanu (Chapter 12) invokes in her analysis of different types of aircrew decision (rule-based and knowledge-based). The rule-based decision is identified where 'a single prescribed response is associated with a specific problem configuration' such as in aborting a take-off, going-around on landing, initiating checklists on landing and in what she calls SOP talk – formulaic utterances such as clearances, radio frequencies, system status and so on. It is worth noting, however, that although the concept of these condition–action rules fits neatly with the behavioural analysis concept of antecedent–behaviour connections, much cognitive work, not captured by a behavioural analysis, may be carried out in the processing of antecedent information and in running forward a mental model so that the consequences of alternative behaviours can be anticipated before an act is committed.

The application of reinforcement for compliance with SOPs (see, for example, Geller *et al.*, 1982; Kalsher *et al.*, 1989) or punishment for non-compliance with them (e.g. censure from a supervisor) are important principles because the evidence indicates that where the following of SOPs is not supported by natural contingencies, the control of behaviour becomes transferred from the SOPs to the natural contingencies (Galizio, 1979). Such a discontinuity is clearly more likely to be a characteristic of ground rather than air operations. Thus a ramp driver may ignore a rule

(SOP) relating to ramp speed restrictions when s/he does not experience any naturally occurring undesirable consequence to breaking the speed limit (such as collision or loss of control).

This potential conflict in the control of behaviour between internalized rules (e.g. SOPs) and natural, external contingencies is echoed precisely in the training literature which contrasts traditional 'decontextualized learning' with 'situated practice'. In the former, learning is conceived as a process of knowledge transfer from teacher to student, in the latter individuals 'build robust and economical forms of knowledge shaped by the demands and constraints of their situation through continued situational experience' (Lintern, in preparation). Situated learning is therefore learning in which the contingencies of experience determine what is learned, rather than what is formally transmitted from teacher to learner. But as remarked earlier, those contingencies can seduce the operator into unsafe practices.

In their discussion of why SOPs are sometimes not followed, Degani and Wiener (Chapter 3) propose a diverse range of explanations. However this diversity can be seen as a rich instantiation of the behavioural principles of motivation discussed thus far, identifying the controlling influence of rewarding consequences (e.g. preference for manual over automated control in *individualism* and consideration of passenger comfort in *technique*); and the controlling influence of aversive consequences (e.g. use of humour to escape monotony and boredom in *humour* and, in *frustration*, avoidance of use of the oxygen mask when one pilot of a two-pilot aircraft leaves the cockpit). Furthermore their category of *complacency*, which refers to a condition of degraded vigilance, provides yet another example of a discrepancy between rule-prescribed behaviour (or SOP) and naturally occurring consequences, where the loss of vigilance may not be followed by aversive consequences for a prolonged period.

The strategy of systematically reinforcing rule-following has had demonstrable success with, for example, safety belt wearing in automobiles, where natural contingencies are not very effective at maintaining the desired behaviour (e.g. Geller *et al.*, 1982; Kalsher *et al.*, 1989; Malenfant and Van Houten, 1988). This approach would seem to have potential for a similar problem in general aviation aircraft where the wearing rate of shoulder harnesses in forward-facing seats is estimated at only 16%. The NTSB have concluded that in survivable light aircraft accidents, the wearing of shoulder harnesses could prevent 75% of fatalities and reduce serious injuries to minor or none in 79% of cases.

Rewards for following safe operating procedures on the ramp might include work privileges, exchangeable tokens, chances to win a lottery or more simply, personal attention and verbal praise. Geller *et al.* (1989)

argue that apart from other considerations, the latter should be contingent on the desired behaviour, be delivered in private, be specific to the observed *behaviour* (not the person – this is especially important if the feedback is negative) and give the recipient space to assimilate the feedback in his or her own way.

Active punishment of unsafe behaviour, to have any deterrence effect, requires both detection and the application of penalties, or at least a raised perception of the likelihood of penalty (Epperlein, 1987; Aberg, 1988). For example, drivers are most likely to conform to a speed restriction if they can both see a police vehicle and it is in a position to intercept them (Shinar and McKnight, 1985), even though the effect on the open road may be short-lived (see Makinen, 1988). Such interventions are however extremely expensive and their cost-effectiveness is open to question, particularly in terms of actually achieving enhanced safety rather than a greater compliance with regulations.

Furthermore, the strategy of policing safety reinforces the perception that safety is somehow imposed on the workforce, rather than a value 'owned' by them. The most effective management of safety occurs where each individual polices him/herself and is open to the reporting of safety-compromising incidents, as well as accidents, without fear of dismissal or demotion (Johnston, in press). Failing such an ideal, however, critical but constructive verbal feedback on unsafe behaviour can have a corrective function, particularly if it involves the recipient making a verbal commitment to behaviour change – a commitment to specific behaviour which will be substituted for the original unsafe behaviour. As mentioned earlier it is important that any criticism should be of the behaviour, not the person (see Geller *et al.*, 1989).

If all else fails, a more draconian intervention might involve the implementation of automatic monitoring systems such as represented by the 'Black box' concept, routinely incorporated in aircraft, and also types of machine intelligence dedicated to monitoring, detection and penalization, currently under development in the Autopolis component of the EC Drive programme (Harper, 1990). It is important to note that the application of additional consequences, such as punishment or the withdrawal of reward, to rule-breaking behaviour to suppress or inhibit it, needs to be consistent, otherwise control will tend to revert to the natural environment. It has been shown, for instance, that where a behaviour has been successfully deterred by being founded on *unrealistic* perceptions of getting caught (such as driving on the public roadway under the influence of alcohol), the undesirable behaviour returns once the real (and low) probability of getting caught is discovered (Epperlein, 1987).

## Global incentive programmes

Under a variety of conditions the rewarding of accumulated *records* of safe behaviour over a period has also been shown to be effective in reducing accident frequency (Wilde and Murdoch, 1982, Wilde, 1985, 1988), although there is room for better controlled studies (Friedland *et al.,* 1987). Through the provision of financial incentives, with and without social incentives such as the maintenance of peer approval and personal or group commitment with public pledges, significant improvements in accident records have been reported (Jannuzo, 1993).

Such procedures are attractive because they are non-punitive and allow individuals some sense of autonomy as to how they might carry out their tasks more safely. However, they suffer from a number of weaknesses, not the least of which is that the required behaviour is not specified for any given situation. What is specified as a requirement in global incentive programmes is not a behaviour but a consequence – a safe outcome. This creates problems for the worker who does not know the safe procedure for a particular task. Thus in their review of 'better driving' incentive programmes, for example, Harano and Hubert (1974) conclude that programme effectiveness is in part determined by the availability of the desired behaviour in the driver's repertoire.

Other problems which may arise with outcome-based incentive programmes are that operatives may become reluctant to report accidents because doing so is punishing (i.e. leading to loss of reward). Further-more, in some instances workers may develop a sense of learned helplessness. They perceive the risk behaviour of others as beyond their control and as a consequence feel relatively helpless in attempting to achieve any outcome-based reward. Geller (1993) recommends that rewards be related to what people *do* for safety, rather than to reductions in accident statistics. This procedure, he argues, attacks the root cause of most work accidents and is also achievement oriented.

## Some implications of behaviour analysis for safety management

As an example of the practical application of behaviour analysis to safety management, Geller *et al.* (1989) published details of a course concerned with behaviour analysis training for occupational safety based on over 45 years of combined behaviour management experience and developed out of a series of occupational safety projects, particularly in the automobile manufacturing industry. In this they argue that there should be a focus on accident prevention through the identification and analysis of near-misses and the design of interventions to avoid them in the future. The analysis of near-misses (or near hits) should include a review of person (knowledge, motivation, human factors), environment (including equipment) and behaviour factors; their emphasis on the latter focusing on the behaviour of the person who experienced the near-miss; the contributing behaviour of others; preventive behaviour which did not occur and the required change in behaviour for safe completion of the task under consideration.

The design of behavioural interventions (as opposed to modifications to equipment, for example) arises out of the answers to three fundamental questions:

1  What are the target behaviours (desired or undesired)?
2  What factors support the target behaviours?
3  What factors can be changed to increase desired behaviour and decrease undesired behaviour?

Geller *et al.* recommend that this entire accident prevention process should be continuous and dynamic, rather than occasional or a 'once-off', and that it should be responsive to feedback regarding its success. Accident prevention then, whether in the air or on the ground, involves a continuing process of observation, intervention, evaluation and programme refinement.

## Safety as intrinsic reward

Much of what has been expounded earlier places the controls on behaviour outside the individuals or group engaging in it, such as the rewarding consequences of praise or monetary bonus. Here the rewards are extrinsic to the individual. However, a potentially more economical and reliable procedure is to make the rewarding consequences intrinsic by enabling the individual to *internalize* those consequences. In the context of aviation safety the implication of this is that a safe outcome should become intrinsically rewarding for every task.

But how does one create this internalized value (reward) or attitude so that safe behaviour is never questioned or compromised? According to Geller (1991) the key point is that attitudes and values (internalized rewarding consequences) arise from the corresponding behaviour. Thus behaviour management techniques can be used to 'act a person into safe thinking'. The continuous practice of safe work procedures contributes to the development of the internalized value of safety.

## The control of behaviour and the behaviour of control

By and large, people do not like to feel controlled but cherish a sense of autonomy. Because of this Geller *et al.* (1989) argue that commitment to a safety policy will be the stronger the more individuals perceive their cooperation to be voluntary. To facilitate commitment to a safety programme they suggest:

1   avoiding the use of imposed 'canned' materials or recommendations emanating solely from top management or outside consultants (unless there is a concurrent strategy that will facilitate employee ownership of the programme);
2   establishing a safety committee, with majority representation of line workers, to be responsible for the design, implementation and monitoring of accident prevention programmes;
3   evaluating programme effectiveness which in turn should
     i)      provide feedback to the workforce;
     ii)     enable modifications to the programme where indicated (i.e. effort must have beneficial consequences to be sustained);
     iii)    include assessment of employees' perception of programme 'ownership' and commitment to its objectives;
4   promoting safety programmes on a continuing basis (e.g. with use of newsletters, posters, videos) and in such a way as to illustrate line-worker commitment to and ownership of the programme. The success of implementing this kind of strategy on the airport ramp, for example, is becoming progressively more evident (Spagnoli, 1993).

## The place of behaviour analysis in a cognitive world

If we pose ourselves the question 'On what is an action based?', most human factors psychologists would probably accept, at least as a working hypothesis, Rasmussen's broad distinction between behaviours that are knowledge-based, rule-based or automatic (Rasmussen, 1987). This sequence also characterizes the progression of skill development, such that a skilled action can ultimately be executed with greater automaticity and fluidity, with greater speed and using less mental capacity.

The issue at hand is where, if at all in this broad classification of action, does the A-B-C concept of behaviour analysis fit? The answer must be at all levels. Behaviour analysis situates behaviour in a framework of associated antecedent events and consequent events, thereby specifying the conditions in which particular acts will achieve particular consequences. It thus captures at one and the same time situations which trigger particular actions and the functional meaning of those actions by relating them to goals or motives. Thus both the stimulus control of behaviour (a kind of *push*) and control by consequences (a kind of *pull*) are both encompassed. And this is achieved with the added bonus of a dynamic mechanism for behaviour change: actions that are followed by rewarding consequences under particular conditions are more likely to occur (i.e. are reinforced) given a repetition of those conditions.

Behaviour analysis fits precisely the concept of rule-based action because rule-based action is precisely what behavioural statements specify: the rule linking action to antecedent events and desired outcomes. The A-B-C paradigm is a paradigm for conditional or production rules of the 'if. . . then' form, where the 'then' term may represent both behaviour and its consequences.

Automatic actions, unless hard-wired in the nervous system, such as in the case of an unlearned reflex response, are examples of the execution of rule-based actions. However, automatic actions have little or no requirement for conscious cognitive activity (working memory) and are the destined status of very many overlearned and sometimes highly complex actions (such as those involved in vehicle driving).

Knowledge-based actions at first sight appear most remotely related to a behavioural analysis. Actions in this category arise in situations where there are no established condition-action rules and the actor has to search long-term memory to explore the properties of the elements of the current situation to create a solution. Here the role of cognitive and indeed metacognitive processes is clearly paramount. Nevertheless, the exploration of the properties of elements must surely include at times the evocation of subsets of production rules of the 'if . . . then' variety concerned with the consequences of action. In other words, the internalized representation of the relationships amongst events described by the A-B-C framework.

Whatever cognitive operations are involved in a task, at some point the behaviour of the operative will involve discrimination of particular events and acting in a particular way to obtain particular consequences. What triggers a behavioural response will typically depend on different investments of cognitive activity and even simple stimulus-response relationships will require some level of pattern recognition. However, more sophisticated forms of cognition, what we think of as mental operations, often occur when the actor involved reflects on diagnostic alternatives or action consequences (as in the process of naturalistic decision making, articulated by Kaempf and Klein in Chapter 11). Where the meaning of a stimulus configuration is not immediately evident or where the consequences of response options are not immediately clear, some cognitive activity of mental simulation, recalling previous experience or derived from a more generic knowledge-base becomes of potential value. But as suggested above, even these activities may involve the exploration of actions situated in their antecedent and consequent associative networks, networks formed through a history of functional engagement with the properties of the experienced world. Furthermore, the very act of reflection itself may be potentially interpretable within the same behavioural paradigm. Thus behaviour analysis, far from representing an alternative to cognitive theory, may be regarded as complementary to it.

## Conclusion

Behaviour analysis does not pretend to provide a complete explanation of behaviour in general or the causes of accidents in particular. But what it does offer is a functional account of why particular behaviours happen or do not happen at points critical for safety. By constructing a model in which the roles of stimulus conditions, behavioural consequences and the previous experiences of the individual are meaningfully integrated, behaviour analysis provides a coherent framework for the identification and analysis of these potentially crucial variables in accident genesis and a theoretical basis from which interventions may be designed. It is to be hoped that its potential contribution to aviation safety will be more fully exploited in the future.

## References

Aberg, L. (1988), 'Driver behaviour and probability of detection on roads with temporary 30 kmh speed limit', in J.A. Rothengatter and R.A. de Bruin (eds), *Road user behaviour: theory and research*, Assen, Van Gorcum, pp. 572–577.

Epperlein, T. (1987), 'Initial deterrent effects of the crackdown on drinking drivers in the State of Arizona', *Accident Analysis and Prevention*, **19**, 285–303.

Friedland, M., Trebilcock, M. and Roach, K. (1987), *Regulating traffic safety: a survey of control strategies*, Report WS 1987-88-(2), Faculty of Law, University of Toronto, October.

Fuller, R.G.C. (1989), 'Behavioural adaptations to potential hazards in a simulated driving task', in A. Coblentz (ed.), *Vigilance and Performance in Automatised Systems*, Dordrecht, Kluwer, pp. 207–217.

Fuller, R.G.C. (1991a), 'Ergonomics of ramp safety', *IATA Ground Handling Council Ramp Safety Seminar*, Frankfurt Airport Conference Centre, Frankfurt, September.

Fuller, R.G.C. (1991b), 'The modification of individual road user behaviour', in M.J. Koonstra and J. Christensen (eds), *Enforcement and Rewarding: strategies and effects*, Leidschendam, SWOV, pp. 33–40.

Fuller, R.G.C. (1991c), 'Behaviour analysis and unsafe driving: warning – learning trap ahead!', *Journal of Applied Behaviour Analysis*, **24**, 73–75.

Galizio, M. (1979), 'Contingency-shaped and rule-governed behavior: instructional control of human loss avoidance', *Journal of the Experimental Analysis of Behavior*, **31**, 53–70.

Geller, E.S. (1991), 'Don't make safety a priority', *Industrial Safety and Hygiene News*, October, p. 12.

Geller, E.S. (1993), 'Are you ready for the revolution?', *Industrial Safety and Hygiene News*, September, 21–22.

Geller, E.S., Lehman, G.R. and Kalsher, M.J. (1989), *Behavior Analysis Training for Occupational Safety*, Newport: Make-A-Difference Inc.

Geller, E.S., Patterson, L. and Talbot, E. (1982), 'A behavioral analysis of incentive prompts for motivating safety belt use', *Journal of Applied Behavior Analysis*, **15**, 403–415.

Grandjean, E. (1988), *Fitting the Task to the Man*, (4th edn), London, Taylor and Francis.

Hale, A.R. and Glendon, A.I. (1987), *Individual Behaviour in the Control of Danger*, Amsterdam, Elsevier.

Harano, R.M. and Hubert, D.E. (1974), *An evaluation of California's 'good driver' incentive program*', Highway Research Report BO146-NTIS CAL-DMV-RSS-74-46.

Harper, J.G. (1991), 'Traffic violation detection and deterrence: implications for automatic policing', *Applied Ergonomics*, **22**(3), 189–197.

Hawkins, F.M. (1987), *Human Factors in Flight*, Aldershot, Gower.

IGHC/APRG (1992), Ground Incident Report, 2 September, 1992.

Jannuzzo, J.T. (1993), 'A handling agent's view of ramp safety', *ACI Seminar Ramp Safety: The Ongoing Challenge*, Rome, Italy, January.

Jensen, R.S. (1989), *Aviation Psychology*, Aldershot, Gower Technical.

Johnston, A.N. (1993), 'Blame, punishment and risk management', in C. Hood, D. Jones, N. Pidgeon and B. Turner (eds), *Accident and Design*, London, UCP.

Kalsher, M.J., Geller, E.S., Clarke, S.W. and Lehman, G.R. (1989), 'Safety belt promotion on a naval base: a comparison of incentives versus disincentives', *Journal of Safety Research*, **20**, 103–113.

Lintern, G. 'On enhancing training effectiveness', in preparation.

'Looking behind the error' (1990), *Flight International*, 31 October–6 November.

Makinen, T. (1988), 'Enforcement studies in Finland', in J.A. Rothengatter and R.A. de Bruin (eds), *Road user behaviour: Theory and research*, Assen, Van Gorcum, pp. 584–588.

Malenfant, J.E.L. and Van Houten, R. (1988), 'The effects of nightime seat belt enforcement on seat belt use by tavern patrons: a preliminary analysis', *Journal of Applied Behaviour Analysis*, **21**, 271–276.

McKelvey, R.K., Engen, T. and Peck, M.B. (1973), 'Performance efficiency and injury avoidance as a function of positive and negative incentives', *Journal of Safety Research*, **5**(2), 90–96.

*National Plan to Enhance Aviation Safety Through Human Factors Improvements* (1989), New York, ATA.

Oborne, D.J. (1987), *Ergonomics at Work*, (2nd edn), New York, Wiley.

Rasmussen, J. (1987), 'The definition of human error and a taxonomy for technical system design', in J. Rasmussen, K. Duncan and J. Leplat (eds), *New Technology and Human Error*, Chichester, Wiley, pp. 23–30.

Shinar, D. and McKnight, A.J. (1985), 'The effects of enforcement and public information on compliance', in L. Evans and R.C. Schwing (eds), *Human Behavior and Traffic Safety*, New York, Plenum Press, pp. 385–419.

Skinner, B.F. (1988), *Beyond Freedom and Dignity*, London, Peregrine.

Spagnoli, O. (1993), 'New training methods', *ACI Seminar Ramp Safety: The Ongoing Challenge*, Rome, Italy, January.

Wiener, E.L. and Nagel, D.C. (1988), *Human Factors in Aviation*, San Diego, Academic Press.

Wilde, G.J.S. (1985), 'The use of incentives for the promotion of accident free driving', *Journal of Studies on Alcohol*, Supplement 10, pp. 161–167.

Wilde, G.J.S. (1988), 'Incentives for safe driving and insurance management', in C.A. Osborne, *Report of enquiry into motor vehicle accident compensation in Ontario, Vol II*, Ontario, Ministry of the Attorney General and the Ministry of Financial Institutions, pp. 464–511.

Wilde, G.J.S. and Murdoch, P.A. (1982), 'Incentive systems for accident-free and violation-free driving in the general population', *Ergonomics*, **25**, 879–890.

# 10 Cognitive Task Analysis in air traffic controller and aviation crew training

*Richard E. Redding and Thomas L. Seamster*

## Introduction

In theory, if not in practice, task analysis is the most important, resource-intensive phase in training development, providing the basis for subsequent design decisions. Advances during the last several decades in cognitive science provide a new method of task analysis called Cognitive Task Analysis (CTA). By providing training models that represent the cognitive structures (e.g. mental models and memory organization) and processes (e.g. decision making and problem solving) underlying expert job performance, CTA is a revolutionary new approach to training and expert systems design that identifies how to expedite learning, and the optimal job performance models, skills and strategies. CTA differs substantially – in terms of its purposes, techniques used and results obtained – from traditional task analysis methods such as Instructional Systems Development (ISD). ISD and similar methods are behavioural rather than cognitive and are concerned mainly with training objectives, materials and behavioural performance, providing little insight into the learning process and the cognitive processes underlying skilled performance. Cognitive Task Analysis has received considerable attention in the professional and academic training literature (e.g. Cannon-Bowers *et al.*, 1991; Edwards and Ryder, 1991), often cited as one of the most promising new technologies for developing training for the complex cognitive tasks confronting today's worker – tasks such as air traffic control and aviation crew decision making (see Edwards and Ryder, 1991), the topics of this paper. Cognitive Task Analysis has been applied in aviation settings for part-task training, air traffic controller curriculum redesign, and the development of computer-based trainers for combat aircrews. In discussing the evolving role of CTA in the development of aviation training programmes, we provide a candid assessment of the problems and potential of this innovative new technology.

The ongoing evolution of CTA within aviation training can be characterized as a three phase process currently in transition from the second to the third phase. The first phase, involving basic research efforts using one or two CTA methods, generally did not integrate or actually apply the results to training design. The second phase of CTA has seen a much greater emphasis on multiple analysis techniques with an integration of the results into a validated cognitive model. A good example of this type of CTA is the two-year long comprehensive effort conducted for the US Federal Aviation Administration to analyse the job of air traffic control (Redding *et al.*, 1991), which integrated the results obtained from multiple methods into a cognitive model of the controller. To the degree that the results were used in developing the training curriculum, it also forecasts the third phase of CTA evolution, involving the integration of the results from a CTA into the training development procedure, training curriculum and training organization. This third phase includes integration of CTA and Instructional Systems Development (ISD) in a cooperative and efficient manner (see Ryder and Redding, 1993), to fully account for the knowledge base and skills needed to perform in complex environments.

We discuss CTA by describing the key research efforts from the first phase that made substantive contributions to this evolution. We then present the second phase of CTA research, with a discussion of the progress made in the integration of results across different analysis methods. Lessons learned from the second phase are addressed in the final section on ecological issues in the study of expertise. These issues must be addressed prior to the third phase when CTA will experience increasing integration within the mainstream of aviation training. The integration of a comprehensive CTA with traditional training methods can already be seen in the Advance Qualification Program (AQP). The AQP is an alternative to the traditional qualifying, training and certifying of aircrews, flight attendants, dispatchers, instructors and evaluators. It requires that the cognitive skills, such as those involved in Crew Resource Management (CRM), be trained and assessed along with the behavioural skills (FAA, 1991). These requirements present a real challenge to airline training organizations looking for ways to integrate CTA results with those from traditional forms of task analysis. The issues discussed in this paper must be addressed before the successful integration of CTA into mainstream training methods can be assured.

*What is Cognitive Task Analysis?*

Cognitive science research in the areas of memory structure, mental models in problem solving, attention allocation, skill acquisition and the nature of expertise provide the theoretical framework for CTA, while research methods from experimental psychology and knowledge engineering provide many of the data collection and analysis techniques. Cognitive task analysis uses interviewing and modelling techniques as well as experimental procedures to determine the cognitive structures and processes underlying skilled job performance. Comparisons are made between experts and those with less experience to determine how a task is learned, how to expedite learning, and optimal job performance knowledge and skills. Instructional developers use CTA data to structure the learning process and to provide students with feedback about the discrepancy between the expert cognitive characteristics and their own.

CTA focuses on decision making and mental models, learning and skill development, the interrelationships among job concepts and task elements, and group performance as well as individual differences. Inherent in CTA is a broader definition of task analysis to include analysis of learning and skill development, and not simply an analysis of the target job performance. Most CTAs identify:

1  The key job components.
2  The knowledge and skills required for similar job components.
3  Important knowledge and skill differences between novices (perhaps intermediates), and experts, or between good and poor performers.
4  The conditions which best facilitate learning.

However, cognitive and behavioural methods are not mutually exclusive; indeed, they ought to be complementary (e.g. see Johnson, 1988), with researchers having proposed their integration into a comprehensive task analysis framework (Ryder and Redding, 1993). Jobs that would most benefit from CTA, given its focus on cognition, are those that have significant problem-solving or decision-making components, require large amounts of knowledge to support performance, or place heavy demands on human attention involving substantial mental workload or competing task demands (multi-tasking). Additionally, CTA is perhaps the most appropriate method for analysing jobs which experts have considerable difficulty verbalizing or demonstrating through overt actions, or for which there is considerable variability in individual decision making. Guidelines for conducting CTA are provided in Lesgold *et al.* (1986), Redding (1992a), and Ryder and Redding (1993).

Cognitive Task Analysis is particularly well suited to aviation tasks due to increases in the automation of modern aircraft and air traffic control systems, as well as the increasingly complex environment in which aircrews and controllers must operate. Air traffic controllers must interpret radar and systems data quickly and accurately in an ever-increasing load of air traffic requiring multi-tasking and complex decision making; complex avionics and information display systems place greater cognitive demands upon aircrews; and combat aircrews must divide and prioritize their attention between many competing tasks to make life and death decisions (Edwards and Ryder, 1991). These jobs share common characteristics including time-constrained, multiple tasks, teams of operators that need to coordinate to perform tasks, and the jobs require expertise entailing not only knowledge of multiple domains but also the ability to implement a series of solutions within severe time constraints. Cognitive task analysis is also useful in providing the data needed for the development of part-task trainers which can simulate the cognitive demands and important skill components of aviation jobs, such as training the spatial components of air traffic control tasks (Schneider *et al.*, 1983).

*Cognitive Task Analysis methods*   There are many methods used in CTA, and the following provides an overview of the most common generic techniques: cognitive interviewing, protocol analysis, psychological scaling, neural network modelling, cognitive and performance modelling, and error analysis. The first two categories of methods share the most in common with those used in traditional task analysis. Psychological scaling and neural network modelling use methods from experimental psychology. The other methods, as well as protocol analysis, entail an ecological approach to task analysis in which the actual job performance of workers is observed, reported or discussed. (It should be noted that observation and open-ended interviewing may also be useful in orienting the analyst to the job and in providing initial insights.)

*Cognitive interviewing* techniques are semi-structured or structured interviewing methods which use questions designed to elicit information about mental processes. Frequently, the questions selected are based upon a particular cognitive theory of expertise and/or are designed to elicit answers suitable for the framework selected for representing or using the results. Zachary (1986) uses a detailed interviewing protocol for analysing decision-making tasks which results in identification of the decision maker's problem representation, goals and strategies. One form of interviewing uses probe questions embedded in a structured interview format, after the discussion of past performance. The critical decision method (Klein *et al.*, 1989) uses a semi-structured set of probe questions following expert's recollection of critical or non-routine incidents. The probe questions are designed to elicit goals guiding performance, options evaluated, perceptual cues used and relevant situational factors.

*Protocol analysis* involves having workers think aloud, either concurrent with, or retrospective to, performing a task. Their verbalizations are then used to infer their thought processes (see Ericsson and Simon, 1984). Retrospective reports can be obtained by asking participants to recall their performance either with or without the aid of additional stimuli such as a videotape recording of the session. As with cognitive interviewing techniques, protocol analysis permits relatively direct access to cognition. Probe questions often are asked while the individual is actually performing the task (Lesgold *et al.*, 1986; Means and Gott, 1988). Hoffman (1987) suggests using different types of problems (e.g. limited-information problems, time-constrained problems, difficult problems) to elicit the range of cognitive processing strategies used for the job under a variety of work conditions. Siegler's (1981) method, for instance, uses a rule-based approach which permits determination of the missteps or errors made by constraining the potential solutions to the problems presented. The results of protocol analysis can include sets of production rules, decision trees, heuristics, algorithms and means-ends hierarchies.

*Psychological scaling* techniques are useful for determining knowledge structures, which cannot be derived readily through more direct methods. These techniques involve having workers make judgements through sorting, recalling, rating or ranking task-relevant knowledge. Proximity estimates for all pairs of the concepts are then derived by measuring interresponse times, output orders or similarity ratings. These proximity estimates typically are used to infer the underlying knowledge structure (e.g. how key job concepts are organized in memory). The assumption is that concepts which are more closely related psychologically will have higher proximity estimates. Popular computer statistical analysis techniques include multidimensional scaling, hierarchical cluster analysis and network analysis. The results often include spatial representations showing conceptual clusters or salient distinctions between concepts (e.g. from multidimensional scaling), tree diagrams showing the hierarchical organization of concepts (e.g. from hierarchical cluster analysis), or networks showing the complex interrelationship among concepts.

*Neural network modelling*, a relatively new methodology, provides cognitive models which represent the neurophysiology of the human brain, depicting a large number of neuronal units interacting with each other. These units represent mental hypotheses, goals or characteristics, which interact by sending activating or inhibiting signals to other units, resulting in a pattern of interconnections. The flexibility of neural nets allows them to better cope with erroneous cues when retrieving information, and to better generalize by identifying elements common to the retrieval cues. These characteristics make neural nets realistic and robust models of human memory. Neural networks have been tested recently because of the limitations of rule-based production systems in modelling aeronautical decision making (Schvaneveldt *et al.*, 1992), which do not function well in novel situations since they are limited to the rules and conditions already specified in the rule set. Neural nets also have the capability to learn from new situations while rule-based systems have to be modified through the addition of new rules. Despite their promise, results from neural network models have thus far been mixed, as illustrated by the findings of Schvaneveldt *et al.* (1992), wherein neural net models of a fighter pilot's selection of air combat manoeuvres produced far greater variability than was desirable. In addition, the neural network's learning process was problematic because the researchers were not completely successful at 'training' the neural network models. Accurate output often can be obtained from a neural network, but the model may provide little insight into the structure and organization of the cognitive process being modelled. Thus, neural network analysis may be more valuable in database development for decision-aiding and expert systems than for curriculum development, where it is important to model the process as well as the output.

*Cognitive and performance modelling* is a relatively recent development entailing construction of a process flow or computer simulation model to represent job performance. Knerr *et al.* (1985) used cognitive modelling to analyse two aircrew tasks, dive bomb delivery and air-to-air radar intercept, using micro-SAINT (Systems Analysis of Integrated Networks of Tasks), a software package designed to model human–machine interaction. Micro-SAINT simulated expert pilot performance by using a rule-based model of the pilot's decision making. Once the expert model had been validated, it was systematically degraded (by omitting certain rules) to simulate the performance errors of inexperienced pilots. Models may be varied by altering their parameters or the environmental conditions to determine the effects upon performance, allowing for systematic investigation of ecological variability. These derived models are therefore useful in constructing databases for intelligent tutoring and expert systems. Although models are difficult and time-consuming to construct, modelling techniques may be useful for deriving the knowledge and skill requirements for complex aviation jobs (see Redding *et al.*, 1991).

*Error analysis* is the systematic analysis of operational or performance errors to determine the relationship between errors types and cognitive processing. This analysis can provide insights into decision making, particularly in critical situations, which would not be gained through the study of routine, error-free performance (see Rasmussen *et al.*, 1987). The information obtained is useful not only in determining thought processes but also in designing man–machine interfaces (Norman, 1984) and in developing judgement or critical incident training programmes (Redding, 1992b). Observation, accident data and self-report are three methods that have been used to study error in aviation environments (Nagel, 1988). A number of frameworks for classifying errors and relating them to human–machine mismatches, task models or human processing characteristics have been proposed (Rasmussen *et al.*, 1987); these frameworks differ in the types of errors they address, e.g. whether diagnostic errors, planning errors, monitoring errors or control errors. Reason (1987) relates errors to various stages of human information processing and maps error types to cognitive processing failures. Aviation errors specifically involve display, control, communication and decision-making errors (Nagel, 1988), which frequently can be traced to faulty mental models.

## Applications in air traffic control training

Air Traffic Control (ATC) presents a special challenge for CTA. The early CTA research analysed the domain knowledge underlying jobs involving primarily sequential, deliberate reasoning as opposed to those such as air traffic control involving task prioritization and workload management. Air traffic control occurs in a complex, time-constrained, multi-tasking, team-work environment. It involves a special form of expertise including not only domain knowledge but also efficient problem-solving and workload management strategies that must be implemented within the time-critical demands of the task. As such, ATC must be studied not through simplified laboratory tasks, but under ecologically valid conditions (such as using simulators) which capture the real-time aspects of the job.

### Early research

A number of early studies provided the basic research on how controllers perceive and organize radar data. This early research provides specific examples of CTA techniques, illustrating how their results provide support for some of the central findings of the more comprehensive analyses discussed later. We do not present all the research; rather, we provide a representative sampling of previous research to demonstrate the evolution of CTA.

Whitfield and Jackson (1982) used several different techniques, including interviews with student controllers, protocol analysis of controller briefings, and retrospective protocols collected through a video replay of actual control sessions. Because it is unreasonable to obtain concurrent protocols from controllers actively working a live sector, they used volunteer controllers to look over the shoulder of the working controller and to discuss the sector 'picture.' Controllers grouped the aircraft based on inbound and outbound aircraft for the different airports and identified the unusual or non-routine events in the sector. They would describe the sector events in relation to abnormal aircraft that were outside the usual routes and posing a conflict with the normally-routed aircraft. These sector events generally were related to sector-specific knowledge which required controller familiarity with the traffic patterns in the sector to be able to identify the normal groupings such as aircraft arrivals, as well as the abnormal events. These findings support those of the Redding *et al.* (1991) research, which found that expert controllers organize information according to sector events, and the similar findings of Schlager *et al.* (1990) that controllers organize information by grouping together aircraft involved in the same traffic patterns. Several valuable research tips also resulted from the Whitfield and Jackson (1982) study. First, it was noted that controllers had difficulty describing their mental picture in verbal terms. Second, the researchers suggest that presenting controllers with an unfamiliar sector may be useful in determining how strategies emerge.

Roske-Hofstrand (1989) introduced the use of eye-movement video data to determine aircraft grouping. By supplementing visual recordings of job performance with eye movement data, it is possible to tell what a user is looking at when performing specific actions. Controllers also were asked to view their actions and to provide a retrospective protocol; the controllers were asked to indicate the boundaries between major events and to identify and comment on each major action. Results from the combined protocol and eye-movement data show that controllers do not organize aircraft by proximity, which supports the findings of Schlager *et al.* (1990). Instead, they organized aircraft into sector-specific flow corridors based on aircraft performance within the groups. Thus, there is strong evidence that the experienced controller's mental model of a specific sector's traffic is not based on aircraft proximity but rather on a set of events related to the sector and the tasks being performed.

## Mental models and knowledge structures

A primary reason for conducting CTA of air traffic control is to identify the mental models and knowledge structures expert controllers use to achieve efficient and superior performance. From a training perspective, there is evidence that individuals with an appropriate mental model learn more quickly and are better able to retain procedures (Kieras, 1988). First, it is useful to distinguish between mental models and knowledge structures. A variety of terms have been used to describe the mental 'pictures' of controllers: mental organization (Roske-Hofstrand, 1989), controller's 'picture' (Whitfield and Jackson, 1982), mental represent-ations (Schlager *et al.*, 1990), conceptual structures (Harwood *et al.*, 1991), and mental model (Redding *et al.*, 1991). Although these are all related concepts, they can be divided into two fundamentally different constructs – mental models and knowledge structures – which have substantially different implications for training system design. Most controller CTAs have focused on mental models. Mental models are dynamic, functional representations which support deductive reasoning and problem solving for the job, as constrained by whatever relevant information is contained in working memory. They are important in maintaining situation awareness about an evolving aviation job situation (Sarter and Woods, 1991). By contrast, knowledge structures are more stable forms of representation relating to how the controller organizes job-related concepts in long-term memory.

Specific job knowledge can be linked to individual controller tasks, and it is in this way that knowledge structures and mental models are related: the knowledge structures are a basic form of organization used to store job knowledge, while the mental model is used by the controller in the real-time environment to access that knowledge in order to perform controller tasks. Mental models and knowledge structures are closely interrelated, and because they are derived through similar data collection and analysis procedures, they often are not clearly distinguished by those conducting Cognitive Task Analysis. Although the distinction between mental models and knowledge structures is not firmly established, a meaningful distinction has been implied by a number of researchers. For example, in discussing device mental models, Kieras (1988) makes the distinction between the 'how it works' and the 'strategic' knowledge. The 'how it works' knowledge (i.e. knowledge structure) consists of information about how device components operate. The 'strategic' knowledge (i.e. mental model) is made up of information and strategies for utilizing the 'how it works' knowledge to more efficiently perform a task.

Although a mental model may include both types of knowledge (Kieras, 1988), with the knowledge structures providing the building blocks for the functional mental models (Johnson-Laird, 1983), 'it is the dynamic computational ability of a mental model beyond that . . . background knowledge that provides the notion with its theoretical utility' (Wilson and Rutherford, 1989, p. 625). We maintain that it is essential to distinguish between these two types of knowledge when performing task analysis. This is because, as discussed below, the two types of knowledge often require different types of data analysis and validation procedures, and have different training applications. In discussing the application of CTA to air traffic controller training, we first present the analysis of knowledge structures followed by a discussion of the analysis of mental models. All the studies utilized similar methods, with most using more than one method to obtain a measure of convergent validity. The distinction between knowledge structures and mental models is important, especially in understanding the evolution of Cognitive Task Analysis. A significant advance is seen in the current phase of CTA evolution which combines the analysis of knowledge structures with the analysis of dynamic mental models. If CTA could provide a comprehensive accounting of knowledge structures as well as mental models, it would make a substantial contribution to training development.

*Analysis of controller knowledge structures*

The air traffic controller must learn about a large number of topics including aircraft, airports, controller facilities, radar systems, radio and communication systems, arrival and departure operations, flight and approach operations, clearance instructions, weather and traffic management. The way controllers organize their knowledge was investigated by Harwood *et al.* (1991). They collected relatedness data about 17 ATC knowledge topics from 11 controllers, and converted it into proximity estimates to generate memory networks. The controllers' networks resulted in substantially fewer links between the concepts as compared to those of a group of non-controllers. (It would have been more meaningful had they compared novice controllers at various stages of training with the more experienced controllers.) These results are consistent with those in other domains where expert rating data tends to generate simpler networks with fewer links (Schvaneveldt *et al.*, 1985). In addition, the experienced controllers had a greater number of common conceptual links, another characteristic of expertise. This analysis of knowledge structures exemplifies an important type of analysis, not frequently performed with air traffic controllers, that could provide the organization for presenting the diverse controller knowledge topics.

Harwood and colleagues propose that controllers may have a number of different cognitive models which can be placed on a concrete to general continuum, and point out that the challenge is to select the proper model for training design. Our proposed distinction between mental models and knowledge structures may provide a partial solution to this problem. Mental models can be used in the design of user interfaces and training systems to represent the dynamic aspects of controller tasks, and to teach the efficient organization of real-time sector- and task-specific data needed to produce superior controller problem solving. The knowledge structures, on the other hand, are more useful for diagnosing conceptual errors and for providing a structure for organizing the vast amount of background knowledge that must be learned before one can efficiently control aircraft.

## Cognitive Task Analysis of controller mental models

The first systematic CTA of air traffic control was conducted by Schlager *et al.* (1990). In addition to protocol analysis, they used several recall techniques including having controllers draw the location of aircraft from specific scenarios, circle the aircraft that they grouped together, and then write in the relevant flight strip information for a subset of aircraft that had been in the scenarios. The recall data was analysed both for accuracy and for the groupings of items recalled together. (Flights strips are paper strips, each containing a record of each aircraft's flight information, including the call number, expected route, altitude, etc. The flight strip is provided before the aircraft enters the controller's sector and currently is manually updated by the controller as the aircraft progresses through the airspace.)

The findings can be categorized according to the data collection method. The protocol analysis provided specific information about control strategies used in particular situations. The traffic drawing task showed that controllers aggregate aircraft into groups according to important sector features to increase efficiency and reduce memory load. The recall task showed that more information was recalled about aircraft requiring control and that the type of information recalled varied as a function of the control actions taken. The most important finding is the 'chunking' effect found in the traffic drawing task: expert controllers reduce their cognitive workload by organizing the radar data into meaningful groups according to aircraft patterns or events. These patterns, in turn, inform the controller about what strategies to use and facilitate the prediction of future traffic events. The researchers concluded that 'what appears to the uninitiated eye . . . as chaos on the radar screen is actually a small number of chunks of information' (Schlager *et al.*, 1990, p. 1312). They suggest that training time could be reduced by teaching students to identify traffic patterns, group aircraft, and to associate certain control strategies with particular traffic patterns.

A major two year study by Redding and colleagues (1991) was designed as a comprehensive CTA of the *en route* air traffic control job to produce data that could be readily used in the Federal Aviation Administration (FAA) curriculum redesign effort. The FAA embarked on this effort to improve the training efficiency of air traffic controllers, with the previous training programme taking too much time and producing graduates insufficiently prepared to work in control centres. This study is best characterized as applied research aimed at producing definitive training models, representing perhaps the first use of CTA to support the development of an entire curriculum for the training of a high-performance task. It illustrates the need for a transition from an emphasis on declarative knowledge structures in the study of expertise, to the study of dynamic mental models and real-time strategies. The research analysed both how experts represent domain knowledge as well as how they put their knowledge to work under real-time conditions.

The methodology for the study was unique in several respects, with a variety of methods (including actual error data for ecological validity) used to obtain convergent validation for the findings and models. The study used specific methods for identifying strategies, mental models and heuristics, and a cognitive modelling procedure designed to model real-time, man–machine interactions in a multi-tasking environment. Eight different data collection and analysis techniques (structured and unstructured interviews, critical incidents interviews, two tests of cognitive style, paired paper problem solving, simulator performance modelling, and simulator problem solving) were used. As in the Schlager *et al.* (1990) study, simulator problem solving was used to capture performance data. Concurrent verbal protocols were obtained from controllers as they performed problems of varying levels of complexity and workload, resulting in the identification of 40 different cognitive strategies grouped into planning, monitoring and workload management categories. Performance modelling was also conducted on simulator problem solving, but this technique utilized a retrospective reporting procedure whereby the control session was replayed to each controller who was asked to describe his intentions, goals and outcome expectations for each control action taken. The data were subjected to the COGNET modelling procedure (see Redding *et al.*, 1991), resulting in an expert mental model and task decomposition of 13 key controller tasks.

This research produced a number of important new findings concerning controller strategy use, goal priorities, problem solving, and critical cues of work overload, which are described in Redding (1992b), Redding *et al.* (1991, 1992) and Seamster *et al.* (1993). We have chosen to limit this discussion to the mental model and associated controller tasks, because these results provided the foundation for the training curriculum and overall model of expertise that was developed. Developing the mental model involved three analytical processes – decomposing the videotaped controller problem solving into a set of tasks, modelling the global problem representation (the mental model), and modelling the individual tasks to the level of task subgoals. This involved an iterative process of refinement, in which each iteration provided greater detail, corrections of

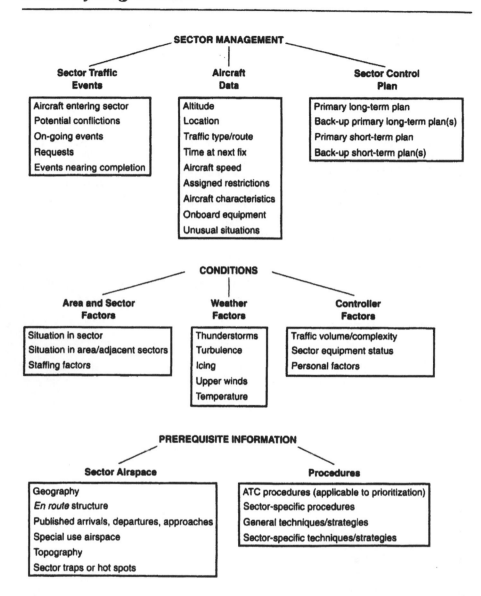

**Figure 10.1   Expert mental model of *en route* air traffic control**
(Source: Redding *et al.*, 1991)

model components, and refinement for consistency and conformity between the parts. The observable, behavioural aspects of the model were determined in early iterations because they were explicit in the videotapes of controller actions. The cognitive aspects of the models were added in subsequent iterations, because they were derived from the protocols and inferred from later interviews with subject matter experts. Cognitive operations were added to the tasks to indicate how information in the mental model was updated through the performance of the tasks. Finally, task triggers were added to the tasks, showing what patterns of information in the mental model should trigger the task. This resulted in an integrated model of controller performance including controller tasks, conditions for task initiation (task triggers), and the mental model used by a controller to understand the traffic situation.

The mental model (see Figure 10.1) contains eight panels clustered into three major categories, depicting the knowledge needed to support ATC performance. The *sector management* panels correspond to working memory and contain information about the aircraft approaching or currently in the sector, the events these aircraft are a part of, and the controller's plan for the sector. The *conditions* panels instantiate the strategies accessed from the long-term memory portion of the mental model and the ecological conditions that determine workload. These conditions include information about the present conditions of runways, airports, equipment, weather conditions that impact traffic control, and subjective factors that determine the controller's general level of stress and workload. The *prerequisite information* panels correspond to long-term memory, and contain knowledge of the physical structure of the sector and the procedures by which control is carried out. These panels contain the controller's knowledge structures; thus, knowledge structures are embedded in the long-term memory portion of the mental model.

The core of the controller's mental model – the working memory portion represented by the *sector management* panels – implies a specific decision making process. The controller perceives data from the radar screen, flight progress strips and communication with pilots, and processes these data about individual aircraft by grouping them into traffic events that must be handled. A long-term plan for controlling the sector is devised to handle the events, and is then translated into a detailed plan of specific control actions involving individual aircraft. Thus, decision making involves *events* rather than individual aircraft. By learning procedures and strategies for event types, the amount of information needed to make decisions can be significantly reduced. The schemas for event types include pointers or parameters that reference specific aircraft data, situational conditions, or other prerequisite knowledge.

The Redding *et al.* (1991) mental model is significant in that it depicts both the dynamic, real-time aspects of ATC (represented by the *sector management* and *conditions* categories within the model) and the underlying knowledge structure held in the controller's long-term memory (represented partially by the *prerequisite information* category). The model represents substantial progress in CTA evolution by combining the dynamic mental model and knowledge structure into a unified cognitive model of the controller. The mental model is also consistent with previous research and current theory in a number of respects. Schlager *et al.* (1990) found that the expert controllers recalled aircraft in groups which tended to include two to four aircraft involved in the same traffic event. These groupings were not related to the aircraft's proximity but rather to their relation to a common task. Thus, the controller's mental representation, as depicted in the Redding *et al.* mental model, is not an analogue of the radar display but is organized by events related to the sector features and the task(s) being performed. In a review of the ATC and memory research, Vingelis *et al.* (1990) concluded that controller working memory consists primarily of aircraft data (such as position) that is hierarchically

---

**Maintains situation awareness** – Maintain understanding of current and projected positions of aircraft in the sector to determine events that require controller activities.

**Develop and revise sector control plan** – Develop and revise a plan for controlling the sector that is current, comprehensive, and that handles contingencies.

**Resolve aircraft conflict** – Evaluate potential conflicts and implement means to avoid them.

**Reroute aircraft** – Change aircraft routes in response to requests or situational considerations.

**Manage arrivals** – Establish sequence and routing of aircraft for arrival into an airport.

**Manage departures** – Maintain safe and efficient departure flows integrated with other sector traffic.

**Manage overflights** – Maintain safe and efficient overflights and integration of overflights with other sector traffic.

**Receive handoff** – Accept, delay, or deny handoffs.

**Receive pointout** – Assess and accept or decline a pointout from another controller.

**Initiate handoff** – Transfer aircraft radar identification and radio communications to the receiving controller.

**Initiate pointout** – Initiate and complete pointout of aircraft to the receiving conroller.

**Issue advisory** – Provide information update to a pilot or another controller.

**Issues safety alert** – Provide mandatory safety warning to a pilot.

---

**Figure 10.2   The 13 key tasks of the *en route* air traffic controller**
(Source: Seamster, *et al.*, 1993)

organized (like the mental model panels) in terms of the importance of the information, within a three to five minute tactical window. Workload changes are a significant factor determining strategy selection. ATC procedures and personal preferences, meanwhile, reside in long-term memory and guide data search and retrieval.

The task decomposition, emphasizing the cognitive activities making up each task, identified 13 key controller tasks (see Seamster *et al.*, 1993). The tasks (see Figure 10.2) were identified by their goal structure rather than their behavioural distinctness, a task being defined as a single, goal-directed activity which would continue to completion if uninterrupted. Tasks could be composed of behavioural or cognitive subgoals, or both. The subgoals often include cognitive operations which use information from the mental model and produce new information which is then added to the mental model. Two interrelated tasks are central to all other tasks and are primarily cognitive: 'maintain situation awareness' and 'develop and revise sector control plan'. As changes in the sector situation occur, these tasks trigger a specific task to handle the event. Task triggers were also identified for each task, represented as *and/or* statements specifying the conditions for task execution.

## Training implications of the Redding et al. study

The study indicates that expertise in ATC is best characterized by an efficient mental model combined with a rapid retrieval system for applying the knowledge when needed. The expert controller's mental model has a number of characteristics allowing for this efficiency. Importantly, it categorizes aircraft into sector traffic events, giving priority to aircraft altitude, location and route in doing so. This allows the expert to work with more aircraft at a time, to better formulate a sector plan, and to use fewer control actions and strategies to work with the aircraft. These findings suggest important training implications.

The mental model provides the framework for more efficient training and learning, forming the basis for teaching ATC knowledge and skills. Based on the mental model, controller procedures should be taught by event type, with training emphasizing the integration of sector aircraft information into sector relevant groupings. This event-based approach includes training in the categorization of aircraft into event types and in the scanning for sector events. This type of scanning can improve memory efficiency by organizing sector data into chunks of related information.

The two cognitive tasks ('maintain situation awareness' and 'develop and revise sector control plan') should receive central emphasis in training, with situation awareness and planning being the primary skills distinguishing expert performance. These two tasks should be taught and practised in an integrated manner to support the performance of the other 11 key controller tasks. Moreover, by emphasizing the cognitive subgoals within each task, training can concentrate on the effective integration of the cognitive operations into the procedural and behavioural sequences for controller task performance.

Currently in development are a series of model-building exercises which allow the student to refine his or her mental model until it approximates that of the expert (Redding, 1992b). Instruction begins with mental model development to promote the efficient acquisition and organization of ATC domain knowledge. In each stage of learning, the mental model is used to guide and organize learning activities, and is elaborated throughout training. Planning skills are then taught in relation to the mental model, followed by procedural knowledge. Once students have developed their mental model and mastered basic planning and controller tasks and skills, students are taught situation awareness skills in relation to the mental model and controller tasks.

*Error analyses*

Other studies have taken a true ecological approach by using real-world accident and incident data to evaluate controller expertise, particularly as related to strategy use. Errors generally relate to a controller's selection or implementation of strategies, to communication with pilots and/or other controllers, or to the controller's mental model. Studies of controller errors provide further, albeit indirect, support for the notion that the grouping of aircraft into meaningful traffic patterns forms a key aspect of the controller's job. Langen-Fox and Empson (1985) observed military controllers' performance, obtained self-reports, and classified their errors. (The self-report data had to be largely discarded because controllers were reluctant to report their errors.) They found that the total number of errors varied according to the number of speaking units on the communication frequency. Since aircraft within the same group are generally assigned the same communication frequency, the error rate relates to the number of aircraft groups. These findings provide ecological validation for the results of the Redding *et al.* (1991) analysis.

One component of the Redding *et al.* (1991) CTA was an analysis of the summary statistical data maintained by the FAA for operational errors occurring during 1989, and a gross-level content analysis of 46 controller incident statements obtained from control centres throughout the United States (Redding, 1992b). The data indicate that a failure to maintain adequate situation awareness is the likely cause of most errors. Errors typically occurred during traffic levels of only moderate 'complexity' (an FAA rating of workload and problem difficulty based on factors such as

206

traffic volume, weather, staffing levels and emergencies), with an average of only eight aircraft under control, and immediately following a break period. According to self-reports, controllers were generally unaware that a critical situation was developing. Communication and coordination errors were the most common. Failures to interpret radar data accounted for the second greatest source of error; these errors could have been avoided through effective monitoring of the radar screen and flight progress strips, and by using the incoming information to update situation awareness.

These findings are consistent with those of the other data analyses conducted as part of the Redding *et al.* (1991) CTA, revealing the central importance of the 'Maintain Situation Awareness' task. Moreover, previous studies using various methodologies for assessing controller workload have also found error to be more frequently associated with average or light workload conditions. Using a variety of workload measures, Stager and Hameluck (1990) analysed 301 operating irregularities in the Canadian air traffic system and found that about 80% of all errors occurred during periods of average or below-average workload. They found attentional factors to be one of the greatest sources of error.

The Redding (1992b) findings based on actual incidents provide additional ecological validation for the new FAA training curriculum developed from the CTA, which emphasizes situation awareness. The subgoals within the 'Maintain Situation Awareness' task direct the controller to the information which is most important for updating the mental model. The task also provides information which often serves to trigger other key controller tasks, with attention frequently switching back to 'Maintain Situation Awareness' (Redding *et al.*, 1991). The results have implications for training situation awareness in the complex aviation environment. Perhaps contrary to common assumptions, simply reducing workload or increasing perceptual vigilance is insufficient to prevent error. Rather, avoidance of critical errors requires the operator to maintain situation awareness by actively processing incoming information and assimilating it into his or her mental model of the evolving situation.

Redding (1992b) has developed a situation awareness training curriculum, taught through computer simulated air traffic scenarios, consisting of six objectives taught in the following sequence:

1  Update the mental model as situational changes (i.e. perceptual events) occur at the workstation.
2  Update the mental model after execution of each task subgoal (i.e. goal-driven events).
3  Return to the 'Maintain Situation Awareness' task whenever other tasks are not being executed.

4   Recall all ongoing traffic events, relevant data on the key aircraft involved, key sector conditions, and sector control plans.
5   Maintain comprehensive situation awareness during a critical situation or job bottleneck.
6   Determine when in danger of decreased situation awareness due to work overload.

For example, the lesson for the third objective, ('return to the "Maintain Situation Awareness" task whenever other tasks are not being executed') presents students with scenarios requiring periodic execution of controller tasks. The tasks are interspersed throughout the scenarios so that there are varying time periods *between* the tasks during which unexpected situational changes may occur in the scenario. At different points during some of these intervening time periods, the scenario is stopped and all displays blanked out. Students must then answer questions concerning aspects of the scenario that changed due to the intervening situational changes. Feedback is provided through comparisons with the actual scenario and accompanying expert descriptions. This lesson provides practice in maintaining situation awareness during periods when other controller tasks are not being performed.

*Automated aiding*

Results from the error studies suggest interesting problems for the Advanced Automation System (AAS) scheduled for introduction into the air traffic control system in the mid 1990s. The AAS will provide a new, more automated controller workstation differing in several significant respects from the present one. Paper flight progress strips will be replaced by display data which can be completely reorganized with a simple data entry, and auxiliary displays and automated aids for separation and planning will be provided. Such changes will reduce the workload on controllers and eliminate many of the kinaesthetic and spatial cues upon which controllers currently rely. This may cause problems because research has shown increased error rates under conditions of low workload and the importance of spatial and kinaesthetic cues in controller performance. Hopkin (1992) and others have pointed out that automated assistance tends to eliminate many routine tasks and to cast humans in a passive role as a system monitor; this can be dangerous given human fallibility as monitors and the importance of routine tasks in maintaining vigilance and situation awareness. Perhaps the most effective automation system is a 'human-centred' one that performs many of the monitoring functions but still allows the human operator to maintain active physical involvement in the system and active cognitive involvement in terms of problem solving and decision making. The paradox here, of course, is that the benefits obtained through the enhanced reliability and efficiency inherent in automation may come at the price of adequate situation awareness on the part of the controller – it remains to be seen how this problem is worked out as the AAS becomes operational.

A *cognitive* task analysis is required to determine the impact of the AAS upon current training. Researchers at CTA, Inc. (1990) have conducted a traditional task analysis of ATC aimed at determining the differences between controller performance requirements for the present system as compared to the AAS. They began by defining the universe of events to which the ATC system must respond and the controller responses to them, looking at the inter-relationships between tasks. Display, control and communication tasks and task elements were analysed according to their cognitive and performance requirements. Although the analysis did determine the facially apparent cognitive requirements underlying the job tasks, it did not use CTA to investigate mental models, expertise differences, decision making, and so forth. Instead, the analysis was behavioural in nature (i.e. mapping inputs to outputs) involving a micro-level analysis of each task from a system functionality perspective. In contrast, the analysis by Redding *et al.* (1991) derived 13 generic controller tasks as units of goal-directed behaviour using a cognitive modelling procedure, emphasizing relationships between the tasks and the mental model, with a focus on planning and maintaining situation awareness.

Of the 373 tasks identified in the CTA, Inc. study, 48 were identified as likely to be very different under the new AAS. Although this research provides invaluable data at the micro-level, it also suggests the need for a comprehensive CTA aimed at determining how controller mental models and strategies might differ between the present system and the AAS. Since so many differences were found at the micro-level, it is likely that differences also exist at the higher cognitive levels, having important implications for training.

*Controller resource management and team communication*

Air traffic controllers, like aircrews, must function successfully as a team to manage their workload and perform their jobs. Aircrew resource management has been a fertile area of research, and training in crew resource management (CRM) is well established at the major air carriers. More recently, research has been undertaken to better understand, from a cognitive perspective, the resource management aspects of air traffic control (see Seamster *et al.*, 1992), by analysing the team communication that occurs between the radar controller and the radar associate controller as they perform their related tasks in the management of air traffic. Seamster *et al.* (1991) used an inductive approach to look at communication frequencies across different types of controller teams and workload levels. The team communications were coded and analysed at the speech turn level, a level that concentrates on one or more conti₃uous speech acts uttered by one controller for one primary purpose.

This research is interesting and important in several respects. The data collection occurred at the behavioural level by recording speech turns, but analysis of this behavioural data in relation to the 13 key controller tasks identified in the concurrent CTA effort (Redding *et al.*, 1991) resulted in useful information about the cognitive aspects of controller team communication, and provided further validation for the CTA findings. Thus, the study demonstrated the complementary nature of behavioural and cognitive analyses. The CTA analysis identified 'Maintaining Situation Awareness' and 'Developing and Revising Sector Control Plan' as the primary cognitive tasks of the *en route* controller. Support for this from the communication study is found in the team members' frequent queries about the sector situation. The importance of the 'Developing and Revising Sector Control Plan' task is supported by the radar controller's frequent statements of intent and the radar associate controller's frequent inquiries about intent. The importance of the cognitive tasks can also be seen in the analysis linking team communication to controller tasks: the two cognitive tasks accounted for more than 80% of all team communication. Additionally, the better performing controller teams had fewer disagreements, suggesting they shared a common frame of reference such as an expert-typical mental model. Thus, mental model instruction may improve controller team communication, which has been a frequent source of error in the past.

The results from Seamster *et al.* (1991) confirm that communication analysis is central to a better understanding of the cognitive aspects of teamwork in aviation, and these findings share a number of points in common with the communication between captains and first officers in the flight deck environment. In that environment, frequencies of certain communication categories have also been associated with better crew performance. Detailed communication analysis can provide an objective dimension to the analysis of resource management, which has been considered too subjective and difficult to quantify by those in the operational environment.

## Applications in aviation crew training

The evolution of CTA can also be seen in its application to air- and ground-crew training. We first present a brief study which demonstrates how the use of just one set of methods produced useful information for training aircraft mechanics. Presented next is a more comprehensive CTA of a more complex aviation job – military electronic warfare. Although this research resulted in a number of training insights and recommendations, it was a project of limited scope which did not achieve integration of CTA within the military training organization. We conclude with a discussion of aircrew decision-making research, which parallels the evolving role of CTA in aviation training.

## Aircraft repair

Eastman (1984) used relatively simply sorting procedures to investigate the proper sequencing of instruction for teaching jet engine repair to military aircraft mechanics. He presented pages from existing repair manuals in random order (each page denoted a particular procedure in the repair process) to expert and non-expert mechanics, who were asked to put the pages in the correct order and note the criticality of each procedure. None of the repairmen reproduced the procedures as they were sequenced in the repair manual, showing that the organization of the manual did not match with their cognitive organization of this procedural task. A cluster analysis of their page orderings revealed the mechanic's knowledge to be hierarchically organized. These findings show how traditional training sequencing, which sequences training according to behaviourally distinct units, may not always comport with the ideal cognitive organization of the job tasks. The study demonstrates the utility of CTA in providing data for designing training according to the architecture of cognition.

## Electronic warfare officer training

A two-year CTA was conducted on the electronic warfare officer (EWO) task in the United States Air Force (Redding and Lierman, 1990). As 'backseaters', electronic warfare officers sit behind the jet fighter pilot or in the back of a bomber, and are responsible for detecting, evading and/or jamming enemy radar. They must acquire an enormous knowledge base about the characteristics of various radar and weapons systems, about countermeasures against threat radars, and about counter-counter-measures possessed by the threat radars. The EWO must also perform a number of critical perceptual and decision-making activities requiring rapid response. The purpose of the CTA was to support the development of part-task training software and to provide recommendations for improvement in the curriculum. Experts (instructors), students and new graduates were given a sorting task requiring them to sort together related threat radar and radar concepts. They were also presented with a free recall task asking them to list specific countermeasures techniques for each category of threat radar, with a prioritization task asking them to rate the relative importance of various factors in deciding which countermeasures to use, and with paper problem scenarios to work which were analysed by protocol analysis.

A number of interesting lessons were learned from this research. It was evident that much of the data obtained were highly dependent upon the individual's own interpretation of the problem scenarios or questions as well as their prior mission experience. This finding is consistent with our hypothesis that mental models, in contrast to knowledge structures, are less stable forms of representation that vary according to environmental and situational factors, and with other research reporting considerable variability in mental models even among experts (Hanisch *et al.*, 1988). Moreover, the interactive computer-based part-task trainer which was developed failed to improve student test scores significantly because students did not make use of this voluntary training system. Students probably felt the system would not enhance their test performance since the multiple-choice course examinations tended to reward short-term rote memorization rather than mastery of cognitive training objectives such as long-term retention, conceptual understanding, mental model development and decision quality. This points to the need for more extensive testing procedures that measure these objectives when evaluating cognitive skills training programs. Cognitive skills development may not be readily evident on an immediate, multiple-choice test. Unfortunately, although the technology exists for doing so (for example, the training programme described above for teaching situation awareness skills could be used to test situation awareness), most testing procedures do not evaluate these higher-level cognitive skills.

Although they did not use the trainer, the class of students that participated in the CTA performed much higher on the examinations than did any of the other classes. Redding and Lierman (1990) speculated that the CTA procedures facilitated the students' conceptual organization and mental model development, and highlighted the importance of cognitive skills. This suggests that training should provide model-building activities such as association tasks and coaching in effective mental model development. Instructors also spontaneously reported using cognitive approaches to instruction in their teaching as a result of participating in the earliest phases of the CTA. These spinoffs demonstrate how an early front-end cognitive analysis can produce many benefits, both expected and unexpected.

## *Aircrew decision making*

Aircrew decision making has been analysed in a wide range of aviation contexts including aircrew team performance, CRM and team training. Researchers have recently addressed the lack of a unified framework in the study of ADM, with a significant problem being that all pilots and crews do not share a common decision-making process. One approach to overcoming this problem of diverse ADM processes is to study experienced crews within the framework of research on expertise (Ericsson and Smith, 1991), focusing on the identification of heuristics used to efficiently reduce the problem space and cognitive load. It is through the study of *expert* aircrew cognition where ADM research is most likely to see progress through the modelling of the common elements of expert crews, rather than by trying to account for the large degree of variance that exists with inexperienced crews. In studying expertise, ADM researchers have recognized the important role of shared mental models in team communication, situation awareness and decision making. Orasanu and Fischer (1991; see also Chapter 12) examined the relationship between crew communication and ADM. Orasanu and Fischer derived their findings by combining decision-making analyses or CTA with the lower-level results produced by the communication analysis. Their results show the value of combining communication analyses with a CTA, as was done in Seamster *et al.* (1991).

The future of ADM research parallels that of CTA in several key respects. Three 'essential' ADM research and development products were specified at a recent ADM research workshop (Lofaro *et al.*, 1992). Seen as the first essential product is a cognitive task analysis, structured by the phases of flight in a commercial aviation mission. The second product is a taxonomy of decision events. Most importantly, the third product is the specification of how the CTA and decision taxonomy would actually be used in the development of aircrew training. The group concluded that past ADM research (like CTA) lacked a common theoretical direction, and had not been integrated with training organizations at the airlines. While producing significant results, ADM analyses had not been conducted within the broader framework of the training system development process. For CRM and ADM research and development to succeed, the results from research in these areas must be integrated at all levels within the training organization. These conclusions point not only to the future direction of ADM but to that of CTA as well.

## Studying expertise in aviation environments

The analyses discussed have revealed a number of methodological problems that deserve special consideration, particularly if CTA is to meet Ulrich Neisser's (1976) challenge that cognitive science be beneficial beyond the laboratory

## Measuring expertise

A basic problem is the selection of experts as participants. Ericsson and Smith (1991) describe expertise as outstanding behaviour attributed to the expert's stable characteristics which can be reproduced under standardized conditions. Based on this definition, it is possible that some of the cognitive analyses in aviation did not properly identify expert performers and that differences in the resulting models may be partly due to this problem. A primary measure used to distinguish experts has been years of experience, seniority or rank. However, years alone are not a sufficient measure of expertise since many individuals will reach an acceptable level of performance without ever achieving a truly superior performance level. Expert selection procedures should therefore include actual measures of competence for the job, although Salthouse (1991) suggests that judgement of expertise by peers or supervisors should be avoided since such judgements are often influenced by extraneous social factors. This makes the selection of expert air traffic controllers, for instance, a particularly difficult task because researchers have traditionally relied on peer or supervisory input for the selection of experts. It is difficult to measure expertise for a complex job such as ATC, which has a number of idiosyncratic problem-solving strategies. There is, however, an urgent need to identify objective measures of controller expertise and to use those measures in the selection of experts for future studies. Possibilities include using outside, independent expert controllers, or automated assessment systems that use quantitative measures, to provide performance ratings.

## Basic domain knowledge versus situational or applied knowledge

Another problem is the identification of appropriate tasks to capture an expert's superior performance under laboratory conditions (Ericsson and Smith, 1991). Use of simulators in the analysis of controller expertise provided laboratory conditions, but sector-specific characteristics of the tasks may have prevented the experts from providing consistent superior performance since their performance was no better than that of the non-experts (Redding *et al.*, 1992). Although experts had significantly more experience in controlling, those who had been trained more recently had more experience with the airspace used in the simulated environment. Thus, the results may have been partially confounded by the fact that the tasks may have been a better measure of how well one knew the simulated airspace and/or adapted to a new sector rather than a measure of controller expertise, with its strong sector-specific knowledge component.

These findings also point to the possibility of a competence-performance distinction in controller expertise. Although the experts performed no better than the non-experts, they demonstrated extensive domain knowledge and problem-solving approaches characteristic of expertise (Redding *et al.*, 1992). While the older, more experienced controllers may have degraded psychomotor performance, they have retained a highly refined and elaborated knowledge base. Similarly, some less experienced controllers may not yet have achieved superior performance but may have already acquired the expert knowledge base. This suggests that domain knowledge may be somewhat distinct from psychomotor and perceptual skill in a task, and that experience may be a relatively good measure of expertise in the knowledge base but not of the actual, real-time job performance.

Additionally, there is evidence from other cognitive analyses that indicates at least two kinds of knowledge. Patel and Groen (1991) have identified two categories of medical expertise: scientific knowledge and the situational knowledge learned in the clinical setting. The basic scientific knowledge, usually learned in the academic environment, is necessary, but the experienced physician is not specifically an expert in this type of knowledge. Rather, the physician's expertise resides with the clinical knowledge. At present, it is not clear how the basic scientific knowledge interrelates with the clinical knowledge. Patel and Groen (1991) hypothesize that this categorization of knowledge into academic components and clinical/situational components may generalize to other domains as well. Aviation tasks may also have this type of knowledge grouping. The pilot's and controller's procedural and aviation knowledge maps well to the basic scientific knowledge, whereas the sector- or cockpit-specific knowledge is analogous to the clinical knowledge.

Future CTAs should include an evaluation of the general background knowledge, as well as the situation-, problem-, or sector-specific expertise developed in a specific cockpit, air traffic control centre, or aviation environment. We contend that both categories of knowledge, and the manner in which they interact, must be analysed to obtain a more complete understanding of expertise. This would entail a systematic analysis of the interrelationship between the mental models and knowledge structures we have identified. There is good reason to avoid analyses of only situational knowledge since the results may not generalize to the entire population of interest. But, situation-specific knowledge cannot be overlooked since it may be one of the major contributors to outstanding performance. For instance, air traffic controllers estimate that the sector-specific knowledge accounts for about 50% of all the required job knowledge.

## Predictive validity

Mental models and performance models for jobs performed in real-time, multi-tasking environments such as air traffic control are relatively complex and variable. Unlike knowledge structures, they cannot be fully validated based on their structure and content alone. A mental model may appear to have a high degree of structural fidelity with that of the expert yet lack functional fidelity because of the complexities of the aviation environment. Thus, models need to be validated based on their outputs or on the prediction of operator performance. This type of validation would be applied mainly to cognitive models such as production systems or neural nets which have the ability to actually generate performance output, and which are also the type of models most useful in modelling jobs like air traffic control and aircrew decision making.

In the complex aviation environment it is often difficult to determine which tasks are being attended to at a specific point in time, and there are often multiple ways to solve a problem. These factors make validation of the models a more difficult and yet essential step in the analysis process. For example, Redding *et al.* (1991) obtained construct validation for the mental model based on data collected from two experienced controllers solving a work overload problem. The resulting data were used to construct timelines for the problem including controller actions, situation changes and controller cognitive processes derived from the protocols. These timelines were examined within the context of the mental model and controller tasks to determine whether all the data could be accounted for by the tasks and mental model panels. Although this procedure allowed us to validate the content and structure of the mental model, it did not provide evidence that the model can actually predict controller cognitive processes or responses.

To obtain predictive validity, cognitive models should be programmable, and their predictive outputs need to be compared with the actions of the operators being modelled. A validation method is needed along the lines of that described in Ryder and Zachary (1991), conducted for a mental model that had been developed for air antisubmarine warfare. Validation was obtained through comparisons between the problem solutions of the actual operators with the solutions generated by a programmed version of the model. To test or validate the model, the operator's lower-level actions were entered into the computer model. Based on that input, the program would predict the current and next task. The model was able to predict 94% of the task instances with an average lead time of 5.38 minutes; thus, the mental model was shown to be a valid functional representation of the operator's cognitive processes.

*Toward the validation of cognitive fidelity*

Predictive validation is not sufficient alone, because a cognitive model useful in training development has the dual requirement that it produce correct results *and* have cognitive fidelity. By cognitive fidelity, we mean fidelity to the actual human cognitive structures and processes, rather than being merely a programmable model which may produce the correct results, yet mischaracterizes the cognitive model of the human operator. Cognitive fidelity exists along two continua (Anderson, 1988). One continuum is pedagogical effectiveness, and the greater the cognitive fidelity, the greater the effectiveness. The second continuum is implementation effort, defined as the effort directed towards actually putting the model to use in job performance. Again, the greater the cognitive fidelity, the greater the implementation effort. This explains in part why cognitive fidelity has not been widely achieved in modelling for training systems. The degree of implementation effort required to achieve cognitive fidelity has limited the number of successful attempts in this area because CTA has not yet been adequately integrated into the training organization.

A model, high in cognitive fidelity, decomposes relevant knowledge into meaningful units and applies that knowledge in ways that are analogous to those used by the experts being modelled. There are at least three approaches to encoding knowledge in cognitive modeling (Anderson, 1988), and these approaches help to explain the continuum from low to high cognitive fidelity. The first, referred to as the 'black box' approach, uses mathematical or neural network techniques. These models produce the correct results, but the way in which the knowledge is structured does not conform to the human learning or performance process, resulting in relatively low cognitive fidelity. The second approach, the mid-point along the continuum of cognitive fidelity, is that used in expert system development where the knowledge is extracted from human experts and is decomposed and organized in meaningful ways. However, the way in which the knowledge is applied does not correspond to the expert's mode of application. The third approach incorporates both the way that humans decompose knowledge with the ways they apply knowledge. This is the most demanding approach to cognitive modelling, but the resulting high degree of cognitive fidelity is ultimately what is needed for cognitive models used in the development of training systems.

Thus, the validation of cognitive fidelity should involve both the validation of the model's knowledge decomposition as well as validation

217

of the model's knowledge application. The former is better established and generally is based on the methods discussed in the analysis of controller knowledge structures, often involving psychological scaling. To validate the knowledge decomposition component, it is important that the researcher work with at least two different data sets and use different forms of data collection and/or analysis. For example, if a knowledge structure analysis was based on sorting data, validation of those structures could be achieved by analysing recall data for the same knowledge concepts. If agreement is obtained across those two data sets, then that knowledge decomposition would be validated. The second form of validation, that of the knowledge application, is less established, and involves a determination that the model's knowledge has been deployed in ways that mirror those of the expert or individual being modelled. One approach to this second form of validation is to have the experts verify the architecture of the model's data flow. That data flow represents the way in which the knowledge is applied, and, by verifying that it is representative of the expert's process, it is possible to achieve a degree of validation of the model's knowledge application. A more difficult but precise validation method might entail constructing constrained problems which yield different solutions depending upon the data flow (or decision) process used (see Siegler, 1981), and then seeing which solutions the experts actually provide.

Cognitive fidelity should be validated in conjunction with predictive validity. Although a model may mirror the expert's knowledge and its organization, the methods used to process that knowledge may be selected for computation efficiency rather than the degree to which they model the expert. These 'black box' models are of little value in training development because they do not specify how the trainee needs to organize and apply domain specific knowledge. The real challenge in the validation of a cognitive models lies in striking a balance between obtaining predictive validity and cognitive fidelity.

*Convergent validity and integrating results into a coherent model of expertise*

Because of the inference and translation involved in CTA, it is important to use complementary methods to produce results that can be integrated and that independently verify the key elements of the derived model(s), resulting in convergent validation. Unfortunately, the problem of integrating results from different methods has been difficult in past CTAs, since it is often one of the less systematic aspects of the entire analysis process. Most analyses have used several different methods, but a careful review of the research suggests that the methods selected for a particular analysis often did not offer independent confirmation for the critical elements of the resulting models. Redding *et al.* (1991) worked with the results from a variety of data collection and analysis procedures, and integration proved to be the most difficult aspect of the entire effort. As new results from CTA converge to establish more robust relationships between the cognitive constructs, it will be easier to integrate results and define the interrelationships between the various constructs. For the present, it is important that analysts establish relationships among the constructs and data collection procedures at the earliest stages of CTA planning and design.

## Conclusion

The outlook for future CTA applications appears bright. It is evident from the studies we have reviewed that CTA has produced research findings that provide insights into the cognitive structures and processes underlying skilled performance in aviation jobs. This has led to the development of exciting, innovative programmes for aviation crew and air traffic controller training. Challenges remain, however, if CTA is to realize its full potential. The cognitive constructs must be better defined and distinguished and, at the same time, more effectively integrated into unified models of expertise. Knowledge structures and mental models must be distinguished and better techniques developed for measuring each. Procedures to validate mental models and performance models require further development, particularly procedures for obtaining predictive validity. For CTA to become fully integrated with training system development, it is essential that the methodologies be made more efficient by automating analysis procedures to the extent possible. (One promising software program for analysing verbal protocol data is SHAPA; the analyst defines the coding scheme which the program then uses to analyse the protocols (Sanderson, 1990).)

In sum, cognitive task analysis must move beyond the stage of being an *ad hoc* collection of techniques and constructs to the production of fully integrated models based upon a unified theoretical and methodological framework (see Ryder and Redding, 1993). It is also critical that CTA continue to evolve so that is becomes fully integrated into existing training development procedures (such as ISD), and become a part of the training organization and culture. We hope that, in proposing applicable distinctions between mental models and knowledge structures and in defining some of the important issues in ecological validity, we have made a modest contribution in advancing this exciting new technology – a technology which gives us a valuable picture of human cognition supporting job performance in the complex aviation environment.

# References

Anderson, J. R. (1988), 'The expert module', in M. C. Polson and J. J. Richardson (eds), *Foundations of Intelligent Tutoring Systems*, Hillsdale, NJ, Lawerence Erlbaum, pp. 21–53.

Cannon-Bowers, J. A., Tannenbaum, S. I., Salas, E. and Converse, S. A. (1991), 'Toward the integration of training theory and technique', *Human Factors*, **33**, 281–292.

CTA Inc. (1990), *FAA Air Traffic Control Operations Concepts Volume VI: ARTCC – Host En Route Controllers* (DOT/FAA/AP-87–01), Washington, DC, Federal Aviation Administration.

Eastman, R. W. (1984), 'Investigation of airmen's conceptual understanding of a complex task', Paper presented at the Annual Meeting of the American Educational Research Association, New Orleans, LA.

Edwards, B. J. and Ryder, J. M. (1991), 'Training analysis and design for complex aircrew tasks', in R. F. Dillon and J. W. Pellegrino (eds.), *Instruction: Theoretical and Applied Perspectives*, New York, Praeger, pp. 150–162.

Ericsson, K. A. and Simon, H. A. (1984), *Protocol Analysis: Verbal Reports as Data*, Cambridge, MA, MIT Press.

Ericsson, K. A. and Smith, J. (1991), 'Prospects and limits of the empirical study of expertise: An introduction', in K. A. Ericsson and J. Smith (eds), *Toward a General Theory of Expertise: Prospects and limits*, New York, Cambridge University Press, pp. 1–38.

Hahn, H. A. (1988), 'Model for measuring complex performance in an aviation environment', *Proceedings of the Human Factors Society 32nd Annual Meeting*, Santa Monica, CA, Human Factors Society, pp. 875–878.

FAA (1991), *Advisory Circular 120–54: Advanced Qualification Program*, Washington, DC, Federal Aviation Administration.

Hanisch, K. A., Kramer, A. F., Hulin, C. L. and Schumacher, R. (1988), 'Novice-expert differences in the cognitive representation of system features: Mental models and verbalized knowledge', *Proceedings of the Human Factors Society 32nd Annual Meeting*, Santa Monica, CA, Human Factors Society, pp. 219–223.

Harwood, K., Roske-Hofstrand, R. and Murphy, E. (1991), 'Exploring conceptual structures in air traffic control (ATC)', *Proceedings of the Sixth International Symposium on Aviation Psychology*, Columbus, OH, Ohio State University, pp. 466–473.

Hoffman, R. R. (1987), 'The problem of extracting the knowledge of experts from the perspective of experimental psychology', *AI Magazine*, **6** (3), 53–67.

Hopkin, D.V. (1992), 'Human factors issues in air traffic control', *Human Factors Society Bulletin*, **35** (6), 1–4.

Johnson, S. D. (1988), 'Cognitive analysis of expert and novice troubleshooting performance', *Performance Improvement Quarterly*, **1** (3), 38–54.

Johnson-Laird, P. N. (1983), *Mental models*, Cambridge, UK, Cambridge University Press.

Kieras, D. E. (1988), 'What mental model should be taught: Choosing instructional content for complex engineering systems', in J. Psotka, L. D. Massey and S. A. Mutter (eds), *Intelligent tutoring systems: Lessons learned*, Hillsdale, NJ, Lawrence Erlbaum, p. 85–112.

Klein, G. A., Calderwood, R. and MacGregor, D. (1989), 'Critical decision method for eliciting knowledge', *IEEE Transactions in Systems, Man, and Cybernetics*, **19**, 462–472.

Knerr, C. M., Morrison, J. E., Mumaw, R. J., Stein, D. J., Sticha, P. J., Hoffman, R.G., Buede, D. M. and Holdings, D. H. (1985). *Simulated-based research in part-task training*, (FR-PRD-85–11), Alexandria, VA, Human Resources Research Organization.

Langen-Fox, C. P. and Empson, J. A. C. (1985), 'Actions not as planned in military air-traffic control', *Ergonomics*, **28**, 1509–1521.

Lesgold, A., Lajoie, S., Eastman, R., Eggan, G., Gitomer, D., Glaser, R., Greenberg, L., Logan, D., Magone, M., Weiner, A., Wolf, K. and Yengo, L. (1986), *Cognitive Task Analysis to Enhance Technical Skills Training and Assessment*, Pittsburgh, PA, University of Pittsburgh, Learning Research and Development Center.

Lofaro, R. J., Adams, R. and Adams, C. A. (1992), *Workshop on aeronautical decision making*, (DOT/FAA/RD–92/14), Washington, DC, Federal Aviation Administration.

Means, B. and Gott, S. P. (1988), 'Cognitive task analysis as a basis for tutor development: Articulating abstract knowledge representations', in J. Psotka, L. D. Massey and S. A. Mutter (eds), *Intelligent Tutoring Systems: Lessons learned*, Hillsdale, NJ, Lawrence Erlbaum, pp. 35–57.

Nagel, D. C. (1988), 'Human error in aviation operations', in E. L. Wiener and D. C. Nagel (eds), *Human Factors in Aviation*, San Diego, CA, Academic Press, pp. 263–303.

Neisser, U. (1976), *Cognition and Reality*, San Francisco, CA, Freeman.

Norman, D.A. (1984), 'Stages and levels in human-machine interaction', *International Journal of Man-Machine Studies*, **21**, 365–370.

Orasanu, J. and Fischer, U. (1991), 'Information transfer and shared mental models for decision making', *Proceedings of the Sixth International Symposium on Aviation Psychology*, Columbus, OH, Ohio State University, pp. 272–277.

Patel, V. L. and Groen, G. J. (1991), 'The general and specific nature of medical expertise: A critical look', in K. A. Ericsson and J. Smith (eds), *Toward a General Theory of Expertise: Prospects and limits*, New York, Cambridge University Press, pp. 93–125.

Rasmussen, J., Duncan, K. and Leplat, L. (eds) (1987), *New Technology and Human Error*, New York, Wiley.

Reason, J. (1987), 'Generic errors modeling systems (GEMS): A cognitive framework for locating common error forms', in J. Rasmussen, K. Duncan and L. Leplat (eds), *New Technology and Human Error*, New York, Wiley, pp. 63–86

Redding, R. E. (1992a), *A standard procedure for conducting cognitive task analysis*, Report to the Federal Aviation Administration, McLean, VA, Human Technology Inc. (ERIC Document Reproduction Service No. ED 340 847).

Redding, R. E. (1992b), 'Analysis of operational errors and workload in air traffic control', *Proceedings of the Human Factors Society 36th Annual Meeting*, Santa Monica, CA, Human Factors Society, pp. 1265–1269.

Redding, R. E., Cannon, J. R. and Seamster, T. L. (1992), 'Expertise in air traffic control (ATC): What is it, and how can we train for it?', *Proceedings of the Human Factors Society 36th Annual Meeting*, Santa Monica, CA, Human Factors Society, pp. 1326–1330.

Redding, R. E. and Lierman, B. (1990), 'Development of a part-task, CBI trainer based upon a cognitive task analysis', *Proceedings of the Human Factors Society 34th Annual Meeting*, Santa Monica, CA, Human Factors Society, pp. 1337–1341.

Redding, R. E., Ryder, J. M., Seamster, T. L., Purcell, J. A. and Cannon, J. R. (1991), *Cognitive task analysis of en route air traffic control: Model extension and validation*, Report to the Federal Aviation Administration, McLean, VA, Human Technology Inc. (ERIC Document Reproduction Service No. ED 340 848).

Roske-Hofstrand, R. J. (1989), 'Video in applied cognitive research for human-centered design', *SIGCHI Bulletin*, **21** (2), 75–77.

Ryder, J. M. and Redding, R. E. (1993), 'Integrating cognitive task analysis into Instructional Systems Development', *Educational Technology Research & Development*, **41** (2), 74–96.

Ryder, J. M. and Zachary, W. W. (1991), 'Experimental validation of the attention switching component of the COGNET framework', *Proceedings of the Human Factors Society 35th Annual Meeting*, Santa Monica, CA, Human Factors Society, pp. 72–76.

Salthouse, T. A. (1991), 'Expertise as the circumvention of human processing limitations',

221

in K. A. Ericsson and J. Smith (eds), *Toward a General Theory of Expertise: Prospects and limits*, New York, Cambridge University Press, pp. 286–300.

Sanderson, P. M. (1990), 'Verbal protocol analysis in three experimental domains using SHAPA', *Proceedings of the Human Factors Society 34th Annual Meeting*, Santa Monica, CA, Human Factors Society, pp. 1280–1284.

Sarter, N.B. and Woods, D.D. (1991), 'Situation awareness: A critical but ill-defined phenomenon', *International Journal of Aviation Psychology*, 1 (1), 45–57.

Schlager, M. S., Means, B. and Roth, C. (1990), 'Cognitive task analysis for the real (-time) world', *Proceedings of the Human Factors Society 34th Annual Meeting*, Santa Monica, CA, Human Factors Society, pp. 1309–1313.

Schneider, W., Vidulich, M. and Yeh, Y. (1983), 'Time compressed components of air intercept control skills', *Proceedings of the Human Factors Society 27th Annual Meeting*, Santa Monica, CA, Human Factors Society, pp. 161–164.

Schvaneveldt, R. W., Benson, A. E., Goldsmith, T. E. and Waag, W. L. (1992), *Neural Network Models of Air Combat Maneuvering*, (AL-TR-1992–0037), Williams Air Force Base, AZ, Human Resources Directorate, Aircrew Training Research Division.

Schvaneveldt, R. W., Durso, F. T., Goldsmith, T. E., Breen, T. J. and Cooke, N. M. (1985), 'Measuring the structure of expertise', *International Journal of Man–Machine Studies*, **23**, 699–728.

Seamster, T. L., Cannon, J. R., Pierce, R. M. and Redding, R. E. (1992), 'The analysis of en route air traffic controller team communication and controller resource management (CRM)', *Proceedings of the Human Factors Society 36th Annual Meeting*, Santa Monica, CA, Human Factors Society, pp. 66–70.

Seamster, T. L., Redding, R. E., Cannon, J. R., Ryder, J. M. and Purcell, J. A. (1993), 'Cognitive task analysis of expertise in air traffic control', *The International Journal of Aviation Psychology*, **3** (4), 257–283.

Siegler, R. (1981), 'Developmental sequences within and between concepts', *Monographs of the Society for Research in Child Development*, **46** (2), (Serial No. 189).

Stager, P. and Hameluck, D. (1990), 'Ergonomics in air traffic control', *Ergonomics*, **33**, 493–499.

Vingelis, P. J., Schaeffer, E., Stinger, P., Gromelski, S. and Ahmed, B. (1990), *Air traffic controller memory enhancement: Literature review and proposed memory aids*, (DOT/FAA/CT-TN90/38), Atlantic City International Airport, NJ, Federal Aviation Administration Technical Center.

Whitfield, D. and Jackson, A. (1982), 'The air traffic controller's picture as an example of a mental model', *IFAC Conference on Analysis, Design, and Evaluation of Man-Machine Systems*, Baden-Baden, Federal Republic of Germany.

Wilson, J. R. and Rutherford, A. (1989), 'Mental models: Theory and application in human factors', *Human Factors*, **31**, 617–634.

Zachary, W.W. (1986), 'A cognitive based functional taxonomy of decision support techniques', *Computer Interaction*, **2**, 25–63.

# 11 Aeronautical Decision Making: The next generation

*George L. Kaempf and Gary Klein*

Over the past 15 years, Aeronautical Decision Making (ADM) has received increasing attention both as a topic of study and as a topic for training. Several facts became apparent during the 1980s: that a pilot's decision making affected safety within the aviation system, that decision requirements change as cockpits become more automated, and that decision skills can be improved through training. People now view good judgement as a malleable skill that can be enhanced through training. ADM training has become commonplace in the aviation industry. Training modules that focus on decision skills can be found in all major sectors of the aviation industry including commercial, general and the military.

ADM training has had a positive impact on pilot performance. Numerous studies report that decision training reduces pilot errors. However, there is still room for improvement. We believe the most profitable avenue for advancing the field of ADM is to revisit the basic models of how people make decisions. Early investigators based their concepts of pilot decision making on decision models that reflected the level of thinking some 20 years ago. These models remain the underlying premise of contemporary models of ADM and ADM training programmes.

Similar to ADM, decision research has advanced considerably during the past 15 years. In fact, an entire field of decision research, Naturalistic Decision Making (NDM), has emerged. NDM researchers have moved the study of decision making out of the sterile, predictable laboratory and into the messy realm of normal human behaviour and the operational environments in which people work. They study how people behave rather than how they should behave. NDM provides insights into how people make decisions in both their work and domestic environments. NDM provides us the means to identify the types of decisions that people make, how they make those decisions, and the cognitive skills that enable people to operate effectively in dynamic, stressful environments.

It is time to integrate the two fields of ADM and NDM. We believe that the NDM perspective can be used to take ADM to another level. To train people to make better decisions, we must understand the decisions they make and how they make them. In addition, we must understand the cognitive processes that underlie decision making and the cognitive skills required to successfully implement those processes. NDM provides the methods and models to accomplish these goals.

The purpose of this chapter is to describe how these two fields may merge and to identify future avenues of research. We begin by describing the history and current state of ADM. Next, we describe what NDM is and some basic NDM concepts and models. Finally, we describe the implications that a naturalistic approach has for ADM.

## An abbreviated history of ADM

Many authors have contributed to this literature. The biennial International Symposium of Aviation Psychology has presented a clear image of the evolution of ADM concepts and training through the 1980s (e.g. Jensen, 1991). However, we do not intend this chapter to serve as a thorough review of the existing literature; rather it is to serve as a springboard to the future of ADM. Therefore, we will touch on only a few historical points here. We hope not to have misrepresented or minimized anyone's contributions to the field.

For many years, aviators held the belief that good pilots made good decisions and that individuals gained good judgement through experience. By 'surviving' a variety of situations, pilots learned which responses were correct under some circumstances and not under others. In addition, aviators held the belief that people are intrinsically motivated to select the safest, most correct course of action. Thus, when confronted with all of the alternative courses of action, a pilot with good judgement would choose the safest option. Obviously, pilots have been making decisions for decades. The concept of ADM is not new. Good pilots make good decisions. The radical shift in thinking occurred when aviators, researchers and trainers realized that decision skills could be enhanced through training.

Jensen and Benel (1977) provided the impetus for changing this perspective and for developing training designed specifically to improve the decision-making skills of pilots and flight crews. In a research effort sponsored by the FAA, Jensen and Benel investigated general aviation accidents that occurred in the US between 1970–1974. They classified errors that contributed to these accidents into three error types: procedural, perceptual-motor, and decisional. Decisional errors led to the majority of fatal accidents (52%) and contributed to many non-fatal accidents (35%). Jensen and Benel concluded that the number of human-error accidents could be reduced by improving pilot decision making, and that decision-making skills could be taught. Diehl (1991) reported similar findings for military and airline accident data for the period 1987–1989.

As a result of Jensen and Benel's findings, the FAA contracted Embry-Riddle Aeronautical University (ERAU) to develop and evaluate judgement training materials for general aviation. Berlin *et al.* (1982) reported the results of this work. They analysed pilot-error accidents to identify underlying concepts that might serve to focus judgement training. Berlin *et al.* developed training materials that familiarized student pilots with the risks inherent in specific flight activities and the consequences of poor judgement. The researchers assumed that exposure to such training would produce attitudinal changes causing students to avoid unnecessary risk as pilots. Berlin *et al.* evaluated the effectiveness of the training materials, and found that they improved the decision making of student pilots when compared to a control group.

Since this original evaluation of judgement training, other researchers have conducted similar evaluations, but in a variety of flight environments. Buch and Diehl (1984) reported that judgement training produced significantly better decisions among civil aviation student pilots in Canada. Diehl (1991) reported a summary of the findings of six government-sponsored evaluations of ADM training programmes that covers the period 1982–1987. Diehl's study includes the findings of Berlin *et al.* (1982) and Buch and Diehl (1984). In each of the six studies, Diehl found that ADM training improved decision-making skills.

The ERAU judgement training materials serve as the foundation for most ADM training courses. The FAA standardized this approach by sponsoring manuals that describe ADM concepts and ADM training programmes for an array of pilot audiences: student and private (Diehl *et al.*, 1987), instructor (Buch *et al.*, 1987), instrument (FAA, 1986), commercial (Jensen and Adrion, 1984), helicopter (Adams and Thompson, 1987), multi-pilot crews (Jensen, 1989), and air ambulance (FAA, 1988). The FAA also published an advisory circular (FAA, 1991) that describes the basic concepts of ADM and recommends methods for training pilots in these concepts.

## Current views of ADM

This section presents an overview of ADM as people view it today. It includes brief discussions of the concepts essential for understanding decision making by flight crews, of the ADM training modules, and of the limitations of the current understanding of ADM and ADM training. This material will serve as a basis for discussing how we can expand our understanding of decision making in the cockpit and how we can apply this understanding.

### ADM concepts

Judgement training programmes currently in use vary to some degree, but most contain the core materials described in the FAA manuals. The core elements include a definition of ADM, a model of ADM, and a discussion of pilot attitudes hazardous to aviation safety. The following paragraphs describe each of these elements. More complete discussions can be found in other references (e.g. Diehl *et al.*, 1987; Telfer, 1989; O'Hare, 1992). We end this section with a brief description of existing ADM training programmes.

*Definition of ADM*    Jensen (1982) proposed a two-part definition of ADM that remains the centre point of many ADM courses. The first component describes a headwork aspect of ADM as 'the ability to search for and establish the relevance of all available information regarding a situation, to specify alternative courses of action, and to determine expected outcomes from each alternative' (p. 64). The second component describes ADM as the 'motivation to choose and authoritatively execute a suitable course of action within the time frame permitted by the situation' (p. 64).

The headwork component describes the purely rational and probabilistic decision maker. Jensen (1982) argues that 'if it were possible to remove this part of human judgment from the second part, people would solve problems in much the same way as computers' (p. 64). Headwork is the perceptual and intellectual ability of the pilot to detect, recognize and diagnose problems, to determine available alternatives, and to determine the risk associated with each alternative (Jensen, 1989).

The second component describes the motivational aspects of the decision maker as 'an attitude or tendency within the pilot to resist non-safety related decision factors and to choose the best alternative consistent with the goals of society within the time frame permitted' (Jensen, 1989, p. 16). Biases and other 'irrational' processes affect the outcomes of the 'rational' decision-making process. Thus, people do not always choose the optimal alternative course of action.

*Model of ADM*   Current ADM training concepts are built on a very basic model of decision making (see Diehl *et al.*, 1987). It describes the decision-making process in three steps: monitoring the environment, recognizing change in the environment, and selecting a response. The decision maker actively monitors the environment as a change in the environment occurs. The decision maker then perceives the change, determines that the change requires some response, generates a list of alternative courses of action, and evaluates the relative values of each alternative. Finally, the decision maker selects and executes the most desirable course of action to compensate for the original change in the environment.

The outcome of this sequence is affected by the pressures exerted by biases and attitudes held by the decision maker. Biases and attitudes explain how a rational person may disregard important information and select a less than optimal alternative course of action.

Finally, execution of the chosen course of action is affected by established procedures and the technical skills of the crew. The crew may choose the correct option, but success of the course of action depends on the crew's ability to carry it out.

*Hazardous attitudes*   Berlin *et al.* (1982) investigated why pilots would choose a course of action that seemed counter-intuitive under the circumstances. Why would a pilot violate the 'rules of good sense' and choose to do something that in hindsight was obviously unsafe? Berlin *et al.* attributed many of these decisional errors to attitudes held by the pilots involved. These attitudes caused pilots to select inappropriate courses of action in the light of information that should have convinced them to take a different action. Berlin *et al.* identified five such attitudes: anti-authority, external control, impulsivity, invulnerability, and macho. Since the early work conducted at ERAU, several authors have investigated and elaborated on the basic attitudes (e.g. Jensen, 1989; Lester and Bombaci, 1984; Lubner *et al.*, 1991). The following describes briefly the five attitudes:

- Anti-authority is an attitude found in people who resent any authority controlling their actions. They do not like to be told what to do. They may resent someone who tells them what to do or they may regard rules, regulations and procedures as inane or unnecessary.
- External control is an attitude or belief that events and outcomes are unrelated to one's own actions. Individuals who hold such attitudes see little connection between their own behaviour and events that happen. They feel that they are not able to make a

great deal of difference in what happens to them. External control is manifested in attributions of outcomes to luck, fate, chance, or someone else's actions.

- Impulsivity is an attitude found in people who do not carefully consider various alternatives before acting. They are apt to act immediately. They do not stop to think about what they are doing or to select the best alternative. Impulsive people do the first thing that comes to their mind.
- Invulnerability is an attitude found in people who act as though nothing bad can happen to them. They feel that accidents happen to other people, but not to them. People who feel invulnerable tend to take chances and increase their exposure to risk.
- Macho is an attitude found in pilots who continually try to prove themselves better than others. People with this attitude take risks to impress others. They tend to be overconfident and attempt difficult tasks solely to gain admiration.

## ADM training

The FAA (1991) describes six steps for good decision making:

- Identify personal attitudes that are hazardous to safe flight.
- Learn behaviour modification techniques to counter the hazardous attitudes.
- Learn how to recognize and cope with stress.
- Develop risk assessment skills.
- Use all resources available in a multi-pilot flight crew.
- Evaluate the effectiveness of one's ADM skills.

ADM training addresses each of these points by providing a structured approach to decision making, recognizing and minimizing the effects of hazardous attitudes, and recognizing and overcoming the effects of stress. Most training programmes provide instruction and practice designed to affect both components of the ADM definition. The remainder of this section describes how ADM curricula present these topics.

ADM training programmes present a structured method for making decisions. The most commonly used model, DECIDE (Benner, 1975), comprises six steps to organize the pilot's thoughts and prevent overlooking anything that may be important:

- Detect that a change that requires attention has occurred.
- Estimate the significance of the change to the flight.
- Choose a safe outcome for the flight.

- Identify plausible actions to control the flight.
- Do or execute the chosen action.
- Evaluate the effects of the action taken on the change and progress of the flight.

Student pilots are taught to consciously consider the DECIDE model as they make decisions in training. As they gain experience, the pilots will gradually implement the model without this forced attention. DECIDE is not the only structured approach to decision making. Some curricula employ other models (e.g. Maher, 1989).

Identifying and assessing risks are viewed as integral parts of decision making. Thus, ADM training programmes include segments that describe the sources of risks and provide opportunities for students to practice incorporating risk assessment into the structured decision-making process. Students learn to assess the risks that originate from the pilot, aircraft, environment, operation and situation.

A second major component of ADM training concerns hazardous attitudes that contribute to decision errors. Students are taught how to recognize each attitude, how these attitudes affect judgement and safety, and how to combat their effects. The curricula include survey instruments designed for the pilots to recognize the degree to which they possess each of the hazardous attitudes. In addition, instructors guide students through numerous incidents and accident reports that illustrate manifest-ations of the hazardous attitudes and how these attitudes contribute to unsafe outcomes. These exercises are intended to sensitize the student pilots to these attitudes in themselves.

However, giving students the skills to recognize undesirable attitudes is not sufficient to change behaviour in the cockpit. ADM training programmes also provide students with tools for minimizing the impact of these undesirable attitudes on flight safety. Through a series of classroom and flight exercises, students learn antidotes for each of the attitudes.

Finally, ADM training includes segments that alert students to the effects of stress on decision making and that teach students to recognize stressors and minimize their effects. Students learn about different sources of stress. In addition, they learn techniques for coping with life stressors, for pre-planning go/no-go decisions, and for minimizing the effects of stressors that originate in the cockpit.

## Limitations of ADM

ADM researchers and trainers have achieved two substantial goals: they have dramatically improved pilot performance through training, and they have drawn attention to the importance of cognitive performance to aviation safety. Despite these successes, ADM concepts have several limitations.

ADM concepts have one point in common with earlier concepts of pilot judgement. They describe judgement in terms of outcomes. Good judgement is indicated if the pilot makes the correct choice; poor judgement is indicated if the pilot makes the wrong choice. This focus on situational outcomes emphasizes the factors that affect outcomes, but ignores the processes which people invoke to reach those outcomes.

ADM models do not explain how individuals make decisions. Borrowing a basic premise from classical decision literature, ADM models assume that pilots compare the benefits or liabilities of available options to select a course of action that has a satisfactory outcome. But, the models of ADM do not describe how pilots make these choices. In the ADM model, the decision event is characterized by the block which reads 'Select Response Type' (FAA, 1991, p. 2). This is a gross oversimplication of complex cognitive events. Even as ADM evolved, a considerable literature existed that described the processes that decision makers invoke to compare various options and ultimately to select a single course of action. Numerous researchers studied processes such as conjunction, disjunction, lexicographic, dominance, and others (see Svenson, 1979). Tversky (1972) described the process elimination-by-aspects; von Winterfeldt and Edwards (1986) described multi-attribute utility analyses. Zsambok *et al.* (1992) provide a more contemporary review of these strategies.

The current view of ADM assumes that pilots and crews invoke a rational strategy for making decisions. That is, they use a strategy derived from the classical decision literature where the decision maker generates an exhaustive list of options, evaluates and weighs the relative value of each option, and then implements the option with the most desirable outcome. Furthermore, the model is prescriptive in that it trains students to implement a specific type of strategy for making decisions. ADM training programmes provide a structured method that directs pilots to generate multiple courses of action, evaluate each option, and then select the most valuable (usually the most safe) option to implement. However, the ADM model does not describe how the decision maker should accomplish these tasks. It offers no guidance for how the decision maker might generate a list of options, nor does it describe the process the decision maker might use to select the most desirable option.

ADM is too narrow in scope. First, it only considers decisions that involve selecting a course of action. Although most researchers and trainers acknowledge the importance of situation assessment, the model of ADM does not address decisions concerning situation diagnoses. As we discuss later, in many operational environments decisions regarding situation diagnoses prove to be the most numerous and the most difficult to make.

Second, the ADM model does not account for different decision processes. The current view of ADM provides training for only a single type of decision, one in which the decision maker can generate multiple options and can evaluate each of these on known dimensions. Real-world operators invoke a variety of strategies to make decisions effectively. Different situations require different decision strategies. Furthermore, decisions about courses of action occur rather infrequently in time-compressed environments (e.g. Kaempf *et al.*, 1992). Thus, providing training in only this one type of decision strategy enhances the flight crew's ability to contend with a relatively infrequent situation, but limits their ability to contend with more frequent situations.

Finally, the target of ADM training is the individual pilot. However, not all pilots fly as individuals. Virtually all pilots in commercial air transport function as part of a team, the flight crew. Research investigating Crew Resource Management (CRM) has revealed the importance of teams to aviation safety and that teams often perform the task of flying aeroplanes better than individuals. In fact, flight crews may be considered as decision-making entities. Thus, we need to refocus ADM training to enhance the decision-making skills of flight crews as well as individual pilots.

Existing ADM training addresses issues that have an impact on decision making and decision quality. However, they do not directly address the cognitive processes invoked in making decisions. We believe that decision makers must possess certain cognitive skills and that these skills may be acquired, enhanced and sustained through training. Furthermore, different decision strategies may require the decision maker to employ different cognitive skills. For example, the cognitive skills required to generate several options, project the outcome of each option, assign probabilities to the likelihood of each option's success, and select the optimal option would be substantially different from the cognitive skills required to recognize a situation as similar to one that has occurred in the past, and identify an appropriate course of action based on this recognition.

We believe that to maximize the benefits of decision training we should not treat cognitive processes as a black box, but should focus directly on the cognitive skills required to make decisions. Decision training will have its highest payoff if we develop training modules designed to acquire, enhance and sustain these cognitive skills. The tough questions now involve identifying the relevant cognitive skills and developing training modules that effectively enhance these skills.

Early proponents of ADM lacked the capability to identify these skills. However, recent advances in cognitive sciences provide a basis for investigating and identifying the decisions that people make in their natural environments and the cognitive processes that they invoke to make these decisions. In the past decade, researchers in Naturalistic Decision Making have developed research methods and models of decision making that allow us to make these determinations. The following section describes the field of NDM and some of its findings.

## Naturalistic Decision Making

Many of the existing limitations of ADM are due to the lack of a comprehensive model that describes how pilots and flight crews make decisions and that can serve as a basis for enhancing decision skills. The field of NDM provides such a model. NDM provides an evolving perspective that allows us to examine these issues and subsequently to apply our knowledge of decision processes to support quality decision making and to reduce operator errors through the design of better systems.

NDM is the study of how people make decisions in their workplace and in their personal lives. A major objective of NDM is to improve decision quality and operator performance through understanding naturalistic decision processes. To learn how decision makers handle the complexities and confusion of operational environments, NDM researchers have moved their research away from the highly controlled and predictable laboratory and into the field to study domains that are complex and challenging.

NDM research is not solely a product of aviation. NDM researchers have studied a wide variety of domains that include aviation. Investigators report findings from domains as diverse as firefighting, anti-air warfare command and control, and power plant control. These domains share a number of characteristics that affect how decision makers make decisions. Thus, much of what is learned in one domain may be applicable to another.

We will discuss a few important NDM concepts that have direct application to the piloting of aircraft. Several other resources provide excellent and thorough reviews of NDM including Klein (1993) and Klein *et al.* (1993). The following paragraphs present a brief history of NDM, several basic NDM concepts including one model of how people make decisions in natural environments, and some of our findings about the types of decisions people make and the strategies they invoke to make those decisions.

## Short history of NDM

NDM is a recent approach. Decision researchers Payne (1976) and Beach and Mitchell (1978) pointed out that the heavily analytical strategies prescribed by classical decision researchers are not practical for many tasks, and that under conditions such as time pressure and uncertainty, people are more likely to invoke simpler strategies. Similar to the classical strategies, however, these contingency models still focused on how people select the best course of action from a set of several alternatives. Several years later, Rasmussen (1985) and Wohl (1981) formulated more detailed descriptions of NDM, and linked the two functions of diagnosing a situation and selecting a course of action. Neither Rasmussen nor Wohl are academic researchers; they are engineers working to resolve design problems in real-world domains. Thus, it may have been easier for them to perceive the relationship between diagnosing a situation and selecting an appropriate course of action for that situation. The importance of considering situation diagnosis will become more clear as we describe one model of NDM later in this chapter.

During this same period, a few researchers investigated naturalistic settings. Hammond *et al.* (1987) showed that highway engineers made effective use of analytical decision strategies for tasks such as estimating traffic load. However, the engineers did better using intuitive strategies for other tasks, such as estimating accident rates. Shanteau and Phelps (1977) found that expert judges were able to make reliable and accurate decisions without following analytical procedures. Their work stands in sharp contrast to the earlier research that emphasized strategies for selecting one course of action from many.

The critical events for NDM occurred in the late 1980s. There had been a growing realization that decision making was more than picking a course of action, that decision strategies had to work in operational contexts, that intuitive or nonanalytical processes must be important, and that situation assessment had to be taken into account. Then a number of researchers presented models showing how decision makers could use their experience to handle operational contexts. Klein and his colleagues (Klein, 1989; Klein *et al.*, 1986) reported on fireground commanders, tank platoon leaders and design engineers. Noble *et al.* (1986) reported on Naval command-and-control personnel. Pennington and Hastie (1981) reported on jurors. Beach (1990) and Beach and Mitchell (1978) studied business decisions. Lipshitz (1989) reported work with Army officers. This research went beyond pointing out the limitations of the classical models of decision making; it presented models of how people make decisions in real-world, operational settings. With the emergence of these models, NDM research achieved coherence as an approach for studying basic and applied issues.

233

## Characteristics of natural environments

What makes natural environments particularly challenging for decision makers? What characteristics of natural environments cause the classical decision literature to lose its relevance for decision makers in the real world? Research has identified the essential characteristics of naturalistic decision environments (Klein, 1993; Orasanu and Connolly, 1993). Table 11.1 lists nine features that are particularly interesting. Not every domain includes these variables, and some naturalistic reasoning strategies apply even when most of these features are missing. Nevertheless, the features in Table 11.1 cover the most challenging aspects of operational settings.

#### Table 11.1  Characteristics of naturalistic domains

| Characteristic | Description |
| --- | --- |
| Time pressure | Decision makers have limited time in which to make decisions and implement responses. |
| Dynamic settings | Situations are not static; they evolve over time. |
| High risk | The consequences of errors are high for either the decision maker or others in the situation. |
| Shifting goals | Dynamic conditions change what is important. As situations evolve the decision maker must modify goals. |
| Feedback loops | Actions taken will alter the situation, and thus may dramatically affect the subsequent goals and actions. |
| Ambiguous, missing, and questionable data | Available data rarely paint a clear picture. Pieces of information may conflict with each other, be missing altogether, or be of unknown quality. |
| Cue learning | Experienced decision makers associate meaning with constellations of cues and with changes in cue clusters. These meanings are not available to novices. |
| Experienced decision makers | Most decision makers have some experience, ranging from journeyman to expert. Decision makers in real-world settings are rarely novices. |
| Teams | Decision makers often work together as teams. |

To help people think clearly under pressure, we must understand how people make decisions under the conditions listed in Table 11.1. However, classic decision models avoided the features listed in Table 11.1. The classical decision theories (Baron, 1988; von Winderfeldt and Edwards, 1986) grew out of mathematics and game theory (Keeney and Raiffa, 1976). These models showed how decision makers should use their estimates and judgements to make optimal choices. The models were formulated for straightforward, well-defined tasks. They were not intended for cases where time was limited, goals were vague and shifting, and data were questionable. Therefore, the classical models are not useful to describe how people work in dynamic, time-compressed settings.

## One model of NDM

The most important finding that has emerged from NDM research is that in operational settings people rarely compare options to select a course of action; that is, they do not decide what to do by comparing the relative benefits and liabilities of various alternative courses of action. For example, Klein *et al.* (1986) investigated how fireground commanders make decisions about deploying their crew members during difficult urban fires. The commanders insisted that they never tried to determine whether one option was better than another. Quite often, they implemented successfully the first course of action that came to mind. For researchers trained to expect that decision making necessarily involved comparison between options, this was totally unexpected. How can skilled decision makers select effective courses of action without comparing options?

NDM research has produced extensive evidence indicating that decision makers can use their experience to size up the situation, recognize it as typical in some ways, and identify the typical way of responding. Therefore, skilled decision makers may never have to consider more than one option when making decisions. The different strategies for contrasting options rarely come into play. Of course, there are times when it is important to contrast optional courses of action, particularly for individuals who do not have sufficient experience. But for most cases, including very difficult incidents, the critical step for the experienced decision maker is to assess the situation. Once the decision maker understands the situation, an appropriate course of action is easily identified.

An incident reported by Kaempf *et al.* (1992) illustrates this point. The commander of an AEGIS cruiser needed to decide whether to shoot down a pair of F-4s that threatened the cruiser. On the surface, the decision was about two different courses of action: to engage or not to engage. On a deeper level, the decision was about assessing the situation by determining the intent of the fighter pilots.

On this particular day, the cruiser was escorting an unarmed ship through the Persian Gulf when two Iranian F-4s took off and began to circle near the end of a nearby runway. Each successive orbit brought the fighters closer to the two ships. The aircraft activated their search and fire control radars and acquired the ships. By the rules of engagement, in effect this was a hostile act. The AEGIS commander would have been justified in engaging the aircraft, but he chose not to. The AEGIS commander needed to defend his ship, but he decided that the two aircraft were not going to attack. How did he form this assessment?

The captain tried to imagine that the F-4s were hostile. He could not imagine that a pilot preparing to attack would be so conspicuous. The pilots had been flying in plain view. They announced their presence by activating their radars. They even used their radars unnecessarily by keeping them on when travelling away from the ships. The captain just could not imagine how pilots planning to attack would behave in this way.

In contrast, the captain could imagine how the pilots were trying to harass him. All of their actions appeared consistent with this hypothesis. Therefore, the captain inferred that the F-4 pilots were simply playing games. Once the captain reached this decision about the situation, then determining a course of action was simple. He would take action to prepare his ship, but would not engage the aircraft.

This incident illustrates several insights derived from NDM research. First, people most often try to satisfice by finding the first workable solution rather than optimize by finding the best solution. Simon (1955) was the first to make this distinction. In operational settings, it is very difficult to determine what the best course of action is, even with hindsight. Decision strategies that try to calculate the optimal course of action only work when time is plentiful and the goals are clearly defined. For example, no one can say that the AEGIS commander was right or wrong in not firing at the F-4s as soon as they illuminated their fire control radar. In this case it worked out, because he avoided an incident by increasing his level of risk while retaining the ability to defend his ship.

Second, situation assessment decisions are distinguishable from course of action decisions. Sometimes, decision makers need to diagnose what is happening, and, perhaps, select one diagnosis from among several. At other times, the decision maker must determine which action to take. In the F-4 example above, the commander was faced with a diagnosis decision. Several studies have shown that diagnostic decisions are predominant in natural settings (e.g. Kaempf *et al*, 1992).

Third, in operational settings, people use their experience to arrive at situation assessments. They can use context to help them draw inferences from the available cues.

Fourth, in most cases, the situation assessment makes the appropriate course of action obvious. Many operational domains have extensive standard operating procedures and preplanned responses. Their purpose is to aid the decision maker by identifying the appropriate course of action. Planners spend considerable effort anticipating contingencies and identifying the responses for each. This removes from the decision maker the burden of generating courses of action, but increases the burden of correctly assessing the situation.

This situation is particularly true for the flight crew of an aircraft. They have extensive checklists, company policies and government regulations that dictate what actions the crew should take in most situations. However, the crew has the responsibility of diagnosing the situation. Once this is accomplished the checklists and regulations tell them which actions to take. The tough task for the pilot or flight crew is to make the diagnosis decision.

Finally, decision makers frequently must act with incomplete and often conflicting information. Decision makers often do not receive the information that would make a decision easy. This may be due to a variety of causes: poor communications, inadequate sensors, malfunctioning equipment, mistakes by others, poor environmental conditions. These factors may also lead to conflicts among the information received from different sources. Experienced decision makers expect these problems and learn methods for handling situations in which they receive inadequate information.

These insights from NDM research portray decision makers as capable of using experience to handle difficult situations without having to evaluate different options. This stands in direct contrast to current ADM training that teaches students to generate many different options, carefully list the strengths and weaknesses of each, and calculate the best. Anything less is seen as deficient. According to the NDM framework, this advice may be useful for novices, but is incompatible with the way that proficient operators make decisions. The available data (Isenberg, 1984; Klein, 1989; Soelberg, 1967) clearly show that decision makers are quite successful, yet they do not follow the classical advice. Furthermore, departing from the classical advice is what experts are able to do to succeed. Thus, it is a model to emulate, not correct. Clearly, there are times to compare options, particularly for the less experienced. But, in highly procedural jobs, these times are relatively rare.

*The Recognition-Primed Decision model*    Several researchers have presented models of NDM. This section describes briefly one of these models, the Recognition-Primed Decision (RPD) model developed by Klein and his colleagues (Klein, 1989; Klein, Calderwood, and Clinton-Cirocco, 1986).

The RPD model (Klein, 1989, 1992) was developed to describe a phenomenon observed in operational domains: people making good decisions without comparing possible courses of action. The initial studies were done with fireground commanders. We expected that they would use

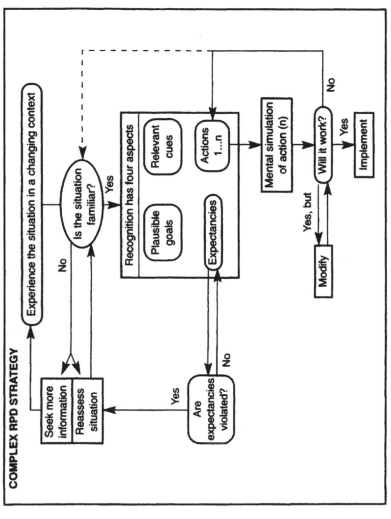

**Figure 11.1  Recognition-primed decision model**

their experience to cut down the number of options they compared, maybe just looking at two. We were wrong – they insisted that they hardly ever compared options. In our interviews with them about how they made tough decisions, we kept hearing about the same type of strategy. We derived the RPD model from what they told us.

There are two components of the model: situation assessment and option evaluation. The RPD model asserts that decision makers recognize the dynamics of a situation, enabling them to identify a reasonable course of action. This course of action is evaluated by imagining how it will be implemented. Experienced fireground commanders can size up a fire quickly. By assessing the type of fire, the type of structure, and the length of time the fire has been active, the appropriate response is usually obvious to an experienced fireground commander. How do you evaluate a course of action if there are no others to compare it with? One strategy fireground commanders use is to imagine carrying out the action. They run it through in their minds. If the risks are great or the course of action is complex, they may run it through several times. We call this process 'mental simulation', because they are simulating the course of action in their heads. This process of mental simulation – mentally enacting a sequence of events – can also appear in the situation assessment phase of the model.

Figure 11.1 shows two versions of the RPD model. The simple version appears in the panel on the left. Here, the decision maker confidently identifies a situation as familiar. This recognition enables the decision maker to know several important things. You know what goals make sense. You know what cues are relevant. You know what to expect so that you can be prepared. Finally, you know the typical ways of reacting.

The panel on the right shows a more complex RPD strategy. Here, the situation assessment was not so easy. The decision maker may need to acquire more information, or there may be several hypotheses about what is occurring. For instance, the commander in the F-4 example had to choose between two hypotheses. Either the aircraft were harassing, or they were preparing to attack. Decision makers use several strategies to arrive at a situation assessment or to choose among different situation assessments. One is to check the hypotheses against the features of the situation. The other is to build a mental simulation, or story, to explain the events. In the F-4 incident, the commander tried out one mental simulation, and judged that it did not make sense. He could not construct a plausible story of how a pilot would make such an attack. The other story did make sense; he based his actions on this diagnosis. Once the decision maker settles on an interpretation of the events, the same functions are accomplished as in the simple RPD model: specifying plausible goals, highlighting critical cues, generating expectancies, and identifying reasonable courses of action.

In complex cases, the expectancies can be violated, leading the decision maker to seek more information and to reassess the situation. Complex cases can also call for evaluation of a course of action. From the interview data we have collected, there seem to be two primary ways of evaluating options: checking them for necessary features and using mental simulation. Sometimes people evaluate a course of action by checking to see if it has the required features, and do not mentally simulate at all. Decision makers may consider a number of courses of action, without ever comparing one to another, by evaluating the options one at a time until they find one that works. This is called a singular generation/evaluation process to distinguish it from situations where people compare options directly to each other.

In more complex cases, the decision maker may try to imagine how a course of action will work in context. This is still a singular generation/ evaluation of options. If you are concerned that F-4s may be preparing to attack your ship, one obvious defence is to use chaff to distract an enemy missile. But if you play this out in your head, you may realize that your ship is between the attacking fighters and the flagship you are defending. So, firing chaff may divert the missiles away from you, but directly towards your flagship. In this case, the electronic warfare coordinator mentally simulated the problem and rejected the option of using chaff. In other cases, mental simulation helps to strengthen a course of action by revealing problems that can be overcome.

This discussion about RPD illustrates the importance of two types of decisions in naturalistic settings: diagnostic decisions and course of action decisions. The RPD model points out the relative importance of diagnostic decisions for people to succeed in dynamic, time-compressed situations. Previous literature focused on the phenomena of choosing a course of action. NDM research demonstrates that these course of action decisions play a relatively minor role in operational settings. Success in dynamic, time-compressed settings requires that people make accurate diagnostic decisions. The value of the RPD model is to:

- explain how people can use experience to make decisions;
- describe how decision makers use situation assessment to identify courses of action;
- describe how decision makers settle on a course of action without considering any others;
- show how people using mental simulation can strengthen a course of action rather than choosing only from the set of original options;

- describe how decision makers can be poised to act, rather than having to wait to complete their comparisons and analyses.

Since its development, the RPD model has been verified in many domains. For example, it describes the decision strategies of tank platoon leaders (Brezovic *et al.*, 1987), commanders and anti-air warfare officers of AEGIS cruisers (Kaempf *et al.*, 1992), critical care nurses (Crandall and Calderwood, 1989; Crandall and Gamblian, 1991), commercial pilots (Mosier, 1990), design engineers (Klein and Brezovic, 1986), and firefighters (Klein *et al.*, 1986).

## NDM strategies

The RPD model describes two types of decisions that are important in natural settings. Now we turn our attention to how people make these decisions, to identify the strategies that people invoke to make these types of decisions. Research has identified several decision strategies for making diagnostic and course of action decisions. The following sections describe some of these findings.

Our use of the term strategy does not mean that a person selects or is necessarily aware of a particular strategy to use. Usually, people invoke a strategy naturally; they may or may not be aware of how they made the decision. Thus, when we say that someone used strategy 'A', we do not mean that the person consciously chooses to use that strategy. Strategy 'A' simply describes how the person solved the particular problem or made the decision. The person may or may not be aware of the process used.

*Strategies for diagnostic decisions* One of the most difficult tasks is to diagnose a problem, sifting among the symptoms and clues to infer what is happening. Our interest is in the strategies that people use in diagnosing conditions where data are missing and ambiguous, parameters keep changing, actions affect the cues, and relationships are complex and uncertain. The available research indicates the use of three primary strategies under these conditions: feature matching, analogical reasoning, and mental simulation.

The most frequently used strategies are feature matching and pattern matching. These strategies allow the decision maker to judge whether the given events are so close to a given hypothesis that the hypothesis can be accepted as true. Often, this judgement is made without awareness, but there are times when the decision maker deliberately reviews the features to see where they match and where they miss. Feature matching also becomes a deliberate strategy when there are alternate hypotheses about the situation, and the task is to determine if one fits the data better than the other. Noble (1989) has been most responsible for showing that feature matching is used for situation assessment in natural settings.

Feature matching is quite appropriate for situations where time to respond is limited, the decision maker is well trained, and the situations can be anticipated or are familiar. Pilots use feature matching quite often to arrive at a situation diagnosis. For example, a pilot may match the existing cues associated with increasing yaw, decreasing rate of climb, and changes in engine noise to the model of engine failure that s/he holds in memory. Based on this match, the pilot would determine that s/he experienced an engine failure.

Reasoning by analogy is another important strategy for diagnostic decisions. Sometimes, a decision maker will retrieve an analogy without thinking about it. At other times, s/he will deliberately search for an analogue and carefully map it onto the current situation. Klein and Weitzenfeld (1978) have described the importance of analogical reasoning for diagnosing a problem. Several projects have clarified different aspects of analogical reasoning (e.g. Gentner, 1982; Gick and Holyoak, 1980).

Perhaps the most interesting strategy for situation diagnosis is mental simulation. To construct a situation diagnosis through mental simulation, the decision maker imagines a sequence of events that would explain the pattern of cues that are observed. Tversky and Kahneman (1974) were the first to point out the importance of mental simulation as a heuristic (Einhorn and Hogarth, 1981). In their description of mental simulation, Klein and Crandall (in press) suggest that a decision maker can use an initial state and build the simulation forward in time (as in evaluating a course of action) or build the simulation backwards in time to determine how the situation evolved. This is how mental simulation is used to diagnose a situation.

Mental simulation is particularly important when the decision maker lacks sufficient information to recognize the situation as familiar. The decision maker must build a story or explanation that links the disparate pieces of information in a causal chain. For the pilot, this is particularly true when the situation involves events that the aircraft cannot sense or present to the flight crew through its instruments.

Situation assessment includes more than diagnosis. In some cases, decision makers do not have to infer underlying causes and dynamics; sometimes they must make simple judgements. For example, a pilot might need to calculate the aircraft's maximum flight range under given conditions or identify the top of the descent point that would allow him/her to meet a certain restriction. The strategies of feature matching, analogical reasoning, and mental simulation seem to apply to this type of judgement as well as to diagnostic decisions. But, there is a difference between forming a situation assessment to reflect the different parameters and a diagnostic decision to imagine what is happening and why certain events have occurred.

*Strategies for course of action decisions*  Course of action decisions may be classified by the function that they serve and by the strategy that the decision maker invokes. Kaempf *et al.* (1992) identified three functions for these decisions. Their most dramatic function is to change or resolve an event. For example, a decision to divert to an alternate airport is intended to eliminate the risk of running out of fuel.

Another decision function is to prepare for resolving an incident. These decisions do not alter the problem, but they enable the decision maker to contend with the problem as it evolves. For example, a ship's captain may activate the weapons systems, not to engage a target, but to be ready to engage if a subsequent diagnosis dictates.

The most common course of action decision is intended to manage resources. Again, these decisions do not alter the problem itself, but they enable the decision maker to contend with the problem. A flight crew must make frequent decisions about allocating resources to address their dynamic environment.

Strategies for making course of action decisions fall into two categories: singular and comparative evaluations. Singular evaluation is a satisficing approach rather than an optimizing approach. Simon (1955) coined the term to describe situations where decision makers looked for the first alternative that would work, rather than going through the trouble of finding the best alternative.

Naturalistic models usually include a satisficing criterion. They describe people trying to find a workable solution quickly. Therefore, NDM strategies for selecting a course of action are singular strategies that evaluate options one at a time until an acceptable one is found. If the first alternative considered is acceptable, the search ends. Often, experience allows skilled decision makers to generate acceptable alternatives as the first ones they consider. A simple singular evaluation strategy is to check the course of action for specific features. This strategy is similar to the feature-matching strategy used for diagnostic decisions.

A more complex and powerful strategy for evaluating a course of action is mental simulation. Klein (1989) reported that decision makers in a variety of domains evaluate options by playing them out in their minds, looking for flaws, making adjustments if necessary, and implementing the option if workable. Mental simulation enables the decision maker to play out the option and see how it will unfold in the given situation. This allows the decision maker to consider the option within the context of the situation and to improve it if necessary.

Most course of action decisions made in natural settings are made using a satisficing strategy. For example, Kaempf *et al.* (1992) reported that AEGIS decision makers used a satisficing strategy to make approximately 95% of their course of action decisions. However, there are times when a decision maker needs to choose among different options, when the decision maker's experience or training does not extend to the current situation. At these times, a comparative strategy for selecting the course of action will prevail.

In contrast to satisficing strategies, comparative evaluation strategies have been discussed in the decision literature for a long time. Svenson (1979) has identified 17 different ones. These all break down the decision into components and features and analyse these features. More recently, Beach (1993) has identified a small set of strategies that people would likely use in natural settings. Beach eliminated strategies from Svenson's list that required computation, high quality data, or that took excessive time to complete.

These strategies typically require the decision maker to identify features of interest, identify the different options, determine the extent to which each option accomplishes the features of interest, and use these data to compare the options. Beach has identified several typical strategies (e.g. elimination-by-aspects (EBA) and conjunction). The following paragraphs briefly describe two of these.

EBA is a method of establishing successive hurdles. The most important feature becomes the first hurdle. All options are evaluated on this feature; any that fail to meet a criterion are eliminated. Then, the decision maker moves to the next most important feature. The process continues until one option is left; this is the option the decision maker implements.

For instance, a pilot has detected a fuel leak and determined that s/he will not be able to make it to the planned destination. The crew must select an alternate airport. The most important feature is that the alternate must be within range of the current fuel status. The crew eliminates all airports beyond that range. Second, the crew eliminates all airports that are not long enough to accommodate the aircraft. Finally, the crew eliminates all airports that do not have company facilities. Thus, the decision makers go through a series of steps to pare the list of options down to one (Orasanu, 1993).

Conjunction requires consideration of several features. The decision maker identifies several features that are equally important and establishes a criterion for each. Each option is reviewed; those options that do not meet the criteria for all of the important features are eliminated. Conjunction may be a one-time process or require more than one iteration to produce a single option. For successive iterations, the decision maker will have to select additional features for comparison or revise the criteria. For example, a person may select a new car by choosing one that is both inexpensive and has an airbag.

The two strategies described above enable the decision maker to select a viable course of action by evaluating options comparatively, but they do not necessarily identify the best option. They are likely candidates for situations where the decision maker is relatively inexperienced.

We have chosen not to present strategies designed to identify the optimal alternative. These often require a great deal of time, high quality data, expertise, and some form of decision support. These conditions are not conducive to use in natural environments. Several authors provide more complete reviews (Beach and Mitchell, 1978; Svenson, 1979; Zsambok *et al.*, 1992).

## Implications for Aeronautical Decision Making

We have spent some effort to describe the current state of the field of Naturalistic Decision Making. NDM describes how people make decisions in operational environments. But, what does this have to do with ADM?

NDM applies to piloting modern aircraft; cockpits are natural environments. Pilots experience many, if not all, of the characteristics of natural decision making domains listed in Table 11.1. They fly planes under time pressure and risk. They must interpret ambiguous, missing, and questionable data. Flight regimes are dynamic, thus goals shift. In most cases, pilots are experienced operators, not novices. And, pilots work in teams. The existence of these characteristics restricts the applicability of the more traditional decision literature to a relatively small subset of cases. Classical analytical decision strategies simply do not work in this kind of environment. However, NDM researchers have identified a variety of decision types and strategies that people use to make decisions effectively under these environmental conditions. But, how can we use this knowledge to improve pilot performance and safety in the aviation system?

The primary objective of the field of ADM has been to improve pilot performance through training. This training is designed to improve the outcomes of a limited set of pilot decisions. NDM can be used as a basis for expanding and improving the effectiveness of decision training.

However, training is not the only means of enhancing decision performance. A second and equally important application for ADM is in the design and implementation of better information displays and human-computer interfaces. Frequently, the quality and difficulty of decisions are based on the ambiguity and completeness of information that the decision maker has at hand. Improving the content and presentation of this information can improve and facilitate the decision-making process. These two applications of ADM are discussed in more detail below.

## Decision-centred training

In the past, ADM training has proven effective in reducing pilot errors. However, as we learn more about human cognition and as aircraft technology changes, the field of ADM must change to assimilate the new findings and technology. The FAA continues to stimulate discussion about the nature of ADM and the optimal methods for enhancing ADM skills. For example, in 1991, the FAA sponsored a workshop about ADM that brought together representatives of academia, commercial and military aviation communities, and government agencies (see Adams and Adams, 1992; Lofaro *et al.*, 1992).

Currently, ADM training addresses performance on a small subset of decisions that flight crews make – course of action decisions that require a comparative evaluation of options. Thus, existing ADM training courses do not impact the larger body of decisions made in the course of a flight: singular evaluations of courses of action and all diagnostic decisions. ADM should be expanded to include training that enhances performance on these decisions, but saying that we should expand decision training is much easier said than done.

We envisage using a hierarchical analysis in which we first identify the types of decisions that flight crews make and when they make them. In several domains, we have identified two types: diagnostic and course of action decisions. Next, we identify the strategies used to make these decisions and the functions that they serve. We have already discussed a variety of decision strategies for both diagnostic and course of action decisions. Decision functions may include ending the event, diagnosing, resource management and obtaining more information. Methods already exist for accomplishing these two steps of the analysis (e.g. Kaempf *et al.*, 1992; Rasmussen, 1986; Woods and Hollnagel, 1987).

The third and final step of the analytic process is to identify the cognitive skills required for the decision maker to implement the decision strategies. Training will be designed to acquire and enhance these basic skills. Once we know what cognitive skills are required, then we can design training objectives and curricula to address these specific skills. Identifying the cognitive skills required to use certain decision strategies is one of the most difficult aspects of NDM and one of the least explored. However, several researchers are actively pursuing this area.

At this point, we can only offer some ideas about the skills that we are investigating. These concern primarily the use of feature matching and mental simulation strategies. Feature matching involves the comparison of existing environmental cues to a model that is held in memory. Thus, the decision maker must hold some mental models of specific situations and be able to recognize specific patterns of cues in the environment. Skill training for feature matching, then, may involve a module that enables the student to build a library of models of specific situations and another process that enhances the speed with which the student can match features or patterns.

Mental simulation is extremely important for flying an aeroplane. It is a cognitive tool that enables a pilot to stay ahead of the aeroplane. Mental simulation allows the pilot to anticipate upcoming events and to prepare responses for those events. In addition, mental simulation enables a pilot to reason into the future about his/her current actions, to think about the effects that these current actions will have on subsequent events. Mental simulation requires that the decision maker build an ordered chain of events either forward or backwards in time. The decision maker must be able to reason about how the current situation came to be or how the existing situation will play out in the future. Thus, the decision maker must understand the mechanics of his/her environment and be able to imagine the effects of specific actions through time. Thus, the training may focus on developing these two skills.

A final point about decision training is that it should not be developed or conducted outside of the context of the physical task. Behavioural task analyses portray an incomplete picture of pilot performance; they do not describe the cognitive elements of performance. Similarly, cognitive analyses do not portray an adequate picture of performance. The two types of analyses must be integrated to fully describe operator requirements.

Too often, flight training programmes appear to be a conglomeration of stand-alone modules developed in response to problems that arise over time. This leads not only to inefficient training, but also to ineffectiveness. For example, many of the existing ADM training modules and Crew Resource Management modules are taught as plug-in units to either the ground training or flight training curricula. We believe that a better approach is to integrate learning the cognitive, team and technical skills required to fly aircraft. Training objectives for cognitive skills should be achieved concurrently with objectives for the technical skills. To accomplish this, training developers must use analytic techniques that consider all components of performance and integrate these components as they develop training objectives.

## Decision-centred design

During recent years, we have witnessed increased concern over the cognitive functioning of systems operators and increased efforts by researchers and designers to consider the cognitive components of performance as they design new systems (Woods and Roth, 1988). To some degree, this increased concern for cognitive task requirements has been generated by well-publicized catastrophic incidents that include a variety of aircraft accidents, the shootdown of an Iranian airliner by an American warship, and the total destruction of the Piper Apha oil production platform in the North Sea. Post-mortem analyses of each of these catastrophes indicated that the systems involved failed to adequately support the operators' decision-making requirements and that these failures exacerbated the severity of the incidents.

Furthermore, aircraft technology has changed dramatically over the last 20 years and promises to change as dramatically over the next 20 years. As technology changes, task demands change. The important elements of these changes in task demands are not physical in nature; they are cognitive. Increased automation in the cockpit has altered and expanded the cognitive requirements of flying aircraft (e.g. Kaempf *et al.*, 1991; Sarter and Woods, 1991; Wiener, 1989). In modern aircraft, proficiency on cognitive skills is of paramount importance to success. Yet, our methods of systems design fail to take the cognitive elements of performance into account. Again, most approaches to systems design rely heavily on behavioural task analyses that portray only the overt physical actions taken by the operator. They do not take into account what the operator is thinking, what decisions s/he must make, or what information the decision maker needs.

Cockpit displays are generally intended to provide information to the flight crew to enable them to maintain current and accurate situation assessments and to make effective decisions. Yet, the systems designers do not consider these components of pilot performance as they build these systems. Producing overwhelming amounts of information for the flight crew is no longer a technological problem. It is a reality. In modern cockpits, flight crews must integrate and interpret an extraordinary amount of information. Under conditions of time pressure and risk, this becomes an impossible task. Another design approach is clearly needed.

Aeronautical Decision Making integrated with NDM can provide a vehicle for incorporating considerations of pilot cognitive performance into the design of cockpit displays and interfaces. Decision requirements are the anchor for assessing cognitive performance and the cognitive requirements of specific tasks.

The objective is to develop interfaces that help crews build rapid, accurate situation diagnoses and to identify the viable courses of action. The approach is to build interfaces that support the natural cognitive processes of the flight crew and present information in a timely and intuitive manner. Systems should not present a vast array of information because it is possible, then leave it to the pilot to find and interpret the needed information.

Different research groups have developed methods for decision-centred design and validated these methods in several domains. We found that a decision-centred design approach made significant improvements to interfaces for the AWACS aircraft (Klinger *et al.*, 1993) and for the AEGIS Combat System (Miller *et al.*, 1992). Woods and his colleagues have used cognitive analyses for developing decision supports in process industries and nuclear power plants (Woods, 1986; Woods and Hollnagel, 1987). In addition, Pew *et al.* (1981) identified control designs based on critical operator decisions.

Thus, the methodological structures exist for incorporating decision requirements into the design of system interfaces. Our task now is to use our expanded knowledge of ADM to affect the design of aircraft systems and cockpit displays. By focusing on critical decisions and decision processes we will be able to design cockpit interfaces that support the needs of the flight crew.

## Summary

In this chapter, we have described the history and the current state of Aeronautical Decision Making, the more general field of Naturalistic Decision Making, and how we may incorporate the findings of one into the other. Previous research has demonstrated the importance of ADM by showing that decision training improves safety in the aviation system. More importantly, ADM has helped to emphasize that cognition is a primary component of pilot performance. Thus, descriptions of pilot performance are not complete without incorporating the cognitive elements of performance. We can describe these cognitive elements by focusing on the decisions that pilots make and the processes they invoke to make these decisions.

The success and findings of NDM research can be used to take ADM to a new level. These findings provide a basis for improving both the scope and the effectiveness of ADM, and to improve aviation safety through better training and improved systems design.

The basis of these applications is a hierarchical analysis of the decision requirements performed for specific aircraft and types of missions. The purposes of behavioural task analysis are to decompose a job into its component elements and then to identify the skills and knowledges required to perform that job proficiently. Training developers can then enhance job performance by designing training modules to enhance these specific skills. We propose a decision-centred analysis that would similarly identify the cognitive skills required for proficient job performance. The analytic anchors for the behavioural task analysis are overt behaviours; the analytic anchors for the analysis of cognitive skills are the decisions that the operator must make to perform the task effectively.

In the first step, we identify the types of decisions that the flight crew must make. Second, we identify how the crew makes these decisions. People employ a variety of cognitive strategies to make decisions in different environments and under different conditions. As we identify these strategies, the cues and factors essential for making the decisions become evident. Finally, we must identify the cognitive skills that people must have to invoke these strategies and make successful decisions.

We propose a new generation of ADM, one that improves flight crew performance in two ways: through expanded and more effective training, and through the design of better human-computer interfaces. Our knowledge of decision making in natural environments provides the capability to do this.

Originally, ADM focused on improving aviation safety by improving the motivational and attitudinal components of pilot decision making. The focus on safety remains, but the scope and depth of ADM has broadened. The field of naturalistic decision making extends our knowledge of ADM by describing the cognitive processes that individuals and teams employ to make decisions. This understanding allows us to move beyond the relatively simple decision cases where a person must choose a course of action from several available alternatives. NDM provides the models necessary to distinguish various types of decisions (e.g. diagnostic versus course of action), and to describe how well-trained individuals make good decisions in dynamic environments under time pressure.

# References

Adams, R. J. and Adams, C. A. (1992), *Workshop on aeronautical decision making: Volume 2 – plenary session with presentations and proposed action plan*, (DOT/FAA/RD-92/14,11), Washington, DC, Federal Aviation Administration.

Adams, R. J. and Thompson, J. L. (1987), *Aeronautical decisionmaking for helicopter pilots*, (DOT/FAA/PM-86/45), Washington, DC, Federal Aviation Administration.

Baron, J. (1988), *Thinking and Deciding*, New York, Cambridge University Press.

Beach, L. R. (1990), *Image Theory: Decision making in personal and organizational contexts*, Chichester, UK, Wiley.

Beach, L. R. (1993), 'Image theory: Personal and organizational decisions', in G. A. Klein, J. Orasanu, R. Calderwood and C. E. Zsambok (eds), *Decision Making in Action: Models and methods*, Norwood, NJ, Ablex.

Beach, L. R. and Mitchell, T. R. (1978), 'A contingency model for the selection of decision strategies', *Academy of Management Review*, **3**, 439–449.

Benner, L. (1975), 'D.E.C.I.D.E. in the hazardous materials emergencies', *Fire Journal*, **69**(4), 13–18.

Berlin, J. I., Gruber, E. V., Holmes, C. W., Jensen, P. K., Lau, J. R., Mills, J. W. and O'Kane, J. M. (1982), *Pilot judgment training and evaluation*, (DOT/FAA/CT-82/56), Daytona Beach, FL, Embry-Riddle Aeronautical University, (NTIS No. AD-A117 508).

Brezovic, C. P., Klein, G. A. and Thordsen, M. (1987), *Decision making in armored platoon command*, (AD-A231775), Alexandria, VA, Defense Technical Information Center.

Buch, G. D. and Diehl, A. E. (1984), 'An investigation of the effectiveness of pilot judgment training', *Human Factors*, **26**(5), 557–564.

Buch, G. D., Lawton, R. S. and Livack, G. S. (1987), *Aeronautical decision making for instructor pilots*, (DOT/FAA/PM-86/44), Washington, DC, Federal Aviation Administration.

Crandall, B. and Calderwood, R. (1989), *Clinical assessment skills of experienced neonatal intensive care nurses*, Yellow Springs, OH, Klein Associates Inc. (Final Report prepared for the National Center for Nursing Research, National Institutes for Health under Contract No. 1 R43 NR01911 01).

Crandall, B. and Gamblian, V. (1991), *Guide to early sepsis assessment in the NICU*, Fairborn, OH, Klein Associates Inc. (Instruction manual prepared for the Ohio Department of Development under the Ohio SBIR Bridge Grant programme).

Diehl, A. E. (1991), 'The effectiveness of training programs for preventing aircrew error', in R. S. Jensen (ed.), *Proceedings of the Sixth International Symposium on Aviation Psychology*, Columbus, OH, Ohio State University.

Diehl, A. E., Hwoschinsky, P. V., Lawton, R. S. and Livack, G. S. (1987), *Aeronautical decision making for student and private pilots*, (DOT/FAA/PM-86/41), Washington, DC, Federal Aviation Administration.

Einhorn, H. J. and Hogarth, R. M. (1981), 'Behavioral decision theory: Processes of judgment and choice', *Annual Review of Psychology*, **32**, 53–88.

FAA (1986), *ADM for instrument pilots*, (DOT/FAA/PM-86–43), Washington, DC, Federal Aviation Administration.

FAA (1988), *ADM for air ambulance helicopter pilots – learning from past mistakes*, (DOT/FAA/PM-88–5), Washington, DC, Federal Aviation Administration.

FAA (1991), *Advisory circular: Aeronautical decision making*, (AC Number 60–22), Washington, DC, Federal Aviation Administration.

Gentner, D. (1982), 'Are scientific analogies metaphors?' in D. S. Miall (ed.), *Metaphor: Problems and Perspectives*, Brighton, UK, Harvester Press.

Gick, M. L., and Holyoak, K. J. (1980), 'Analogical problem solving', *Cognitive Psychology*, **12**, 306–355.

Hammond, K. R., Hamm, R. M., Grassia, J. and Pearson, T. (1987), 'Direct comparison of the efficacy of intuitive and analytical cognition in expert judgment', *IEEE Transactions on Systems, Man, and Cybernetics*, **17**(5), 753–770.

Isenberg, D. J. (1984), 'How senior managers think', *Harvard Business Review*, November/December, 80–90.

Jensen, R. S. (1982), 'Pilot judgment: Training and evaluation', *Human Factors*, **24**, 61–73.

Jensen, R. S. (1989), *Aeronautical decision making – cockpit resource management*, (DOT/FAA/PM-86/46), Washington, DC, Federal Aviation Administration.

Jensen, R. S. (ed.) (1991), *Proceedings of the Sixth International Symposium on Aviation Psychology*, Columbus, OH, Ohio State University.

251

Jensen, R. S. and Adrion, J. R. (1984), *Aeronautical decision making for instrument pilots*, (DOT/FAA/PM-86/43), Washington, DC, Federal Aviation Administration.

Jensen, R. S. and Benel, R. (1977), *Judgment evaluation and instruction in civil pilot training*, (FAA Report No. FAA-Rd-78-24), Savoy, IL, University of Illinois Aviation Research Laboratory.

Kaempf, G. L., Klein, G. A. and Thordsen, M. L. (1991), *Applying recognition-primed decision making to man–machine interface design*, (Final Technical Report), Fairborn, OH, Klein Associates Inc. (Prepared under contract NAS2–13359 for NASA-Ames Research Center, Moffett Field, CA.)

Kaempf, G. L., Wolf, S., Thordsen, M. L. and Klein, G. (1992), *Decision making in the AEGIS combat information center*, Fairborn, OH, Klein Associates Inc. (Prepared under contract N66001–90-C-6023 for the Naval Command, Control and Ocean Surveillance Center, San Diego, CA.)

Keeney, R. L. and Raiffa, H. (1976), *Decisions with multiple objectives: Preferences and value tradeoffs*, New York, Wiley.

Klein, G. A. (1989), 'Recognition-primed decisions', in W. B. Rouse (ed.), *Advances in Man-Machine System Research*, Vol. 5, 47–92, Greenwich, CT, JAI Press.

Klein, G. (1992), *Decision making in complex military environments*, (Final Report), Fairborn, OH, Klein Associates Inc. (Prepared under contract N66001–90-C-6023 for the Naval Command, Control and Ocean Surveillance Center, San Diego, CA.)

Klein, G. (1993), *Naturalistic decision making – Implications for design*, WPAFB, OH, CSERIAC.

Klein, G. A. and Brezovic, C. P. (1986), 'Design engineers and the design process: Decision strategies and human factors literature', *Proceedings of the 30th Annual Human Factors Society*, **2**, 771–775, Dayton, OH, Human Factors Society.

Klein, G. A., Calderwood, R. and Clinton-Cirocco, A. (1986), 'Rapid decision making on the fire ground', *Proceedings of the 30th Annual Human Factors Society*, **1**, 576–580, Dayton, OH, Human Factors Society.

Klein, G. A. and Crandall, B. W. (in press), 'The role of mental simulation in naturalistic decision making', in J. Flach, P. Hancock, J. Caird and K. Vicente (eds), *The Ecology of Human-Machine Systems*, Hillsdale, NJ, Lawrence Erlbaum.

Klein, G. A., Orasanu, J., Calderwood, R. and Zsambok, C. E. (1993), *Decision Making in Action: Models and methods*, Norwood, NJ, Ablex.

Klein, G. A. and Weitzenfeld, J. (1978), 'Improvement of skills for solving ill-defined problems', *Educational Psychologist*, **13**, 31–41. (Also released as Wright Patterson AFB Technical Report AFHRL-TR-78-31.)

Klinger, D. W., Andriole, S. J., Militello, L. G., Adelman, L., Klein, G. and Gomes, M. G. (1993), *Designing for performance: A cognitive systems engineering approach to modifying an AWACS human-computer interface*, (Final Report), Fairborn, OH, Klein Associates Inc. (Prepared under contract F33615–90-C-0533 for Armstrong Laboratories, Wright-Patterson AFB, OH.)

Lester, L. F. and Bombaci, D. H. (1984), 'The relationship between personality and irrational judgment in civil pilots', *Human Factors*, **24**(5), 565–572.

Lipshitz, R. (1989), *Decision Making as Argument Driven Action*, Boston University, Center for Applied Social Science.

Lofaro, R. J., Adams, R. J. and Adams, C. A. (1992), *Workshop on aeronautical decision making: Volume 1 – executive summary*, (DOT/FAA/RD-92/14,1), Washington, DC, Federal Aviation Administration.

Lubner, M., Phil, M. and Markowitz, J. (1991), 'Towards the validation of the five hazardous thoughts measure', in R.S. Jensen (ed.), *Proceedings of the Sixth International Symposium On Aviation Psychology*, Columbus, OH, Ohio State University.

Maher, J. W. (1989), 'Beyond CRM to decisional heuristics: An airline generated model to examine accidents and incidents caused by crew errors in deciding', in R. S. Jensen (ed.), *Proceedings of the Fifth International Symposium On Aviation Psychology*, Columbus, OH, Ohio State University, 439–444.

Miller, T. E., Wolf, S. P., Thordsen, M. L. and Klein, G. (1992), *A decision-centered approach to storyboarding anti-air warfare interfaces*, Fairborn, OH, Klein Associates Inc. (Prepared under contract N66001–90–C–6023 for the Naval Command, Control and Ocean Surveillance Center, San Diego, CA.)

Mosier, K. L. (1990), *Decision making on the air transport flight deck: Process and product*, Unpublished dissertation, Berkeley, CA, University of California.

Noble, D. (1989), *Application of a theory of cognition to situation assessment*, Vienna, VA, Engineering Research Associates.

Noble, D., Boehm-Davis, D. and Grosz, C. G. (1986), *A schema-based model of information processing for situation assessment*, Vienna, VA, Engineering Research Associates (NTIS No. ADA163150).

O'Hare, D. (1992), 'The "artful" decision maker: a framework model for aeronautical decision making', *International Journal of Aviation Psychology*, **2** (3), 175–191.

Orasanu, J. (1993), 'Decision making in the cockpit', in E. Wiener, B. Kanki and R. Helmreich (eds), *Cockpit Resource Management*, New York, Academic Press.

Orasanu, J. and Connolly, T. (1993), 'The reinvention of decision making', in G. Klein, J. Orasanu, R. Calderwood and C. E. Zsambok (eds), *Decision Making in Action: Models and methods*, Norwood, NJ, Ablex.

Payne, J. W. (1976), 'Task complexity and contingent processing in decision making: An information search and protocol analysis', *Organizational Behavior and Human Performance*, **16**, 366–387.

Pennington, N. and Hastie, R. (1981), 'Juror decision making models: The generalization gap', *Psychological Bulletin*, **89**, 246–287.

Pew, R. W., Miller, D. C. and Feehrer, C. E. (1981), *Evaluation of proposed control room improvements through analysis of critical operator decisions*, Electric Power Research Institute Report NP-182, Palo Alto, CA.

Rasmussen, J. (1985), 'The role of hierarchical knowledge representation in decision making and system management', *IEEE Transactions on Systems, Man and Cybernetics*, **15**(2), 234–243.

Rasmussen, J. (1986), *Information Processing and Human-Machine Interaction: An approach to cognitive engineering*, New York, North-Holland.

Sarter, N. B. and Woods, D. D. (1991), 'Situation awareness: A critical but ill-defined phenomenon', *The International Journal of Aviation Psychology*, **1**(1), 45–57.

Shanteau, J. and Phelps, R. H. (1977), 'Judgment and swine: Approaches and issues in applied judgment analysis', in M. F. Kaplan and S. Schwartz (eds), *Human Judgment and Decision Processes in Applied Settings*, New York, Academic Press.

Simon, H. A. (1955), 'A behavioral model of rational choice', *Quarterly Journal of Economics*, **69**, 99–118.

Soelberg, P. O. (1967), 'Unprogrammed decision making', *Industrial Management Review*, **8**, 19–29.

Svenson, O. (1979), 'Process descriptions in decision making', *Organizational Behavior and Human Performance*, **23**, 86–112.

Telfer, R. (1989), 'Pilot decision making and judgment', in R. S. Jensen (ed.), *Aviation Psychology*, Aldershot, UK, Gower.

Tversky, A. (1972), 'Elimination by aspects: A theory of choice', *Psychological Review*, **79**(4), 281–299.

Tversky, A. and Kahneman, D. (1974), 'Judgment under uncertainty: Heuristics and biases', *Science*, **185**, 1124–1131.

von Winterfeldt, D. and Edwards, W. (1986), *Decision Analysis and Behavioral Research*, Cambridge, UK, Cambridge University Press.

Wiener, E. L. (1989), *Human factors of advanced technology ('Glass Cockpit') transport aircraft*, (NASA Contractor Report 177528), Moffett Field, CA, Ames Research Center.

Wohl, J. C. (1981), 'Force management decision requirements for Air Force Tactical Command and Control', *IEEE Transactions on Systems, Man, and Cybernetics*, 11(9), 618–639.

Woods, D. D. (1986), 'Paradigms for intelligent decision support', in E. Hollnagel, G. Mancini and D. D. Woods (eds), *Intelligent Decision Support in Process Environments*, New York, Springer-Verlag.

Woods, D. D. and Hollnagel, E. (1987), 'Mapping cognitive demands in complex problem-solving worlds', *International Journal of Man-Machine Studies*, **26**, 257–275.

Woods, D. D. and Roth, E. M. (1988), 'Cognitive systems engineering', in M. Helander (ed.) *Handbook of Human-Computer Interaction*, New York, North-Holland.

Zsambok, C. E., Beach, L. R. and Klein, G. (1992), *A literature review of analytical and naturalistic decisionmaking*, Fairborn, OH, Klein Associates Inc. (Prepared under contract N66001–90-C-6023 for the Naval Command, Control and Ocean Surveillance Center, San Diego, CA.)

# 12 Shared problem models and flight crew performance

*Judith M. Orasanu*

## Introduction

Why do aeroplanes crash? In most cases, it is not because the aircraft is unsafe to fly or the pilots lack technical flying skills. In most cases, problems have been found in the management of crew resources (70%, according to Lautman and Gallimore, 1987). Inadequate communication, poor judgement and decision making, inefficient task management, and absence of leadership have all been identified as causal factors in air transport accidents (Kayten, 1993). However, perhaps we should be less surprised to find occasional cases of poor crew coordination than to see the overwhelming number of cases in which crews work together efficiently and effectively.

How is it that two or three individuals who are complete strangers to each other can climb into the cockpit of a complex jet aircraft and interact smoothly, coordinating their actions to fly the plane safely from one point to another, usually under schedule pressures, often in less than ideal weather conditions, and sometimes with an inoperative subsystem. Several researchers recently have suggested it is because the crew members have *shared mental models* for how to fly aeroplanes (Cannon-Bowers *et al.*, 1990; Orasanu, 1990). These shared models consist of common knowledge and understanding about the aircraft systems, the task, crew member roles, standard procedures and company policy[1]. This shared knowledge, which is grounded in common training and experience, enables crew members to act in predictable and appropriate

[1]Hackman (1986) and Ginnett (1993) have described this shared knowledge as an organizational 'shell'.

255

ways. As cognitive psychologists have pointed out, mental models provide a basis for describing, predicting and explaining events, a point that shall be elaborated later in this chapter.

Flying a large jet aircraft, even one with high-technology automated flight and guidance systems, is a complex task. It is sufficiently complex that two crew members are still required (Wiener, 1993). These crew members are not just performing their own tasks in parallel; their activities are highly interdependent. Monitoring systems and each other's actions, entering data into flight management computers, updating weather and airport information, and communicating with Air Traffic Control (ATC), company operations, cabin crew, and passengers, all require coordinated action (Orasanu and Salas, 1993).

Crew member interdependence is magnified when abnormal or emergency situations arise. While the workload on each individual crew member increases, the need for coordination among crew members increases as well. Crews must determine what the problem is, judge the levels of risk and time pressure, plan for contingencies, and decide on a course of action. In addition, they must determine who will fly the plane and carry out non-routine tasks. Workload may increase enormously when a problem arises in an already high-workload phase, such as take-off or landing. Individuals under stress are likely to make more errors and to make poorer decisions because of psychological factors such as 'tunnel vision', and failure to consider critical information (Broadbent, 1971; Bronner, 1982; Wickens *et al.*, 1991; Wright, 1974). A summary of causes of accidents indicates that crew judgement and decision making were causally involved in 47% of all accidents from 1983–1987 (NTSB, 1991a). With two or three crew in the cockpit and additional resources on the ground, the question remains why crews continue to make judgement errors that lead to accidents.

This chapter will examine the relations between crew communication, problem solving, decision making, and overall levels of crew performance, especially in abnormal and emergency situations (see Figure 12.1). While several aspects of communication will be addressed, the focus will be on communication that fosters the creation of shared mental models among crew members for the problem at hand. These will be referred to as *shared problem models*, to distinguish them from other forms of shared background knowledge. Problem solving and decision making are treated together because in the cockpit decision making frequently is embedded in a broader process of problem solving. Before a decision can be made, the crew must first recognize that a problem exists, determine its nature, and determine the desired outcome (see Orasanu and Connolly, 1993, for a broader treatment of the issue). The third element is crew performance on their primary task, which is flying the plane. While safety is the

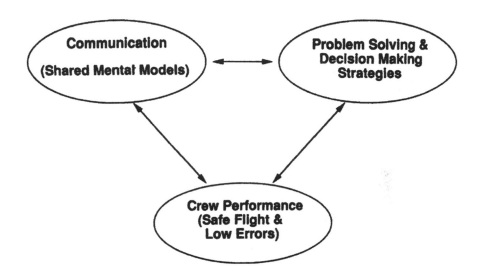

**Figure 12.1   Components of effective crew performance**

primary criterion of task performance, the number of aviation accidents is sufficiently low that surrogate measures may be used to assess crew performance, such as errors, crew coordination, efficiency or economy.

The main contention of this chapter is that communication that builds shared problem models among crew members contributes to the quality of crew performance. However, effective communication is not something that leaps into play only in emergencies; it permeates all aspects of crew behaviour and is a general feature of effecive crews. Evidence for this position will be drawn primarily from research on crews performing in full-mission simulators and from accident and incident reports. Unfortunately, the base of empirical data on this topic is quite meagre, in large part because of the cost and difficulty in obtaining it. Until recently, little research had been conducted on decision making in applied settings (see Klein *et al.*, 1993 and Chapter 11, for a summary of recent naturalistic decision-making research).

## The role of mental models in crew performance

What are mental models? Cognitive psychologists have used this term to refer to mental representations of complex systems, specifically for components, their linkages, and the relations among them as the system operates (Rouse and Morris, 1986; Gentner and Stevens, 1983). Unlike more simple schemas for the structure of a static object like a text

257

(Rumelhart, 1980), mental models are dynamic. That is, by 'running' the model one can see the consequence of an event in one part of the system on performance elsewhere in the system. For example, a trained electrical technician is able to predict the effect of an electrical short in a complex guidance system on the behaviour of the system. Mental models provide the basis for *describing* causal relations among events, for *predicting* what will happen as a consequence of an event in the system, and for *explaining* why a particular outcome occurred. Mental models have been seen as critical to troubleshooting of very complex systems such as electronic diagnostic centers (Lesgold *et al.*, 1986), and much technical training research has focused on techniques to develop mental models for the system of interest. A feature that distinguishes experts from novices is the presence of well-elaborated, richly connected and complete mental models for domains of expertise (Chi *et al.*, 1988; Ericsson and Smith, 1991).

To say that a team or crew has a *shared* mental model means that crew members have similar mental representations of the system with which they are interacting. As pointed out by Cannon-Bowers *et al.* (1990), in addition to shared models for the equipment and the environment within which they are operating, team members also have shared models for team tasks which include knowledge about the roles and responsibilities of each team member (Hackman, 1986). They may also have models about the specific individuals filling each role, which may include knowledge about how a person is likely to behave in a situation and the person's level of competence. These personal expectations contrast with the prescribed behaviour associated with roles, and are more likely to arise from personal experience.

In routine situations shared mental models enable a team to function efficiently with little need for explicit coordination (Kleinman and Serfaty, 1989). Witness the coordination of a jazz combo or a basketball team (Hackman, 1993). Implicit coordination is possible in highly proceduralized tasks where team member roles are well-specified in advance. Flying a plane clearly falls into this category, in contrast to other team tasks that are less well structured, such as designing a system (e.g. Olson *et al.*, 1992). For highly proceduralized tasks, an experienced and well-trained team may not need to talk except to pass on information necessary for performing the task (see Hutchins' 1989 analysis of information transfer in navigating a large ship into San Diego harbour, and Heath's 1991 analysis of a 'Little League' baseball team's communication).

Unlike the string quartet or baseball team that practises together extensively, cockpit crews may not have the advantage of personal knowledge of team members. The importance of familiarity among crew members was discovered inadvertently by Foushee *et al.* (1986) in a study designed to assess the effects of fatigue on crew performance. The operational definition of fatigue used in that study was completion of a normal 3–5 day duty cycle. Crews came directly to the simulator facility upon completion of duty and participated in the study. Non-fatigued crews were those who had been off duty for several days and were assembling to go out on their next duty. Foushee *et al.* found that the so-called fatigued crews actually performed better in the simulated mission than the rested crews. Further examination of the data led to the discovery that many of the 'fatigued' crews were composed of individuals who had just flown together, but few of the 'non-fatigued' crews had recently flown together. Differences in operational errors and co-ordination between the more and less familiar crews highlight the importance of shared personal models for crew performance.

While their knowledge of role-specific behaviours, procedures and systems enables crews to function effectively as a team when situations are routine, it may not be sufficient when non-routine situations arise. Then, explicit coordination may be needed to assure successful crew response. This is where communication comes in. Consider the situation a crew faces when a warning light comes on. Cues signal a problem. What does the cue mean? How much risk is present? What kind of response is required? Is the situation deteriorating? Is time available to collect more information? A shared model specific to the just encountered problem and its management must be developed by the crew. This emergent *problem* model builds on and incorporates shared background knowledge, but goes beyond it. Communication is the mechanism by which the shared problem model is built and sustained.

What does a shared problem model include?

1  Shared understanding of the situation, the nature of the problem, the cause of the problem, the meaning of available cues, and what is likely to happen in the future, with or without action by the team members.
2  Shared understanding of the goal or desired outcome.
3  Shared understanding of the solution strategy: What will be done, by whom, when, and why?

An emergent problem model assures that all crew members are on the same wavelength and are solving the same problem. NTSB accident reports are filled with examples of failures of crews to talk about the problems they faced prior to the accident. For example, while taking off in a snowstorm almost 45 minutes after having been de-iced, the first officer of a B-737 recognized that engine indicators, sounds and feel were not quite right: 'God, look at that thing' (when engine indicators surged);

then, 'That don't seem right, does it?' and 'Ah, that's not right'. The captain replied, 'Yes, it is, there's eighty,' but the first officer persisted, 'Naw, I don't think that's right'. During the 35 second take-off roll the crew did not discuss the meaning of the abnormal indicators or check other engine indicators that would have told them that their power settings were below normal for take-off (NTSB, 1982).

In a second case, the captain had told his first and second officers that he would circle for 15 minutes to diagnose a problem with their landing gear and to allow the cabin crew to prepare passengers for a possible emergency evacuation. The second officer observed, 'Not enough. Fifteen minutes is gonna really run us low on fuel here.' The captain did not take this information into account, circled for more than 15 minutes before requesting clearance to land, and ultimately ran out of fuel short of the airport (NTSB, 1979).

What these examples show is the critical relation between communication, cognitive activity and crew performance. In neither case was there evidence that the crew members held a common understanding of the nature of the problem, despite efforts of one crew member to share his awareness with others. What is evident in these cases and many others is the absence of discussion about the nature the problem, the risks involved, and the need to take action consistent with those perceptions.

This chapter consists of three sections that focus on the three components described above: crew communications, decision making in flight, and their joint contributions to overall crew performance. To provide a context for discussing crew communication, the first section will describe various types of decisions crews face and the cognitive processes involved in making them. The second section will address functions of communication in the cockpit and features of effective communication based on empirical studies. The third section will examine how communication and decision making interact to affect overall flight safety. The chapter will close by discussing some implications for training.

## Decision making

Decision making is discussed first to lay a foundation for understanding what crews talk about in abnormal or emergency situations. This section will consider the various types of decisions that must be made during flight (a fuller treatment of cockpit decision making can be found in Orasanu, 1993). Different types of events make different cognitive demands of the crew. Crews in turn bring different knowledge, skills and strategies to the task. This perspective echoes Hammond's Cognitive

Continuum Theory (Hammond *et al.*, 1987) which holds that task features induce intuitive or analytic decision processes.

On the demand side, situations differ in two major factors: the clarity of the problem and the number and type of response options that are available. Together these factors determine the complexity of the problem and the type of cognitive work that must be done (see Figure 12.2). Two other factors, time pressure and risk, amplify problem difficulties. Problem features also set the scene for the types of errors that are likely to occur (Orasanu *et al.*, 1993).

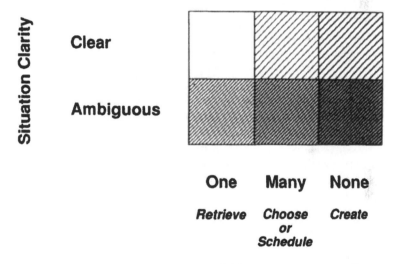

**Number of Responses Available and Required Cognitive Work**

**Figure 12.2   Decision complexity as a function of situational features (darker blocks are more complex)**

Cues that signal problems vary in how clearly they define the underlying problem. Some system indicators are fairly diagnostic, like a low engine oil indicator. Other cues are ambiguous: noises, thumps, vibrations, smells or inconsistent system indicators. Weather is inherently ambiguous and constantly changing. In ambiguous cases diagnostic work must be done to determine what is wrong prior to deciding what to do about it. In general, ambiguous situations will demand more crew work than clear problems.

Problems also differ in the number and type of response options available. In certain cases a single prescribed response is associated with a specific problem configuration. In keeping with Rasmussen (1983), these will be referred to as rule-based decisions. Company procedures, manufacturer's manuals and government regulations often specify exactly what action should be taken when certain conditions are encountered. These rules often apply in situations involving high risk and time pressure, when it is not desirable to have crews pondering about what they should do. Reports submitted to the Aviation Safety Reporting System (ASRS) (1991) indicate that the most common type of decision in this category is aborting a take-off, a decision that must be made very rapidly. Others include a go-around on landing or initiating checklists following a system malfunction. Decisions of this type are what Klein (1993) calls Recognition-Primed Decisions and Hammond calls intuitive decisions.

Other decision situations offer multiple response options or no response options. Both of these cases demand more complex cognitive work from the crew than rule-based decisions. Again, in keeping with Rasmussen, these will be referred to as knowledge-based decisions. Some situations present multiple response options from which a single option must be chosen. Decisions of this type have been termed consequential choice problems (e.g. Lipshitz, 1993) because options are evaluated in terms of their projected outcomes. Examples of choice decisions include diversions (to an alternate destination airport) or departing with an inoperative system component (which may involve projecting future consequences, such as low fuel, especially if weather is bad).

In other cases multiple tasks must be accomplished within a time window. The nature of the decision is not to choose from among a set of options, but to decide which task will be done first, second, or perhaps not at all. These decisions are most difficult when problems cascade or occur during high workload periods like take-off or landing.

If the problem is clearly understood, if response options are readily available, and·if the requirements of an adequate solution are known, the problem is well-defined. When the problem is not understood, and when response options are not available, the problem is ill-defined. Ill-defined problems are the most difficult, least frequent and least predictable. These require both diagnosis of the problem and creation of a solution. If the problem can be diagnosed definitively, it may reduce to a rule-based or choice problem. On the other hand, the crew may diagnose the problem correctly, but then find there is no prescribed solution, in which case they must invent one (as in the case of a DC-10 that lost all hydraulic systems at 37 000 ft which was landed by manipulating the power on the two remaining engines (NTSB, 1990)).

Understanding differences in situational demands is important for defining effective decision making and errors. A description of underlying problem structures permits comparison of different decisions in terms of requisite behaviours and provides a baseline for examining crew performance.

How do these different types of decision tasks affect the crew's decision-making process? A simplified decision process model (Figure 12.3) suggests how crews deal with each type. (This model summarizes a set of similar models in the literature, e.g. Bloomfield, 1992 Kaempf and Klein, this volume; Smith, 1992; Stokes, 1991; and Wickens and Flach, 1988).

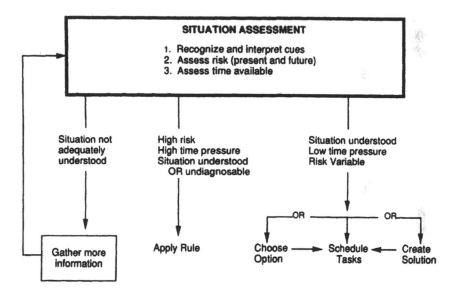

**Figure 12.3  Simplified decision process model**

The two primary dimensions of problem situations are mirrored in the two basic components of decision making in all natural environments: *situation assessment* and *choosing a course of action* (Klein, 1993). We have included *risk assessment* and *time assessment* under situation assessment (Delta Air Lines includes explicit treatment of these two features in its crew resource management course (Byrnes and Black, 1993)).

When crews first note cues that are out of the ordinary, they must try to figure out what the problem is. Even if the cues are ambiguous, the crew must make an initial judgement about risk and time pressure. How serious is the situation? Is there enough time to gather more information or to work out a plan? Perceptual recognition processes play a major role. If risk is high and time is limited, a prepackaged solution will be sought. For most emergency situations (e.g. fire, rapid decompression), condition-action rules are available as standard procedures. The real cognitive work is determining whether the situation is an instance of case *x*, and then retrieving the associated response. Even in time-limited situations, the appropriateness of carrying out the prescribed response must be evaluated: Is there any reason *not* to take the prescribed action? This sequence ccorresponds to Klein's (1993) simple recognition-primed approach.

If the situation is ambiguous and judged to be high risk, the crew may decide to treat it as an emergency and apply an appropriate response, such as landing immediately. That is, a standard procedure is applied even when the situation is not fully understood, because the level of risk warrants it.

If no rule is available that prescribes an action, then the crew must identify a solution that meets their goal and takes into account prevailing constraints. If multiple options are inherent in the situation, the decision is posed as a choice. For example, if a diversion is required because of fog at the destination, and if the plane has suffered a system malfunction, the decision may be: Should the flight try again to land at its original destination, divert to its designated alternate, or go to a different alternate that is more suitable given the consequences of the system failure? The goal is to choose the alternate that best satisfies the constraints imposed by the situation, e.g. weather, fuel, maintenance facilities, passenger convenience, approach path and runway length. A choice is made based on a strategy for weighting these factors and evaluating options (see Zsambok and Klein, 1992, for a discussion of various decision strategies).

In the above case, only one option will be chosen from among several possibilities. In other cases, the real decision facing a crew is to schedule a number of responses, all of which must be performed within a limited window of time. Cognitive work includes assigning relative priorities, estimating the time required to carry out each, determining any sequential dependencies, and scheduling events so that all are accomplished with the available time. For example, when an engine oil temperature indicator moves into the cautionary range, the crew should go through the appropriate checklist (a rule-based decision) and determine whether to leave the engine running, to reduce power to idle, or to shut it down (a choice decision). However, if this event occurs while the crew is coping with a second problem, e.g. a runaway stabilizer trim, the qustion becomes which problem to deal with first and how to allocate resources to managing the problems and to flying the plane (a scheduling decision). Thus, we see a choice decision embedded in a scheduling problem.

In some rare cases, no response option is available and the crew must invent one. This is clearly the most difficult situation, and one which the airline industry tries to avoid by establishing contingencies for every situation they can imagine. However, 'impossible' situations do arise (such as the previously mentioned loss of all flight controls due to a catastrophic engine failure that severed all hydraulic systems (NTSB, 1990). More often, problems cascade or co-occur, which increases the need for problem management and scheduling. Consider the following report to the Aviation Safety Reporting System (ASRS, 1991) (see Chapter 8) involving ambiguous cues and multiple decisions made over the course of a flight in response to an event that occurred on take-off.

> Just after liftoff, an aircraft waiting to take off on 35R reported to tower that they observed rubber leaving our aircraft. We acknowledge the report and continued with the departure. All engine indications were normal. The decision was made to continue to Portland if all other indications were normal rather than risk a heavy landing at a high altitude airport with unknown tire damage. Enroute, after checking weather and NOTAMS, decided to overfly Portland and continue to Seattle, which was the next stop. Seattle was reporting better weather with a strong wind directly down the runway. Seattle also has a 1,000 ft longer runway. In Seattle, flew by tower at 300–400 ft for gear inspection. They advised us that the left inboard tire showed damage (ASRS Report No. 170760).

*First decision: Continue take-off or a return to origin?* The first indication of a problem was the report on take-off of flying rubber, which was assumed to reflect tyre damage. Passengers and flight attendants reported hearing pieces hit the aircraft and smelt hot rubber. They also thought the engine sounded louder than normal during climb. Diagnostic actions included checking engine indicators for evidence of damage and sending an observer to the cabin to check engine sounds and to make a visual inspection. Risk of continuing was assessed. The crew concluded that it was safe to continue: there was no evidence of engine damage, but probable tyre damage.

*Second decision: Given possible tyre damage, continue to destination or divert to an alternate?* Because the plane was heavy and Denver airport is a mile high, the crew decided to continue towards their destination (Portland). The tyre damage posed no problem for flying the plane. Also, they wanted to reduce their weight by burning fuel and to get passengers to their destination.

*Third decision: Land at destination or divert to alternate?* Approaching their destination, the crew checked the weather and runways. They found that poor weather was forecast and runways were short. The tyre damage imposed constraints: the crew was concerned with controlling the plane on landing and thus wanted to avoid poor runway conditions (e.g. wet or cross-winds). They considered the next stop *en route* (Seattle); checked weather and runway. Found good weather, a long runway and a good headwind. As they had sufficient fuel, they decided to divert.

This example illustrates the dynamic nature of a real decision problem and how various cognitive element are woven together in solving it.

## Communication in the cockpit

*Functions of communication*

Communication is both a medium through which crews accomplish their tasks and a reflection of the cognitive process of individual crew members. Three major communication functions support problem solving:

1 Sharing information.
2 Directing actions.
3 Reflecting thoughts.

These map loosely onto two functions identified by Kanki and Palmer (1993) – provide information and management tools. Kanki and Palmer also note that communication contributes to team building and social/emotional support. Given this chapter's focus on problem solving, those other functions will not be addressed here, although they surely contribute to team effectiveness.

The first function, *sharing information*, is perhaps one we think of most readily when we hear the term 'communicate'. Information is passed from one person to another via some medium: speech, writing, codes, graphics, gestures or facial expression. Perhaps most critical to cockpit problem solving is sharing information about system status or progress of a plan, e.g. 'The only thing we haven't done is the alternate flap master switch.' Much routine communication is quite formulaic; that is, the classes of information, their functions and their form are very predictable, which allows unfamiliar crew members to work together readily. Formulaic utterances include clearances (headings, altitudes, speeds), radio frequencies, system status ('V-One'), flight progress ('Out of ten', 'Glide slope's captured'), and information obtained from airport charts or checklists. Much of the information in this category is discrete and numeric. Getting the value precisely correct is important. This type of talk is referred to as 'Standing Operating Procedure' (SOP) talk.

The second function of talk is to *direct action*. This includes direct commands for routine flight actions (e.g. 'Gear up', 'Flaps 15'). It also includes action commands for dealing with problems, like gathering information required for a decision ('Call dispatch and ask them if any other airport is open', or 'Tell· 'em we want immediate clearance to Roanoke'). Commands are also used to allocate tasks, that is, to let crew members know who should do what and when. For example, 'You handle him (ATC) for a second. I'm gonna get you the weather', or 'You hold that lever down and I'll turn it'. Through commands of various types, actions are scheduled and directed and workload is managed.

The third function of cockpit talk is to *share* what one is *thinking*, as opposed to passing on information obtained from an external source. It may have the indirect effect of directing an action, but that is not its immediate purpose. This function is most central to building a shared problem model among crew members. It includes perceiving and interpreting cues that may signal a problem ('Looks like a cat-two up there'), stating overall or intermediate goals ('We need a lower altitude'), suggesting plans for reaching those goals ('Let's see if the weather's improving . . . Like to wait around a few minutes. We'll try it again'), evaluating conditions ('Pretty heavy to be landing there'), warning or predicting what may occur in the future ('Keep your eye on that temperature'), and providing explanations or justifications ('I want to know how far we are *so I can get down there*'). This type of talk reflects thinking about the problem, and is closely related to metacognitive processes, which will be discussed in the third section of this chapter.

## Features of effective crew communication

If it is true that effective crew problem solving and decision making depend on a shared problem model, then we should be able to identify features of communication that distinguish more from less effective crews and find relations between communication and specific problem-solving and decision-making behaviours. Studies conducted by researchers at NASA-Ames Research Center and at the US Naval Training Systems Center over the past decade have addressed this problem, based mainly on analyses of crew performance in full-mission simulators (e.g. Foushee and Manos, 1981; Kanki *et al.*, 1991; Kanki and Palmer, 1993; McIntyre *et al.*, 1988; Mosier, 1991; Orasanu and Fischer, 1992; Oser *et al.*, 1990; Prince *et al.*, 1993; and Veinott and Irwin, 1993). An advantage of collecting data in a simulator is that all crews face the same system malfunctions, weather conditions and problem demands. Crew performance can be assessed both on-line and after the fact based on videotapes, allowing calibration of judges and establishing reliability of observation instruments.

Early studies led to several general conclusions about the relation between crew communication and performance. Crews that made few operational errors showed common communication patterns: they talked more overall, and had higher levels of commands, observations and acknowledgements (Billings and Reynard, 1981; Foushee and Manos, 1981). Several new analyses both confirm and elaborate on the original conclusions.

Recent findings pertain to *what is said* and *how* it is said. In brief, cockpit talk that is associated with effective crew performance is both *explicit* and *efficient*. Explicitness applies to what is said; efficiency applies to how it is

267

said. Explicitness is essential for avoiding ambiguity or lack of information which may lead to serious errors. Being explicit means that you do not assume that your listeners know what you know or think; you tell them. Billings and Reynard (1981) reported that an analysis of reports to the Aviation Safety Reporting System showed that 70% of the incident reports involved some failure of information transfer: 37% of the cases involved inaccurate, incomplete, or ambiguous messages, while in another 37% of the cases, information available to one crew member simply was not transmitted to the other.

A highly publicized accident highlights the contribution of ambiguous communication to a tragic outcome. In this case, the pilots had limited fluency in English, a fact that invariably played a role, but the communication problem apparently went beyond fluency. Approaching Kennedy Airport in New York after flying from Bogota, Colombia, the crew of a B-707 was given three holds totalling over 90 minutes due to bad weather. The first officer, whose command of English was better than the captain's, handled all radio communications. Towards the end of the last hold, the first officer advised ATC that they did not have enough fuel to reach their designated alternate airport. However, because the crew did not declare a fuel emergency and did not state how much longer they could fly, ATC did not clear them to land ahead of other aircraft.

The captain had trouble stabilizing the final approach because of windshear and executed a missed approach. ATC gave the aircraft instructions that put them back into the sequence of arriving aircraft, rather than putting them in front. The controller asked if the proposed routing was 'OK with them and their fuel'. The first officer replied, 'I guess so.' The captain asked his first officer if ATC knew they were 'in emergency' and the first officer mistakenly replied 'yes'. Shortly afterwards, the engines flamed out due to fuel depletion and the aircraft crashed (NTSB, 1991b). The crew seemed unaware of how to communicate the severity of their predicament to ATC. Moreover, it did not appear that the crew itself shared an understanding of the problem.

Cockpit Voice Recorders (CVR) rarely provide positive examples of communication associated with effective performance because they are only reviewed when accidents occur (but see Predmore, 1991, for an analysis of effective communication). Crew performance in simulators can provide more abundant postive cases. Orasanu and Fischer (1992) analysed the talk of two-member and three-member crews flying B-737 and B-727 simulators (studies originally conducted by Foushee *et al.*, 1986, and Chidester *et al.*, 1990). The studies involved similar scenarios (Figure 12.4): bad weather on departure with a forecast for deteriorating weather at the destination, a missed approach at the destination due to weather, and a failure of the primary hydraulic system during climb-out

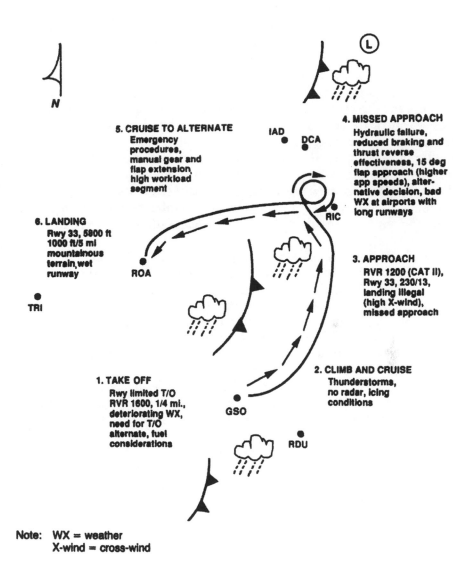

5. **CRUISE TO ALTERNATE**
Emergency
procedures,
manual gear and
flap extension,
high workload
segment

6. **LANDING**
Rwy 33, 5800 ft
1000 ft/5 mi
mountainous
terrain, wet
runway

4. **MISSED APPROACH**
Hydraulic failure,
reduced braking and
thrust reverse
effectiveness, 15 deg
flap approach (higher
app speeds), alter-
native decision, bad
WX at airports with
long runways

3. **APPROACH**
RVR 1200 (CAT II),
Rwy 33, 230/13,
landing illegal
(high X-wind),
missed approach

2. **CLIMB AND CRUISE**
Thunderstorms,
no radar, icing
conditions

1. **TAKE OFF**
Rwy limited T/O
RVR 1600, 1/4 mi.,
deteriorating WX,
need for T/O
alternate, fuel
considerations

Note: WX = weather
X-wind = cross-wind

**Figure 12.4   NASA full-mission simulation screnario**
(Source: Foushee, 1986)

after the missed approach. Crews had to decide where to land after those two events, taking into account constraints imposed by the hydraulic failure. High workload complicated the decision-making task and required planning, scheduling, and task allocation to cope with the circumstances.

Crews were divided into higher and lower performing groups based on error scores. These were determined by check-airmen who were familiar with standard company procedure relevant to the scenarios. Analysis of cockpit talk showed that crews that made fewer errors were led by captains whose talk was more explicit than that of captains of higher error crews. Explicitness pertained to two functions described earlier – sharing thoughts and directing actions. Captains of higher performing crews were more explicit in stating their goals, plans and strategies, making predictions or warnings, and providing explanations. In other words, they let the crew in on their thinking about the problems they faced. In addition, these captains were more explicit in giving commands and in allocating tasks to crew members. This directive function was more evident in three-member crews than in two-member crews; not surprising given the greater possibility for misunderstanding in three-member crews.

The above findings pertained only to captains' talk. What were the other crew members' contributions? Overall first and second officers did more status monitoring and transmitted more information, but these functions did not distinguish more from less effective crews. However, in two-member crews first officers in *lower* performing crews stated more plans and strategies than first officers in higher performing crews. It appears that the first officers were attempting to compensate for lack of direction on the part of their captains.

Based on observations of crews in Line Oriented Flight Training (LOFT), Law (1993) reported that more effective first officers demonstrated higher levels of inquiry and assertion. The FAA's CRM introductory handbook (Driskell and Adams, 1992) defines assertion as 'stating and maintaining a course of action until convinced otherwise by further information' (p. 21). Advocacy is defined as 'the willingness to state what is believed to be a correct position and to advocate a course of action consistently and forcefully' (p. 20).

Status differences and non-responsiveness on the part of captains may suppress advocacy and assertion on the part of junior officers. Their reluctance may be compounded by the ambiguity inherent in a situation. However, vague hints or indirect statements often are ignored by captains, whereas direct statements more likely will grab the captain's attention (Goguen *et al.*, 1986). For example, one crew facing a thunderstorm with lightning over the destination airport acknowledged the storm: The captain commented, '. . . Smell the rain. Smell it?' The first officer replied, 'Yup. Got lightning in it too.' The captain did not pick up on the first officer's observation about the lightning, which could be read as an indirect warning. No discussion followed about whether to continue with the approach, which resulted in a windshear encounter (NTSB, 1986).

Based on NTSB reports, major points on which advocacy seems necessary are (a) problem recognition and definition, (b) risk assessment, and (c) suggestions for immediate action. In fact, Delta Air Lines is training its first and second officers to add a recommended action to their statement of a problem, when possible. Simply noting a problem may not be enough to get the captain's attention.

The above examples illustrate how explicit communication on the part of both the captain and other crew members contributes to building shared problem models. These models enable the crew to function as a single coordinated unit. Having a shared problem model does not mean that all crew members think exactly the same, but it does increase the likelihood that they are all moving in the same direction, with the same understanding of what the problem is, what is to be accomplished, and how.

The second aspect of effective communication is *efficiency*. In efficient communication participants convey their messages briefly, saying as much as needed, but not too much (cf. Grice, 1957). Cockpit talk has a cost in attention and cognitive resources, both on the part of the speaker and on the part of the receiver. Talking or listening may interfere with some other tasks. The 'sterile cockpit' rule of no unnecessary talk below 10 000 ft clearly reflects this awareness. Early analyses suggested that crews that talked more performed better, but Foushee and Manos (1981) pointed out in their analysis of the Ruffel Smith (1979) simulator data, 'The type and quality of communication are the important factors – not the absolute frequency.'

That more effective crews appreciate the need for communication efficiency is evident in their communication strategies during normal (low workload) and abnormal (high workload) phases of flight. Among the two-member crews analyzed by Orasanu and Fischer (1992), captains of higher performing crews talked less during the high-workload phase of flight than during the normal phase of flight. During the high-workload phase, they had to cope with a major hydraulic system failure and decide on an alternate landing site following a missed approach. Despite the increased cognitive demands, they did not talk more, but were more selective in what they said. The number of commands actually dropped, while the relative proportion of plan and strategy utterances increased. Plans and commands differ mainly in that plans are future oriented and provide temporal or event-based guidelines about when something should be done. Commands are immediate: 'Do *x* now'. Therefore, planning statements by captains allow other crew members to organize their own activities to fit within time frames, whereas commands might actually disrupt ongoing activities because of their immediate nature.

By laying out their plans and inviting suggestions or reactions from

271

other crew members, captains establish a shared model of how to deal with the problem. Strategic expectancies are generated concerning how the problem is to be managed and subsequent events are then expected to unfold in a particular fashion; at this stage a shared action-event sequence has been triggered which may not need further discussion. We cannot confirm a causal link, but the fact is that fewer operational or procedural errors were committed by crews that talked more about plans than crews that did less planning (see also Pepitone *et al.*, 1988). It should be noted that in the process of making their decisions, higher performing crews actually collected more information and considered more constraints than lower performing crews, so it is not the case that they simply talked less. They just used talk differently.

Economizing on talk is not the only effective strategy for managing cognitive load, however. Strategy appropriateness depends on the resources available. When three crew members are present, captains of more effective three-member crews actually talked more during the high-workload phase than the low-workload phase, but this was after they had transferred the task of flying the plane to the first officer. This strategy freed up the captains to deal with the hydraulic problem and the choice of a landing alternate. Trying to fly the plane while managing a problem appears to be difficult to do well. In fact, an NTSB accident report (1985) noted that one captain's attempt to both fly the plane and diagnose a noisy vibration just after take-off (while the first officer was on the radio and the second officer was standing by) was a poor use of crew resources and probably contributed to the accident.

The final type of communication efficiency is closed-loop communication. This is communication in which an acknowledgement signals receipt and understanding of a message. Foushee and Manos' (1981) early analysis found a high level of acknowledgements associated with low-error crew performance. Kanki *et al.* (1991) found that low-error crews were characterized by high levels of commands by captains and acknowledgements by first officers. Moreover, the communication patterns of these crews were homogeneous. Kanki *et al.* interpreted this finding to mean that normative communication establishes a basis for predictability among crew members, facilitating coordination. It may also reduce the cognitive load on the speakers.

## The relation between communication, decision making, and crew performance

Do effective communication and shared problem models assure good decision making and safe flight? Unfortunately, not always. The reason is the loose connection between crew performance and outcomes. Some accidents will happen regardless of what the crew does. In these cases, crew behaviour can only improve the outcome, not prevent the event. For example, when United Airlines Flight 232, a DC-10, lost all flight controls due to severed hydraulic lines resulting from a catastrophic engine failure, the crew saved the situation probably as much as was humanly possible. They utilized all available resources in the cockpit, on the ground and in the cabin, taking advantage of a flight instructor flying as a passenger. Despite the fact that they crash landed, lives certainly were saved due to the direct efforts of the crew (NTSB, 1990).

In contrast, the absence of a shared problem model is likely to lead to poor performance in abnormal or emergency conditions. NTSB accident files are full of cases in which a minor problem became a major catastrophe because of the way the crew handled the situation. For example, crews have flown into thunderstorms, encountered windshear and crashed, when they easily could have decided to go around. After losing both generators one crew decided to fly on battery power to its destination rather than return to its departure airport. The destination was 39 minutes away; the departure was 6 minutes away. Battery power gave out before they reached their destination and they crashed (NTSB, 1983). Review of the transcripts of voice communication in these cases shows no discussion of the risks facing the crews nor of the options available to them. In neither case was time pressure so great that no discussion was possible.

What do shared problem models do for a crew? They are most essential for coping with problems and for making good decisions when the unexpected happens. Crews need to have a process in place that will allow them to respond flexibly to diverse types of problems. It is important that the behaviours be so well established that they are practised during normal routine flights, and are not just seen as special skills to be called upon when problems arise. The FAA recognizs this fact, and has recommended embedding CRM training into technical training for crews (FAA, 1993), so that it will not be seen as an optional add-on. *Efficient, explicit communication must become routine so it is a matter of course when problems are encountered.* This is the philosophy behind Aer Lingus's integrated training programme that emphasizes crew management of non-normal events with technical training in the simulator, using a generic non-normal checklist (Johnston, 1992). Evidence of routine use of effective communication by crews in full-mission simulators was found by Mosier (1991): higher performing crews gathered more general information than lower performing crews, thereby keeping themselves better informed and up to date on evolving conditions.

Shared problem models enable crews to exploit available resources and to create a framework for dealing with the problem. They allow crew members to contribute, to interpret each others' actions, and to anticipate others' information needs. They provide the basis for coordination, in conjunction with shared procedural and role knowledge.

When a problem is encountered, the crew needs to talk about how to manage it:

1 What problem do the cues signify? What are their implications for future actions?
2 What level of risk does the immediate situation pose? Future risk?
3 Do we need to act immediately or do we have time to think about a plan?
4 Is a standard response available? If not, what are our options?
5 What information do we need to make a decision?
6 What are the constraints on our options?
7 Who will perform each action?

While this chapter has emphasized the importance of effective communication, the goal of CRM training is to get crews to *think* more effectively, as well. The kind of thinking of concern is what psychologists have called 'metacognition'. In research on individual learning or problem solving, the term metacognition refers to an individual's awareness of the demands of the task in relation to her or his competencies, including skills and knowledge that can be brought to the task, one's own limitations, and what is needed to overcome them (Flavell, 1981; Redding and Seamster, this volume). The same type of awareness is seen in effective cockpit crews. Metacognitive skill includes the ability to discern the nature of the problem and its requirements, awareness of potential limitations of the crew, especially under high workload, stress and time limits, and awareness of available resources (both within the cockpit and outside, including cabin crew, dispatchers, maintenance and ATC). Three components have been identified based on crew performance in the NASA full-mission simulator studies (Fischer *et al.*, 1993; Orasanu, 1990). They characterize effective overall crew performance:

1 Situation awareness.
2 Decision strategies.
3 Task management strategies.

Although conceptually distinct, these three components work together, and comprise what could be called the crew's metacognitive competence.

*Situation awareness* in the cockpit goes beyond simple recognition of cues that signal a problem. It is the interpretation of the significance of those cues that is critical to performance (Sarter and Woods, 1991). The crew must recognize the potential risk inherent in the situation and the implications of conditions. The crew must appreciate the instability and ambiguity present in many dynamic situations (e.g. weather or malfunctions can deteriorate) and prepare for a worst case scenario, if warranted.

For a crew to be effective, situation awareness must be a crew phenomenon, not just an individual one. One crew member may notice a problem, but unless its significance becomes shared by all crew members, the crew is not likely to do anything about it. In the simulator data we found that more effective crews talked more about cues that signalled the *possibility* of future problems, which set them up to monitor the situation closely and to establish contingency plans in case conditions deteriorated.

Good situation awareness sets the scene for development of good *decision strategies*, primarily by influencing the perceived need for certain kinds of information. For example, crews aware that bad weather might mean a missed approach were more aggressive in obtaining weather updates and analysing conditions. More effective crews realized that cross-winds and runway visual range were just borderline, which prompted them to request updated weather. This new information allowed them to decide at an earlier point to abort their landing than crews lacking sensitivity to these conditions. More effective crews are also more sensitive to constraints on decisions. This means being aware of factors that define a desirable solution. For example, in the NASA scenario the choice of an alternate airport following a missed approach and loss of the main hydraulic system was influenced by runway length, final approach, high ground and weather at candidate airports. More effective crews considered more of these factors in making their decision.

The third metacognitive feature of effective crew performance is *task management*. Task management refers to scheduling activities and assigning tasks so that essential tasks are accomplished under manageable workloads. To do so requires awareness of the demands of the task in conjunction with realistic assessment of the capabilities and limitations of the crew. Time required to perform tasks, relative priorities of tasks, and capabilities of crew members to carry out tasks must all be assessed. Several analyses have noted that effective crews are 'ahead of the aircraft', meaning they have anticipated events, planned for them, and briefed or reviewed essential information (Amalberti, 1992; Pepitone *et al.*, 1988). In contrast, less effective crews always seem to be behind and never quite catch up when problems arise that increase the demands on the crew.

The following are specific examples of good task management behaviours that reflect awareness of task demands and available resources (Fischer *et al.*, 1993):

- Planned for contingencies and gathered information early which enabled the crew to make decisions expeditiously.
- Managed workload through assignment of tasks, including the task of flying the plane, working the radios and checklists, and managing problems.
- Managed cognitive load by managing 'windows of opportunity', or the timeframe during which actions should be accomplished, e.g. requested holding to buy time to gather decision-relevant information rather than jumping to a less adequate but quicker decision.
- Scheduled and rehearsed abnormal tasks – such as manual flap and gear extension – to allow quick, smooth and coordinated execution of the tasks.
- Used briefings to prepare mentally for high-workload phases such as landing with a system malfunction. Reviewed approach plates and abnormal landing checklists early and often, allowing crew members to anticipate extra work and what systems would not be functioning.

As Amalberti (1992) and Valot and Deblon (1990) have pointed out in their analyses of mission planning strategies used by single combat pilots, the goal is to 'preprocess the problems' likely to be encountered, considering planned events, potential events, and the means to face them. One way fighter pilots do this is by 'mission impregnation,' analogous to crew briefings, to reduce the need to review materials under time and workload pressures. In general, by talking through strategies, crews prime themselves to be ready for what *might* occur so they can respond quickly and with little effort.

The aviation industry, including manufacturers, government regulators and airline companies, has established standard procedures to reduce the need for crews to make decisions under high workload, time-pressured and risky conditions. For crews to do this for themselves depends on metacognitive processes. Situation awareness is needed to recognize that a special need may exist in the future and to plan for it. Valot and Deblon (1990) point out that fighter pilots establish contingency plans on the *ground*, not during flight, unless essential. Airline crews can do the same in their preflight briefings, in which they establish how they will handle possible events, for example, during take-off (Who will control the plane? Who will decide? What are the triggers?). Ginnett (1993) has shown that more effective crews are led by captains who state their expectations and ground rules literally on the ground, in crew briefings prior to the flight. They expand the organizational 'shell' or shared model for crew interaction that reduces uncertainty later on.

## Training implications

Findings about crew problem solving, communication and flight safety reported in this chapter have implications for crew training. Many are already present in CRM courses offered by airlines, but some are not (see Helmreich and Foushee, 1993). The primary goal of training is to establish robust routine habits of thinking and communicating so they will be available when problems are encountered in flight. Designing any training programme requires two questions to be addressed: *What* is to be taught? and *How* is it to be taught? Specific skills to be taught – the *what* – derive directly from research and fall into four major areas: situation assessment, response selection, task management, and communication. The first two comprise the two major elements of the decision-making process illustrated in Figure 12.2. Overall problem management depends on metacognitive skills, and communication is the medium by which the others are enacted.

Specific training targets within each of these categories could include the following:

- *Situation assessment* Assess risks and recognize uncertainty in ambiguous conditions. Practice may be needed in interpreting ambiguous situations (e.g. bad weather), and crews must consider worst case scenarios. Specific conditions that are triggers for behaviours under emergency and abnormal conditions must be quickly recognized. It is important to assess when fast action is required versus when time is available for more planful responding.
- *Response selection* Crews must discern what kind of decision task they face: a rule-based decision, a choice problem, scheduling, or one that requires creative problem solving. Different strategies are appropriate for each, and oversimplifying a problem increases the risk of a poor decision. Crews must actively generate the constraints and information requirements necessary to make effective decisions.
- *Task management* Metacognitive skills are required to match resources to the demands of the situation. Task priorities must be established and time requirements determined. Crews need to assess the resources available not only in the cockpit, but also in the cabin and on the ground and how they might be used to solve a problem. Resource limitations must be appreciated. When workload is heavy, it is important for the crew to manage its cognitive resources to provide thinking 'space' (either by delegating tasks or deferring unnecessary ones). Planning for contingencies, especially during low-workload periods, allows the crew to be prepared if conditions deteriorate.

- • *Communication* Communication is needed to build shared problem models. Explicit and efficient communication assures that all crew members share an understanding of the problem, strategies for coping with it, and who will do what tasks. The goal is not to increase the level of talk overall, but to assure that critical aspects of problems are discussed, especially interpretation of ambiguous cues, possible consequences, risk levels and possible actions. Talk that is future oriented (plans, briefings, rehearsal of non-normal tasks) seems to benefit crew performance, as does closed-loop communication.

Strategies for *how* to achieve these training targets can be found in recent literature on training. Two general principles dominate: first, if we want crews to perform as crews, they must be trained as crews (FAA, 1993; Hackman, 1993); second, training for problem solving and decision making should be embedded in meaningful task contexts as much as possible. This serves two purposes: context-based training establishes retrieval cues so the desired skills will be accessed and applied when they are needed, preventing the problem of 'inert' knowledge (Bransford *et al.*, 1986; Brown *et al.*, 1989). Context-based training also makes it possible to build in some of the stressors that are likely to be present in real abnormal or emergency situations. These include time pressure, workload, ambiguity and distractions. In addition, target skills need to be generalized, that is, elicited by many different specific situations. Practice in multiple situations that differ in critical features is necessary to establish the applicability of desired skills in many contexts.

Three general strategies for developing desired skills are:

1 Provide positive models of the skills through demonstration, video-tapes, or case study.
2 Provide opportunities for practice of realistic problem solving and decision making in meaningful contexts with feedback on performance.
3 Provide supports for the desired skills.

Airline CRM training programmes already are taking advantage of videotape to present examples of positive and negative behaviours in abnormal and emergency situations, often using re-enactments of accidents. These powerful tools connect immediately with their audience. But it is essential for crews to practise the desired behaviours, not just to be passive recipients of information. Practice can be done in contexts of varying levels of fidelity to the actual cockpit, ranging from small group problem-solving exercises to low-fidelity part-task trainers to full-mission simulation (Butler, 1993).

Aer Lingus has developed a unique technique to support desired behaviours in practice. They have developed a generic non-normal checklist that focuses the crew on critical behaviours (Johnston, 1992). Their checklist items map onto the skills our research has identified as essential. For example, their system malfunction checklist includes the following items:

- Assure that someone is flying the plane.
- Assign tasks and establish coordination.
- Determine the nature of the problem.
- Monitor and cross-check critical actions (like engine shut-down).
- Determine consequences of the problem and deal with them.

The checklist serves as a scaffold to support the type of thinking needed for problem solving. A simple item like assigning the task of flying the plane requires the captain first to assess the demands of the situation and available resources: who is best suited to perform the necessary tasks? The last checklist item (Discuss consequences of the problem) should prompt forward thinking and appropriate discussion of contingencies and constraints. By supporting this form of explicit discussion, the checklist supports the desired type of thinking. As checklists are 'public' and shared procedure prompts, they are more likely to be accepted than other techniques for developing communication skills. Some flight problems are not tied to system malfunctions, but once crews have become used to applying the non-normal checklist, we might expect that the underlying 'habits of thought' would generalize to other classes of problems, as well.

## Conclusion

While effective crew performance depends on the crew's collective skills in responding to the demands of various types of events, communication is the means by which crew performance is enacted. The behaviours we have analysed – communication, decision making, workload management, situation awareness, planning, assertion and advocacy, leadership – are all prescribed CRM behaviours in the FAA Advisory Circular on CRM (1993) and the CRM introductory handbook (Driskell and Adams, 1992), both of which grew out of recommendations at the NASA/MAC CRM workshop (Orlady and Foushee, 1987). What our research has done is to analyse how the pieces fit together in response to specific problems encountered in flight, and how they contribute to overall crew performance.

The intimate relations between thinking, talking and acting are highlighted as crews cope with in-flight problems. It is not enough for one crew member to perceive a problem or to have a good idea. The thought must be shared for coordinated action to occur. In addition, mistakes in thinking can only be corrected when they are made public through talk. Critical review is also the basis for avoiding 'groupthink' (Janis, 1972), or the tendency of a group to go along with a plan proposed by a high status member, i.e. the captain, without critical examination. Its antithesis is a 'devil's advocate' approach, in which the plan is challenged for the purpose of checking its validity. Assertiveness and advocacy recommendations directed to junior officers in the cockpit clearly target this problem.

Talk is the means by which the cognitive power of a crew is unlocked. Having multiple flight crew, as well as ground and cabin personnel, contributing to problem resolution might be expected to enhance performance compared to a solo problem solver (cf. Orasanu *et al.*, 1993). But this does not always happen. If the cognitive resources remain locked up within each individual, no benefit results. The key to unlocking crew resources is training that fosters shared models through communication. Pilots must be 'brought up' within the organization to know that collective crew efforts are the norm, not the exception.

Current airline training programmes prepare crews well for the task of flying modern aircraft. The emerging trend is to embed CRM training with technical training, primarily in full-mission simulation. The challenge of the next generation of training is to prepare crews to manage in-flight problems and decisions as well as they manage their aircraft.

## Acknowledgements

I wish to express my gratitude to NASA (Office of Space Science and Applications) and to the Federal Aviation Administration for their support of my research on which much of this chapter is based. Thanks are also extended to Kimberly Jobe for her assistance in preparing this chapter, and to all the research assistants at NASA-Ames Research Center who diligently transcribed, coded and analysed the data reported here. Finally, special appreciation is expressed to the airline companies and individual pilots who have participated enthusiastically in this research and shared their insights and experiences so generously.

## References

Adams, R. J. and Adams, C. A. (eds) (1992), *Workshop on Aeronautical Decision Making: Vol. II – Plenary Session with presentations and proposed action plan*, Department of Transportation/FAA/RD-92/14, 11, Washington, DC, pp. 146–164.

Amalberti, R. (1992), 'Safety in flight operations', in W. B. Quale (ed.), *New Technology, Safety and System Reliability*, Hillsdale, NJ, Lawrence Erlbaum.

Aviation Safety Reporting System (1991), *FLC Decision-Making Situations*, (ASRS SR No. 2276), Moffett Field, CA.

Billings, C. E. and Reynard, W. D. (1981), 'Dimensions of the information transfer problem', in C. E. Billings and E. S. Cheaney (eds), *Information Transfer Problems in the Aviation System*, (NASA Technical Paper 1875), Moffett Field, CA, NASA-Ames Research Center.

Bloomfield, J. R. (1992), 'Elements of a theory of natural decision making', in R. J. Adams and C. A. Adams (eds), *Workshop on Aeronautical Decision Making: Vol. II – Plenary Session with presentations and proposed action plan*, Department of Transportation/FAA/RD-92/14,11, Washington, DC, pp. 146–164.

Bransford, J., Sherwood, R., Vye, N. and Rieser, J. (1986), 'Teaching thinking and problem solving: Research foundations', *American Psychologist*, **41**, 1078–1089.

Broadbent, D. (1971), *Decision and Stress*, New York, Academic Press.

Bronner, R. (1982), *Decision Making under Time Pressure*, Lexington, MA, D.C. Heath.

Brown, J. S., Collins, A. and Duguid, P. (1989), 'Situated cognition and the culture of learning', *Educational Researcher*, **18** (1), 32–42.

Butler, R. E. (1993), 'LOFT: Full-mission simulation as crew resource management training', in E. L. Wiener, B. G. Kanki and R. L. Helmreich (eds), *Cockpit Resource Management*, San Diego, CA, Academic Press, pp. 231–259.

Byrnes, R. E. and Black, R. (1993), 'Developing and implementing CRM programs: The Delta experience', in E. L. Wiener, B. G. Kanki and R. L. Helmreich (eds), *Cockpit Resource Management*, San Diego, CA, Academic Press, pp. 421–443.

Bowers, J. A., Salas, E. and Converse, S. (1990), 'Cognitive psychology and team training: Training shared mental models of complex systems', *Human Factors Society Bulletin*, **33** (12), 1–4.

Chi, M. T. H., Glaser, R. and Farr, M. J. (1988), *The Nature of Expertise*, Hillsdale, NJ, Lawrence Erlbaum.

Chidester, T. R., Kanki, B. G., Foushee, H. C., Dickinson, C. L. and Bowles, S. V. (1990), *Personality factors in flight operations: Volume I. Leadership characteristics and crew performance in a full-mission air transport simulation*, (NASA Technical Memorandum No. 102259), Moffett Field, CA, NASA-Ames Research Center.

Driskell, J. E. and Adams, R. J. (1992), *Crew Resource Management: An introductory handbook*, DOT/FAA/RD-92/26, DOT-VNTSSC-FAA-92–8, Washington, DC, DOT, FAA.

Ericsson, K. A. and Smith, J. (eds) (1991), *Toward a General Theory of Expertise*, Cambridge, Cambridge University Press.

Federal Aviation Administration (1993), *CRM Training Advisory Circular*, AFS-210, AC No. 120–51A, February 10, Washington, DC, DOT, FAA.

Fischer, U., Orasanu, J. and Montalvo, M. L. (1993), 'Effective decision strategies on the flight deck', in R. Jensen (ed.), *Proceedings of the Seventh International Symposium on Aviation Psychology*, Columbus, OH, Ohio State University.

Flavell, J. H. (1981), 'Cognitive monitoring', in W. P. Dickson (ed.), *Children's Oral Communication Skills*, New York, Academic Press.

Foushee, H. C. and Manos, K. L. (1981), 'Information transfer within the cockpit: Problems in intracockpit communications', in C. E. Billings and E. S. Cheaney (eds), *Information Transfer Problems in the Aviation System*, (NASA Technical Paper 1875), Moffett Field, CA, NASA-Ames Research Center.

Foushee, H. C., Lauber, J. K., Baetge, M. M. and Acomb, D. B. (1986), *Crew factors in flight operations: III. The operational significance of exposure to short-haul air transport operations*, (Technical Memorandum. No. 88322), Moffett Field, CA, NASA-Ames Research Center.

Gentner, D. and Stevens, A. L. (eds) (1983), *Mental Models*, Hillsdale, NJ, Lawrence Erlbaum.

Ginnett, R.C. (1993), 'Crews as groups: Their formation and their leadership', in E. L. Wiener, B. G. Kanki and R. L. Helmreich (eds), *Cockpit Resource Management*, San Diego, CA, Academic Press, pp. 71–97.

Goguen, J., Linde, C. and Murphy, M. (1986), *Crew communications as a factor in aviation accidents*, (NASA Technical Memorandum No. 88254), Moffett Field, CA, NASA-Ames Research Center.

Grice, H. P. (1957), 'Meaning', *Philosophical Review, 67*. (Reprinted in D. Steinberg and L. Jakobovits (eds) *Semantics: An interdisciplinary reader in philosophy, linguistics, and psychology*, Cambridge, Cambridge University Press, pp. 53–59.

Hackman, J. R. (1986), 'Group level issues in the design and training of cockpit crews', in H. H. Orlady and H. C. Foushee (eds), *Proceedings of the NASA/MAC Workshop on Cockpit Resource Management*, Moffett Field, CA, NASA-Ames Research Center, pp. 23–39.

Hackman, J. R. (1993), 'Teams, leaders, and organizations: New directions for crew-oriented flight training', in E. L. Wiener, B. G. Kanki and R. L. Helmreich (eds), *Cockpit Resources Management*, San Diego, CA, Academic Press, pp. 47–70.

Hammond, K. R., Hamm, R. M., Grassia, J. and Pearson, T. (1987), 'Direct comparison of the efficacy of intuitive and analytical cognition in expert judgement', *IEEE Transactions on Systems, Man, and Cybernetics, 17* (5), 753–770.

Heath, S. B. (1991), ' "It's about winning!" The language of knowledge in baseball', in L. Resnick, J. Levine and S. Behrend (eds), *Shared Cognition: Thinking as social practice*, Washington, DC, American Psychological Association.

Helmreich, R. L. and Foushee, H. C. (1993), 'Why crew resource management? Empirical and theoretical bases of human factors training in America', in E. L. Wiener, B. G. Kanki and R. L. Helmreich (eds), *Cockpit Resource Management*, San Diego, CA, Academic Press, p. 3–41.

Hutchins, E. (1989), 'The technology of team navigation', in J. Galegher, R. E. Kraut and C. Egido (eds), *Intellectual Teamwork: Social and technical bases of cooperative work*, Hillsdale, NJ, Lawrence Erlbaum, pp. 191–220.

Janis, I. L. (1972), *Victims of Groupthink*, Boston, MA, Houghton Mifflin.

Johnston, A. N. (1992), 'The development and use of a generic nonnormal checklist with applications in *ab initio* and introductory advanced qualification programs', *The International Journal of Aviation Psychology, 2* (4), 323–337.

Kanki, B. G., Folk, V. G. and Irwin, C. M. (1991), 'Communication variations and aircrew performance', *The International Journal of Aviation Psychology, 1* (2), 149–162.

Kanki, B. and Palmer. M. T. (1993), 'Communication and CRM', in E. L. Wiener, B. G. Kanki and R. L. Helmreich (eds), *Cockpit Resource Management*, San Diego, CA, Academic Press, pp. 99–134.

Kayten, P. J. (1993), 'The accident investigators perspective', in E. L Wiener, B. G. Kanki, and R. L. Helmreich (eds), *Cockpit Resource Management*, San Diego, CA, Academic Press, pp. 283–310.

Klein, G. A. (1993), 'A recognition-primed decision (RPD) model of rapid decision making', in G. Klein, J. Orasanu, R. Calderwood and C. Zsambok (eds), *Decision Making in Action: Models and methods*, Norwood, NJ, Ablex, pp. 138–147.

Klein, G. A., Orasanu, J., Calderwood, R. and Zsambok, C. (eds) (1993), *Decision Making in Action: Models and methods*, Norwood, NJ, Ablex.

Kleinman, D. L. and Serfaty, D. (1989), 'Team performance assessment in distributed decision making', in R. Gilson, J. P. Kincaid and B. Goldiez (eds), *Proceedings: Interactive Networked Simulation for Training Conference*, Orlando, FL, Naval Training Systems Center.

Lassiter, D. L., Vaughn, J. S., Smaltz, V. E., Morgan, B. B., Jr. and Salas, E. (1990), 'A comparison of two types of training interventions on team communication performance', Paper presented at the *1990 Meeting of the Human Factors Society*, Orlando, FL.

Lautman, L. G. and Gallimore, P. L. (1987), 'Control of the crew caused accident: Results of a 12-operator survey', *Boeing Airliner*, April–June, Seattle, WA, Boeing Commercial Airplane Company, 1–6.

Law, J. R. (1993), 'Position-specific behaviors and their impact on crew performance: Implications for training', *Proceedings of the Seventh International Symposium on Aviation Psychology*, Columbus, OH, Ohio State University, pp. 8–13.

Lesgold, A., Lajoie, S., Eastman, R., Eggan, G., Gitomer, D., Glaser, R., Greenberg, L., Logan, D., Magone, M., Weiner, A., Wolf, K. S. and Yengo, L. (1986), *Cognitive Task Analysis to Enhance Technical Skills Training and Assessment*, Pittsburgh, PA, University of Pittsburgh, Learning Research and Development Center.

Lipshitz, R. (1993), 'Decision making as argument-driven action', in G. Klein, J. Orasanu, R. Calderwood and C. Zsambok (eds), *Decision Making in Action: Models and methods*, Norwood, NJ, Ablex, pp. 138–147.

McIntyre, R. M., Morgan, B. B., Jr., Salas, E. and Glickman, A. S. (1988), 'Team research in the eighties: Lessons learned', unpublished manuscript, Orlando, FL, Naval Training Systems Center.

Mosier, K. (1991), 'Expert decision making strategies', *Sixth International Symposium on Aviation Psychology*, Columbus, OH, Ohio State University Press.

National Transportation Safety Board (1979), *Aircraft Accident Report: United Airlines, Inc., McDonnell-Douglas DC-8-61, N8082U, Portland, Oregon, December 28, 1978*, (NTSB/AAR-79-7), Washington, DC.

National Transportation Safety Board (1982), *Aircraft Accident Report: Air Florida, Inc., Boeing 737-222, N62AF, Collision with 14th Street Bridge near Washington National Airport, Washington, DC, January 13, 1982*, (NTSB/AAR-82-8), Washington, DC.

National Transportation Safety Board (1983), *Aircraft Accident Report: Air Illinois Hawker Siddley HS 748-2A, N748LL, Near Pinckneyville, Illinois, October 11, 1983*, (NTSB/AAR-85/03), Washington, DC.

National Transportation Safety Board (1985), *Aircraft Accident Report: Galaxy Airlines, Inc., Lockheed Electra-L-188C, N5532, Reno, Nevada, January 21, 1985*, (NTSB/AAR-86/01), Washington, DC.

National Transportation Safety Board (1986), *Aircraft Accident Report – Delta Airlines Flight Inc., Lockheed L-1011-385-1, N726DA, Dallas–Fort Worth International Airport, Texas, August 2, 1985*, ( NTSB/AAR-86/05), Washington, DC.

National Transportation Safety Board (1990), *Aircraft Accident Report – United Airlines Flight 232, McDonnell Douglas DC-10-10, Sioux Gateway Airport, Sioux City, Iowa, July 19, 1989*, (NTSB/AAR-91-02), Washington, DC.

National Transportation Safety Board (1991a), *Annual Review of Aircraft Accident Data: US Air Carrier Operations Calendar Year 1988*, (NTSB/ARC-91/01), Washington, DC.

National Transportation Safety Board (1991b), *Aircraft Accident Report – Avianca, The Airline of Colombia, Boeing 7707-321B, HK2016, Fuel exhaustion, Cove Neck, New York, January 25, 1990*, (NTSB/AAR-91-04), Washington, DC.

Olson, G. M., Olson, J. S., Carter, M. R. and Storrosten, M. (1992), *Small group design meetings: An analysis of collaboration*, (Technical Report No. 43), Ann Harbor, MI, Cognitive Science and Machine Intelligence Laboratory.

Orasanu, J. (1990), *Shared mental models and crew decision making*, (Technical Report No. 46), Princeton, NJ, Princeton University, Cognitive Science Laboratory.

Orasanu, J. (1993), 'Decision making in the cockpit', in E. L. Wiener, B. G. Kanki and R. L. Helmreich (eds), *Cockpit Resource Management*, San Diego, CA, Academic Press, pp. 137–168.

283

Orasanu, J. and Connolly, T. (1993), 'The reinvention of decision making', in G. Klein, J. Orasanu, R. Calderwood and C. Zsambok (eds), *Decision Making in Action: Models and methods*, Norwood, NJ, Ablex, pp. 3–20.

Orasanu, J., Dismukes, R. K. and Fischer, U. (1993), 'Decision errors in the cockpit', *Proceedings of the Human Factors and Ergonomics Society 37th Annual Meeting*, Seattle, WA.

Orasanu, J. and Fischer, U. (1992), 'Team cognition in the cockpit: Linguistic control of shared problem solving', *Proceedings of the 14th Annual Conference of the Cognitive Science Society*, Hillsdale, NJ, Lawrence Erlbaum, pp. 189–194.

Orasanu, J. and Salas, E. (1993), 'Team decision making in complex environments', in G. Klein, J. Orasanu, R. Calderwood and C. Zsambok (eds), *Decision Making in Action: Models and methods*, Norwood, NJ, Ablex, pp. 327–345.

Orasanu, J., Wich, M., Fischer, U., Jobe, K., McCoy, E., Beatty, R. and Smith, P. (1993), 'Distributed problem solving by pilots and dispatchers', *Proceedings of the Seventh International Symposium on Aviation Psychology*, Columbus, OH, Ohio University Press.

Orlady, H. W. and Foushee, H. C. (eds) (1987), *Proceedings of the NASA/MAC Workshop on Cockpit Resource Management*, (NASA Conference Publication 2455), Moffett Field, CA, NASA-Ames Research Center.

Oser, R. L., Prince, C., Morgan, B. B. and Simpson, S. S. (1990), *An analysis of aircrew communication patterns and content*, Technical Report 90–009, Orlando, FL, Naval Training Systems Center.

Pepitone, D., King, T. and Murphy, M. (1988), *The role of flight planning in aircrew decision performance*, (SAE Technical Paper No. 881517), Warrendale, PA, Society of Automotive Engineers.

Predmore, S.C. (1991), 'Microcoding of communications in accident analyses: Crew coordination in United 811 and United 232', *Proceedings of the Sixth International Symposium on Aviation Psychology*, Columbus, OH, Ohio State University, pp. 350–355.

Prince, C., Chidester, T. R., Cannon-Bowers, J. and Bowers, C. (1993), 'Aircrew coordination: Achieving teamwork in the cockpit', in R. Swezey and E. Salas (eds), *Teams: Their training and performance*, Norwood, NJ, Ablex, pp. 329–354.

Rasmussen, J. (1983), 'Skill, rules, and knowledge: Signals, signs and symbols, and other distinctions in human performance models', *IEEE Transactions on Systems, Man and Cybernetics*, **13** (3), 257–267.

Rouse, W. B. and Morris, N. M. (1986), 'On looking into the black box: Prospects and limits in the search for mental models', *Psychological Bulletin*, **100**, 359–363.

Ruffell Smith, H. P. (1979), *A simulator study of the interaction of pilot workload with errors, vigilance, and decisions*, NASA Technical Memorandum No. 78482. Moffett Field, CA, NASA-Ames Research Center.

Rumelhart, D.E. (1980), 'Schema: The building blocks of cognition', in R. J. Spiro, B. C. Bruce and W. F. Brewer (eds), *Theoretical Issues in Reading Comprehension*, Hillsdale, NJ, Lawrence Erlbaum.

Sarter, N. B. and Woods, D. D. (1991), 'Situation awareness: A critical but ill-defined phenomenon', *International Journal of Aviation Psychology*, **1** (1), 45–57.

Smith, K. M. (1992), 'Decision task analysis for the air transport pilot', in R. J. Adams and C. A. Adams (eds), *Workshop on Aeronautical Decision Making, Vol. II – Plenary Session with presentations and proposed action plan*, Department of Transportation/FAA/RD-92/14,11. Washington, DC.

Stokes, A. F. (1991), 'Flight management training and research using a micro-computer flight decision simulator', in R. Sadlowe (ed.), *PC-based Instrument Flight Simulation: A first collection of papers*, American Society of Mechanical Engineers, New York, pp. 25–32.

Swezey, R. W. and Salas, E. (eds) (1992), *Teams: Their training and performance*, Norwood, NJ, Ablex.

Valot, C. and Deblon, F. (1990), 'Mission preparation: How does the pilot manage his own competencies?', *Troisième Colloque International Interactions Hommes Machine et Intelligence Artificielle dans les Domaines de l'Aeronautique et de l'Espace*, Toulouse, France, 26–28 September.

Veinott, E. S. and Irwin, C. M. (1993), 'Analysis of communication in the standard versus automated aircraft', *Proceedings of the Seventh International Symposium on Aviation Psychology*, (in press).

Wickens, C. D. and Flach, J. M. (1988), 'Information processing', in E. L. Wiener and D. C. Nagel (eds), *Human Factors In Aviation*, San Diego, CA, Academic Press, pp. 111–149.

Wickens, C. D., Stokes, A., Barnett, B. and Hyman, F. (1991), 'The effects of stress on pilot judgement in a MIDIS Simulator', in O. Svenson and J. Maule (eds), *Time Pressure and Stress in Human Judgement and Decision Making*, New York, Plenum Press.

Wiener, E. L. (1993), 'Crew coordination and training in the advanced-technology cockpit', in E. L. Wiener, B. G. Kanki and R. L. Helmreich (eds), *Cockpit Resource Management*, San Diego, CA, Academic Press, pp. 199–230.

Wright, P. (1974), 'The harassed decision maker: Time pressures, distractions and the use of evidence', *Journal of Applied Psychology*, **59**, 555–561.

Zsambok, C. E. and Klein, G. A. (eds) (1992), *Decision-making Strategies in the Aegis Combat Information Center*, San Diego, CA, Naval Ocean Systems Center.

# 13 Stress and crew performance: Challenges for Aeronautical Decision Making training

*Carolyn Prince, Clint A. Bowers and Eduardo Salas*

## Introduction

In a typical duty day, cockpit crew members encounter difficult situations and issues that are considered to be a standard part of the job. These include high workload periods, time limitations, noise which obscures critical flight information, required adherence to regulations or policies which may seem to compromise safe flying practices, possible criticism from (or conflicts with) other crew members, potential mechanical malfunctions or weather problems, and a long work day which may end in disturbed rest away from home (Stone and Babcock, 1989). If stress is considered to be a factor that threatens an individual's well-being and taxes or exceeds that individual's resources (Folkman and Lazarus, 1985), then the flight crew's environment describes an ideal setting for stress.

This operational environment is also where multiple decisions must be made, some routine, and some, life-threatening. A stressor, such as mechanical failure, may force a crew to make a decision. Conversely, a decision, such as one to continue a flight despite certain warning conditions, may increase the stress for the crew members. Thus, because of the persistent presence of both stress and decision-making demands in the cockpit, they are often interrelated. Given the knowledge that stress can affect how decisions are made (Dwyer, 1992), it is evident that to improve aviation performance, attention must be focused on this critical area. Knowledge gained from this attention must then lead to the development of training that will help crew members make good decisions under conditions of stress.

Unfortunately, at present, there is a conspicuous lack of aviation-specific scientific data that can guide training design. Because this is such a critical issue, we will attempt to integrate the results of the available aviation research with that from other relevant areas to establish what is known and what needs to be known before optimal training interventions can be developed. In this chapter, four distinct bodies of literature that might contribute to our understanding of stress effects on Aeronautical Decision Making (ADM) are reviewed:

1  Conceptual approaches to the problem of stress and human performance.
2  Research on stressors in the cockpit.
3  Investigations of the effects of stress on ADM.
4  Studies of the effects of stress upon crew processes.

There are three primary goals that we pursue in this chapter. First, we summarize research findings on stress as it relates to human performance. This is done to uncover questions that remain to be answered by research so that more effective training and assistance can be developed for the aviation crew members who must make decisions in the presence of multiple stressors. These questions are presented in the form of research challenges. Next, we review information on existing stress interventions to assess the degree to which our field is prepared to respond to these challenges and assist operational personnel who must perform under these difficult conditions. Finally, we integrate the information that we had uncovered to describe a preliminary training strategy for reducing stress effects on ADM. The approach to the proposed intervention is based on an existing Crew Resource Management (CRM) training approach, and is taken for reasons of both economy and compatibility. This approach lends itself to incorporation in a CRM programme as well as within existing ADM training, and can be presented also as stand-alone training. This intervention is presented not as an ultimate solution, but as a starting point for training that can be altered and improved as research begins to provide more guidance in training design.

## The conceptual approach

Based on research on team performance in general, Driskell and Salas (1991a) proposed a model to explain the stress/performance relationship. According to this model, there are four stages in reacting to stress:

287

1 The presence of specific stimuli.
2 Appraisal (consisting of evaluation of both the threat and the resources to combat the threat).
3 Development of performance expectations (which then determine the task performance differentials).
4 Psychological, behavioural and physiological effects.

Driskell and Salas (1991a) suggested that this process is affected by perceptions of control, predictability and threat characteristics such as proximity, intensity and duration. Further, Driskell and Salas stated that the stress/performance relationship is also mediated by the type of task, the severity of the stress, the presence of others, and individual differences. The model is illustrated in Figure 13.1.

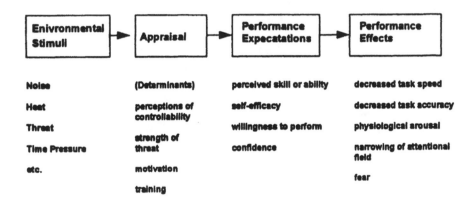

**Figure 13.1  Conceptual model of stress and performance**
(Source: Driskell and Salas, 1991a)

The Driskell and Salas model (1991a) provides a conceptualization of the stress/performance relationship, which may be used as an initial base for explaining that relationship in the cockpit. It is inadequate for a complete understanding of that relationship, however, since there are at least three characteristics of cockpit stress/performance relationships that are insufficiently addressed. First, the model was developed to explain the effects of acute stressors only, but flight operations clearly provide exposure to both chronic and acute stressors. Second, the model is concerned with individual performance, while flight generally requires the performance of more than one individual. Finally, the model is based on research considering less complex performance than ADM. For these reasons, the Driskell and Salas model requires some additional considerations to be adequate for application to ADM.

## The first challenge

There is a need to develop a theoretical model of stress effects on ADM to guide training research. The model should include consideration of both chronic and acute stressors, possible mediating factors, and the effects of stress on the requirements for decision making, such as information processing. Although the development of such a model is beyond the scope of this chapter, we will point out factors which should be included within a conceptualization of stress and ADM.

*Stress in the cockpit*   A dictionary definition of stress identifies it as 'a constraining force or influence' (Merriam-Webster, 1985). Stress is considered to occur when an 'external stimulus is so compelling that a person cannot adequately adjust' (Hayakawa, 1968). These somewhat negative connotations of stress are common when it is used in reference to a person; in fact, synonyms for stress include words such as 'pain, distress, tension, pressure' (Roget, 1977; Rodale, 1978). Although there is some evidence of positive effects of stress (Lazarus *et al.*, 1952) it is the negative effects that will be the focus here. The effects of stressors on individuals can be measured through their physiological arousal, loss of motivation, increased attention on the self, and after-effects.

Several attempts have been made to delineate the stressors that are present in work environments in general, and in the cockpit in particular. Weitz (1970) developed a list of stressors acting on a pilot that included time pressure, group pressure, noxious stimuli, perceived threat, disrupted physical functions, isolation and confinement. Moore (1977) cited temperature, sleep loss, turbulence and fatigue as some cockpit stressors. In addition, Driskell and Salas (1991a) identified five stressors that have been shown to affect task performance on a variety of tasks: crowding, noise, performance pressure, workload and anticipated threat of danger including flight emergencies and combat. All of these can occur in the cockpit.

According to Driskell and Salas (1991a), stress can be ambient or performance contingent. An example of ambient stress is stress that is part of the work background, such as a noisy, hot, fast-paced environment where performance of the task does not relieve the stress. On the other hand, performance contingent stress is removed if the crew performs satisfactorily. The stress that accompanies an engine malfunction in flight will be relieved if the malfunction is successfully identified and corrected and the threat of danger is avoided. Stress in this case may affect the decision-making process, and the decision that is made is likely to affect the stress. To varying degrees, both types of stress are found in the cockpit.

### The second challenge

It is well established that a number of potential stressors exist for crews performing flight tasks. However, knowledge about the specific number, types or nature of these stressors is still incomplete. Thus, the second challenge for training is to identify the types of stressors that can confront flight crews, and to develop methods that crew members can use in operational settings to mitigate these stressors' effects.

*ADM and stress* Expertise in a flight crew may help avoid some emergency situations (Ericsson, 1992) and subsequently reduce the number of decisions that may be imposed by those emergencies, but expertise does not eliminate decision making from the cockpit. There are still decisions which remain that are routinely required in the cockpit, even in the most uneventful flight.

Several researchers have recently addressed decision making in the cockpit and have all recognized that there are a variety of decisions that are required. Smith (1991) specified ten different operational decision points that are part of every transport flight. He included decisions associated with take-off operations, initial climb, departure, *en route* climb, cruise, initial descent, intermediate descent, approach descent, final approach, and landing. Each of these ten operational tasks represents multiple possible decisions, where the number and complexity of decisions is based partially on the result of the previous decision. Smith (1992) noted that not all decision points are equal in the demands they impose on the decision maker and suggested that there are four levels of decision complexity (i.e. least complex, moderately complex, complex, and highly complex). Using this four-level classification, he estimated the complexity of each of the identified decision points for transport flight. For example, Smith rated initial climb decisions least complex, and those made at landing he classified as highly complex. Bloomfield (1992) found decisions could be classified in four different categories based on decision consequences. These categories were described as subordinate, everyday, moderately important, and consequential. Still another scheme for decision classification is based on types of aeronautical decisions (Orasanu, 1992). At least three decision types have been specified: rule-based (i.e. go/no-go decisions); (ii) knowledge-based, where the problem is relatively clear; and (iii) knowledge-based decisions where the problems are ill-structured.

This brief overview of some of the current research in ADM illustrates the inevitability of decision making in the cockpit and the extreme importance of some of those decisions. Despite the co-existence of stress and the requirement for multiple decisions in the cockpit, there is relatively little data regarding the impact of stress on ADM. However, the information on stress and decision making outside of the aviation arena indicates that several negative outcomes are possible. Stress may have a critical impact on the way a crew assesses decision information, the way they make a decision, and the way they carry that decision through to completion. Stressor effects that are especially relevant to decision making are cue restriction and narrowing of the perceptual field, decreased search behaviour, decreased vigilance, degraded problem solving and performance rigidity (Driskell and Salas, 1991a). Also relevant and disturbing for the cockpit environment is the finding that performance stress may increase errors on operational procedures, reaction time, and time to complete manual tasks (Driskell and Salas, 1991a). That is, even once a decision is made, its enactment may be inhibited by these performance problems.

Dwyer (1992), in surveying the stress/performance literature, observed that perceptual narrowing has been found in a wide range of jobs and a wide variety of studies. Similarly, Berkun (1964) found a decrement in cognitive ability when subjects were placed in a stress condition. Keinan (1988) reported that individuals under stress conditions displayed a tendency to make decisions before all alternatives were reviewed, used a non-systematic search of alternatives, and made fewer correct decisions. Further, it has been found that decision makers under stress are prone to biases that serve to expedite the decision-making process. These biases include satisficing, deciding based on the perceived desires of others, limiting the range of alternatives, and over-weighting historical analogies (George, 1980). This tendency towards 'premature closure' can severely limit the quality of the eventual decision, especially during complex operations (Janis, 1988). Thus, it seems that ADM might be at particular risk for these stress effects.

*The third challenge*

Currently, there is no agreement on terminology and classifications for decisions, yet there is clearly a recognition that the requirement for decision making in the cockpit is ever present, that there are diverse decisions that must be made, and that some of these decisions are far more complex than others. In addition, there is sufficient evidence to suggest that the decisions that are made in the cockpit in the presence of stressors are likely to be determined in part by the reactions of the decision makers to the stress. The need for a wide variety of decisions to be made with stressors present sets the stage for the third challenge to be faced in training ADM. The challenge is to design training such that, under stressful conditions, a variety of decisions are addressed and that appropriate strategies for making different types of decisions and coping with the possible effects of stressors are provided.

*Stressor effects and crew processes*    Based on the explanations put forth for the effects of stress for individuals, some explanations for expected team effects can be made. One approach to defining and measuring stress has introduced the importance of individual perception in determining if a particular situation is a stressor. This perception is determined, in part, by the assessment of the individual of his/her own resources in the face of the stressor. In a team situation, where the team is well trained for the task, the resources available to any individual are increased by the resources of the team. That is, each individual assesses the capabilities of other team members as part of his/her resources. This resource may not be considered by a team member if the team has not developed process skills which facilitate team interactions or if the team leader does not encourage the use of those process skills. This would be the case when a crew functions more as a set of individuals working on tasks in close physical proximity to one another, rather than functioning as a team.

Burgess *et al.* (1991) reviewed literature on stress, team performance and leader traits and behaviours. They reported that although performance for teams was poorer under stress, there were certain behaviours performed by the leaders that lessened the impact of the stress. These included: accepting team input, providing feedback, structuring the team, planning and coordinating team task performance, emphasizing goals, preparing the team for upcoming crises, and giving explanations for decisions made and actions taken. Leader traits or characteristics, on the other hand, were found to have no effect on performance of the teams.

Driskell and Salas (1991b) noted that although the effects of stress on individuals have been studied for a number of years, little is known about the effects of stress on group performance. This lack of knowledge is compounded by an incomplete understanding of the processes by which teams accomplish their tasks (Prince, 1991). This has particular relevance to the airlines, where few flights are flown by a single individual. Ruffel Smith (1979), in observing cockpit crews flying a realistic simulator scenario, found that the majority of errors committed by the crews could be attributed to deficiencies in group process, not problems with individual proficiency. Foushee and Helmreich (1988) noted that subordinate group members acquiesce to authority when under stress; they lack the assertiveness to impose their understandings and views of the situation on the captain or other senior members of the crew.

Driskell *et al.* (1986) found a decline in task cue utilization when two-person teams were given increased responsibilities and presented with an environmental threat. They attributed this to the increased demands on attention made by the stressor. In a another study, Driskell and Salas (1991b) tested two competing hypotheses of group decision making under stress using a highly ambiguous task combined with stressors. The first hypothesis, centralization of authority, predicts that as authority is centralized under stress, subordinate group members transfer more responsibility for group decisions to the leader while the leader becomes less likely to accept input from subordinates. The second hypothesis, increased receptivity, says that group members tend to defer to the group leader for decision making while the leader becomes more, not less, responsive to inputs from other team members. Driskell and Salas (1991b) found that performance changes of team members under stress are somewhat dependent on the positions they occupy in the group status structure. Specifically, they found that subordinate group members defer more frequently in decision-making interactions and higher status members seek more inputs from the others in the team. Driskell and Salas (1991b) also noted that this increased reliance of lower status group members on the group leader and increased tendency for the group leader to attend to subordinates' input may result in greater workload for the team leader. This change in the leader's workload could have performance implications. Kantowitz and Casper (1988) stated that an individual is most reliable in conditions where the workload does not change suddenly and erratically.

## The fourth challenge

Given the interest in involving the entire crew in the management of its resources, it is only logical that the entire crew involvement in some of the decisions be recognized and the appropriate training offered. It is then the challenge to training developers to develop ADM training for all the members of the crew, recognizing the effects that stress may have on them based upon their particular crew positions.

## Interventions for stress effects

In trying to overcome the effects of stressors in the operational environment there are three major remedies that can be applied: re-design of the task and/or its environment; selection of the crew members for the personal qualities that withstand the stressors; and training of the crew members. Since the beginning of manned flight, aviation systems have undergone constant and dramatic change. Most of the changes have reduced stressors that were found in the cockpit or the task, but, with progress, other stressors have been added. For example, as navigation systems improved and reduced the uncertainties of location, air traffic increased and added the concern for mid-air collisions. The change in gauges, from steam to glass cockpit, has reduced some workload and increased the information available to the aircrew for decision making, yet they have introduced other potential stressors which may interfere with the decision process. Eventually, as the equipment improvements continue, the potential stressors in the cockpit can be reduced but they can never be eliminated. Flying an aeroplane takes place in a complex environment that includes the features of the aeroplane, the airspace and its traffic, weather and various other considerations. Improving one area may lead to problems in another. This suggests that stress and ADM will be a part of aviation for a very long time.

Selection, the second alternative, has three major drawbacks. The first of these is that it is not practical to select from the existing crew members those that can best withstand stress while making decisions. These individuals have been highly trained for their jobs at considerable expense, in both monetary costs and time. Selecting among them would present additional problems in human terms and in organizational terms. It could cause considerable shortage of well-trained crew members. The second drawback is a problem with measurement. There is not enough information about the meaning of the measures of reactions to stress, and there is not always agreement among the different measures that can be made. We are not yet at a stage of knowledge where we can measure an individual's reactions to stress under one set of circumstances and be reasonably certain that we can predict his/her behaviour under another set of circumstances. The third problem with selection is the difficulty of creating a valid selection test. It would require identifying all possible stressors and their combinations and would need to include aviation-specific decisions. Aviation-specific decisions are required since there is some information that decision-making expertise is tied to a particular subject matter and does not generalize (Lippert *et al.*, 1989). Lippert *et al.* (1989) found that surgeons, who were also pilots, did not use the same decision-making skills that they employ in the operating room, even where the decision processes are relevant. The most reasonable intervention, therefore, is training. Specifically, our goal is to identify methods to augment current ADM training to result in crew decision-making processes that are more resilient to stress effects.

## Training interventions for performance under stress

Despite the reports of negative effects of stress on human performance, there have been relatively few attempts to develop training interventions to mitigate these negative effects. The existing training interventions can be assigned, for the most part, to one of two categories: stress management interventions and stress exposure interventions. Each of these approaches is reviewed below.

*Stress management training*    Stress management training refers to a broad class of programmes which attempt to reduce the subject's anxiety by providing knowledge about human stress as well as by teaching subjects a variety of physiologic and cognitive stress-reduction techniques (Rosch and Pelletier, 1989). The specific interventions employed in stress management techniques typically include relaxation training, biofeedback, cognitive reappraisal, information about diet and exercise, and so forth. Thus, stress management techniques may be conceptualized as global, knowledge-based techniques designed to assist subjects in coping with the variety of stressors they encounter in everyday life.

Stress management training has been utilized in attempting to minimize the impact of a variety of stressors across a number of performance areas. For example, the beneficial effects of stress management training have been demonstrated in athletics (Crocker, 1988), nursing (Johansson, 1991) and SCUBA diving (Griffiths *et al.*, 1981). Yet, much of this research focuses on the comfort of the subject and not on actual performance changes. Therefore, the stress management literature has been criticized due to the lack of empirical research regarding its effectiveness in eliciting performance change (Goldstein, 1980). There is a clear need for continued research to provide a better understanding of the effectiveness of this technique for applied psychology.

*Stress exposure training*    Stress exposure training seeks to reduce the negative effects of stress by providing an opportunity to experience the stressor. Specifically, trainees practise coping responses under conditions of graded exposure in a controlled setting prior to confronting the actual stressful experience. Therefore, this approach attempts to intervene in the stress-performance relationship by altering the subject's perception of his or her ability to cope with the target stressor and by providing an opportunity to practise specific coping behaviours in a setting that will provide feedback and coaching. This intervention is based on a type of psychotherapy used in clinical psychology to assist people in coping with anxiety or anger (Meichenbaum and Jarembo, 1983). However, the

295

rationale underlying stress inoculation training appears to be applicable to occupational psychology as well (Meichenbaum, 1985).

Stress exposure training is similar to stress management training in that training typically progresses through a series of distinct modules. However, the two approaches differ in terms of the focus of the training. While stress management programmes typically attempt to assist subjects in dealing with a broad range of stressors encountered in everyday life, the objective of stress exposure training is for trainees to use particular skills to enhance performance in the presence of a defined, predictable stressor (Hall *et al.*, 1992). Encouraging results for stress exposure training have been obtained with trainees preparing to undergo combat (Keinan, 1988), students preparing to take academic tests (Hussain and Lawrence, 1978), police officers (Novaco, 1977) and oil workers undergoing training for smoke diving (Hytten *et al.*, 1990). It should be noted that stress management and stress exposure paradigms are not mutually exclusive. Rather, researchers have indicated that there might be some benefit in combining the two approaches (i.e. Smith, 1980).

*Stress, training and flight performance*

The training interventions for stress that are described above might not appear to be practical for use in flight situations. One might hypothesize that the use of stress interventions such as relaxation and cognitive restructuring might compete with cognitive resources needed to execute the flight task itself. However, there are a few empirical studies which have demonstrated that these interventions are useful in reducing stress effects in the cockpit. For example, Krahenbuhl, Marett, and Reid (1978) describe a stress exposure experiment in which pilots in undergraduate flight training were divided into treatment and control groups. The treatment group experienced power-on stalls and spins in a high fidelity simulator prior to training while the control group encountered these stressful flight situations without such exposure. Stress levels (assessed by norepinephrine/epinephrin ratios) were collected immediately after the actual flight lesson regarding spins and stalls. The results of the study indicated that the experimental group demonstrated significantly less stress in response to the lessons than did the control group. Hence, these interventions might have a beneficial impact on performance without directly competing for processing resources.

*Stress, decision making and cockpit resource management*

Although there are training programmes dedicated specifically to ADM (i.e. Paul, 1992), it is currently the case that most crew-level decision-making training is provided in the context of Cockpit Resource Management (CRM) or Aircrew Coordination Training (ACT) programmes. As such, we will focus on incorporating the existing stress coping technologies into a current ACT programme. Certainly, however, these concepts are equally valid for programmes designed specifically to train decision-making skills.

CRM was developed in response to reports that many aviation mishaps were attributable to crew error (i.e. Cooper *et al.*, 1979). Because this is a relatively recent observation, scientists have thus far attempted to intervene by applying models and techniques originally developed for use in other areas of human performance.

Using McGrath's (1964) input–process–outcome model as a foundation, Foushee and Helmreich (1991) attempted to develop interventions to improve safety within the aviation system. They adopted the suggestion of Hackman and Morris (1975), who suggested that group performance is changed most efficiently by intervening in input factors. Therefore, CRM development proceeded by attempting to identify those input variables that appeared to place crew members at risk for unsafe behaviour. One such variable, operator attitudes, has received particular attention in this regard. This is due to the fact that, since the history of aviation, pilots have been portrayed as fearless, individualistic risk-takers. Thus, interventions designed to alter these attitudes were believed to have the highest potential for providing safety benefits. These interventions are believed to set the stage for changes in the group's structure, another important input variable. That is, by altering crew members' attitudes about the importance of group interaction, one is able to introduce interventions designed to optimize the contribution of all of the resources in the cockpit.

Many of the existing CRM training programmes have been deployed as 'awareness phase' interventions. These training programmes attempt to satisfy the goals described above by providing knowledge designed to alter attitudes toward cockpit coordination. The majority of these programmes include a discussion of aeronautical decision making. For example, a recent military training programme includes a number of subject areas. Training begins with a discussion of decision making and the responsibilities of the crew members. This is followed by a discussion of frequent 'psychological pitfalls' which include habits or attitudes that might serve to place crews at risk for faulty decision making (i.e. 'get-there-itis' which is defined as the fixation on the target destination without consideration of alternates). Trainees are encouraged to consider pilot, aircraft, environmental and operational factors in the decision-making process. If any of these factors are not optimal, trainees are encouraged to make the most conservative decision rather than taking an unwarranted risk.

Additionally, pilots are provided with instruction in assessing their own attitudes and identifying 'hazardous' beliefs. In this section, pilots are informed about commonly held negative attitudes and are provided with suggestions about how to correct these thoughts. Each section of this training includes exercises and role plays designed to help pilots understand and accept the new information. These exercises frequently include vignettes for trainees to discuss or role play. Finally, the decision-making discussion also includes a discussion of stress effects. The information provided is very similar to that described previously in the stress management section. Trainees are provided with general knowledge about the impact of stress and the importance of being aware of stress effects. This section also includes a description of relaxation procedures and lifestyle interventions in coping with stress.

*Stress and decision making: Future directions for training*

The state of the art in coordination training programmes is similar to that seen in stress management programmes several years ago. In other words, we have amassed a number of methods that appear to be effective in communicating general information which trainees might utilize in attempting to cope with the demands of the flight task. In general, these programmes are well received by trainees (Foushee and Helmreich, 1988). Furthermore, there are some data to suggest that these interventions result in enhanced group processes in the cockpit (Helmreich, 1991). However, it might be the case that coordination training interventions, like the stress reduction programmes previously described, can benefit from integrating specific behavioural techniques designed to assist crews in dealing with stress. Furthermore, it is hypothesized that decision making under stress will improve when crews are provided with an opportunity to practise newly acquired skills under conditions of graded exposure to common stressors. In fact, a review of the decision-making literature suggests that stress exposure training is among the most promising of interventions to improve this aspect of human performance under stress (Janis, 1982). Therefore, the remainder of this chapter will suggest a coordination training programme for air crews that include the components of stress exposure training to result in both prevention of stressful events as well as ability to cope with unavoidable stressors. This approach is hypothesized to satisfy the goals of coordination training while providing the dual benefit of greater decision-making performance under stress.

## *Suggestions for future ADM training programmes*

In attempting to incorporate stress exposure interventions into crew coordination training, we will strive to follow the guidelines described by previous researchers (i.e. Brecke, 1982; Hall *et al.*, 1992; Meichenbaum, 1985). However, it should be noted that existing stress exposure paradigms have been developed exclusively for reducing stress effects in individuals. Because the primary negative effect of stress upon team performance is disrupted team process (Morgan and Bowers, in preparation), we will focus on interventions designed to maintain effective coordination in the decision-making process rather than reduction of individual anxiety.

Hall and her colleagues (1992) suggest that the first step in developing a stress exposure training paradigm should be a thorough needs analysis to identify the typical stressors encountered in task performance. Additionally, it was suggested that the needs analysis procedure should also include an assessment of the knowledge, skills and abilities required to promote technical performance during the period of stress exposure. The majority of these data are available in the existing literature. The stressors encountered in aviation have been documented in discussions of aviation human factors (i.e. Stone and Babcock, 1988). However, an assessment of these stressors for the community of interest is still advisable.

Furthermore, an initial assessment of the coordination skills required for effective flight performance has been provided by researchers attempting to develop a behaviourally-specific coordination training programme (Franz *et al.*, 1990). The result of these efforts was a series of seven dimensions which are believed to comprise the behavioural elements of a thorough air crew coordination training programme. We suggest that behaviours from each of these seven dimensions are required for crews to display effective decision making under stress. Therefore, behaviours from each of these dimensions will be included in the training programme.

The next recommended step in this intervention is to teach the actual target behaviours. The number of behaviours required for this training is much greater than is typical for training programmes of this type. Therefore, we recommend that the training proceed in two distinct phases. We have labelled the first of these phases as *prevention-focused behaviours*. These coordination behaviours are considered useful in observing, communicating, and assessing problems in the cockpit. Therefore, the impact on decision making is to attempt to reduce stress effects by identifying difficult situations as quickly as possible to reduce complications and to allow the crew as much time as possible to solve problems.

The training of prevention-focused behaviours includes three of the skill dimensions identified by Prince and Salas (1989): communication, situational awareness and assertiveness. Each of these dimensions includes specific behaviours to be trained and practised. For example, the target behaviours within the communication dimension include repeating information as required and using standard terminology. The behaviours within the situational awareness dimension are targeted towards main-taining an accurate perception of the aircraft's location as well as detecting situations which require attention and includes behaviours such as noting deviations, maintaining awareness of the performance of others, and identifying potential problems. The final dimension contained within this phase of training is assertiveness. This dimension is included within the prevention-focused phase because it is believed that the willingness to provide one's opinion is an important element in preventing stress effects or in allowing crews the maximum possible time to cope with unavoidable stressors. The specific behaviours that constitute assertiveness are asking questions when uncertain, making suggestions and confronting ambigu-ities and conflicts.

The training of each of these skill dimensions should progress according to the three steps described by Hall and her colleagues (Hall *et al.*, 1992). That is, trainees should be provided with instruction regarding the importance of each skill dimension as well as an explanation of how the skills can be applied in coping with the target stressors. For example, descriptions of aviation mishaps can be used to generate discussion about how the specific behaviours can be applied to prevent the causes of actual mishaps

The second step in the skill acquisition process is to teach the specific behaviours contained within each of the dimensions. A candidate training strategy for this training is behaviour modelling (Hall *et al.*, 1992; Prince *et al.*, 1992). Trainees are thought to acquire these skills most efficiently when the specific behaviours are demonstrated by the instructor, followed by an opportunity to practise and receive feedback from the instructor and fellow trainees about the execution of each skill.

The final step in the skill training process is to allow trainees to practise the newly acquired skills under conditions of gradually increasing stress. This exposure training will most likely take place in a high fidelity, full mission simulator, although there are some data to suggest that low fidelity simulations might provide an adequate opportunity for the training of these skills (Smith and Salas, 1991). However, regardless of the level of fidelity used, the challenge for the training developer is to create scenarios that provide reliable, gradually increasing levels of the stressor of interest. In the case of the prevention-focused behaviours, the goal for scenario development might be to provide increasingly difficult situations to diagnose. Guidelines for this type of scenario development have been provided by Prince and her colleagues (Prince *et al.*, 1993). This training should progress, with feedback following each exposure trial, until performance is maintained under moderately high levels of stress.

The second set of behaviours recommended for training are *problem-focused behaviours*. These are used to assist crews in making decisions in the presence of unavoidable stressors. These behaviours are effective because they facilitate the identification, selection and execution of effective alternatives. Four behavioural dimensions are contained within this phase. The first, mission analysis, refers to the behaviours required in creating and revising the crew's strategy. The behaviours include devising short- and long-term plans and assessing the impact of unplanned events on the mission. A second behavioural dimension is adaptability. This refers to the ability to alter one's course of action to respond to changes in the environment or task. Specific behaviours to be trained include changing flight plans in response to new information and willingness to consider the suggestions of others. A third dimension in the problem-focused phase of training is final decision making. This refers to the behaviours required in collecting pertinent data and choosing the best possible alternative. Specific behaviours in this dimension are gathering and checking information and anticipating the outcome of decisions. The final area of training is leadership, which emphasizes the allocation of the crew's resources in executing the final decision. Specific behaviours include determining the tasks to be assigned and assigning tasks to appropriate crew members.

Training for the elements discussed above should progress in a fashion identical to that described above. That is, trainees will pass through the steps of skill description, modelling and practice, and exposure to stress. In this case, the goal of the stress exposure trials is for the crew to maintain its performance in the presence of increasing levels of stress.

Finally, attitudes, stress levels and performance should be assessed to determine the effectiveness of the training. This information is useful not just to suggest alterations in the training programme, but to form a basis for feedback to crew members.

## Summary

We have identified four distinct challenges for training crews to make effective decisions when confronted with the stressors of the operational environment: to develop a useful theoretical model, to provide skills that are effective in coping with the large variety of stressors that confront aircrews, to provide training to support a variety of decision types in the presence of these stressors, and to provide skills that allow effective utilization of all crew members in the decision-making process. However, as demonstrated by the preceding sections, our knowledge of the effects of stress on ADM is severely limited by the lack of empirical research regarding this area of performance. Thus, there is a need for accelerated research regarding the effects of various stressors on crew decision making in aviation before these challenges can be realized.

In the interim, we have provided a method of augmenting existing training programmes to improve crews' abilities to collectively maintain their decision-making ability when faced with stress. The stress exposure paradigm discussed here has been effective in other areas of human performance, and is likely to satisfy several of these challenges. However, this intervention has thus far been supported by research relevant to relatively simple areas of human performance. There are little data to suggest its efficacy for decision making in general, and multiple-person decision making in particular. There is a critical need to assess the effectiveness of this type of training programme and to identify methods to strengthen it.

## Acknowledgements

The views expressed in this paper are those of the authors and do not reflect the official position of the Department of the Navy, Department of Defense or the US Government.

Dr Bowers conducted this research at the Naval Training Systems Center during a Summer Faculty Fellowship sponsored by the American Society for Engineering Education. He has subsequently returned to the University of Central Florida.

Please address correspondence to Dr Eduardo Salas at the Naval Training Systems Center, Code 262, 12350 Research Parkway, Orlando, FL 32826-3224, USA.

## References

Berkun, M. M. (1964), 'Performance decrement under psychological stress', *Human Factors*, **6**, 21–30.

Bloomfield, J. (1992, May), 'Elements of a theory of natural decision making', *The Federal Aviation Administration Aeronautical Decision Making Workshop*, Denver, COL.

Brecke, F. H. (1982), 'Instructional design for aircrew judgment training', *Aviation, Space, and Environmental Medicine*, **53**, 951–957.

Burgess, K. A., Riddle, D. L., Hall, J. K. and Salas, E. (1992, March), 'Principles of team leadership under stress', *38th Annual Meeting of the Southeastern Psychological Association*, Knoxville, TN.

Cooper, G. E., White, M. D. and Lauber, J. K. (eds) (1979), *Resource management on the flight deck*, (NASA Conference Publication 2120), Moffett Field, CA, NASA-Ames Research Center.

Cox, T. (1978), *Stress*, New York, Macmillan.

Crocker, P, R. (1989), 'Evaluating stress management training under competition conditions', *International Journal of Sport Psychology*, **20**, 191–204.

Diehl, A. E. (1992, August), 'Overview of aeronautical decision making effectiveness', in R. J. Adams and C. A. Adams (eds), *Workshop on Aeronautical Decision Making (ADM), Volume II: Plenary Session with Presentations and Proposed Action Plan*, (Document No. DOT/FAA/RD-92/14, 11), Washington, DC, Federal Aviation Administration.

Driskell, J. E., Moskal, P. and Carson, R. (1987), *Stress and human performance*, (Technical Report TR86–022). Orlando, FL, Naval Training Systems Center.

Driskell, J. E. and Salas, E. (1991a), 'Overcoming the effects of stress on military performance: Human factors, training, and selection strategies', *Handbook of Military Psychology*, 183–93.

Driskell, J. E. and Salas, E. (1991b), 'Group decision making under stress', *Journal of Applied Psychology*, **76**(3), 473–478.

Dwyer, D. (1992), 'Reducing stress-induced performance decrement through stress exposure training', Unpublished manuscript.

Ericsson, K. A. (1992, May), 'Methodology for studying and training expertise', in R. J. Adams and C. A. Adams (eds), *Workshop on Aeronautical Decision Making (ADM), Volume II: Plenary Session with Presentations and Proposed Action Plan*, (Document No. DOT/FAA/RD-92/14, 11), Washington, DC, Federal Aviation Administration.

Folkman, S. and Lazarus, R. S. (1985), 'If it changes it must be a process: Study of emotion and coping during three stages of a college examination', *Journal of Personality and Social Psychology*, **48**, 150–170.

Foushee, H. C. and Helmreich, R. L. (1988), 'Group interaction and flightcrew performance', in E. L. Weiner and D. C. Nagel (eds), *Human Factors in Aviation*, San Diego, CA, Academic Press.

Franz, T. M., Prince, C., Cannon-Bowers, J. A. and Salas, E. (1990), 'The identification of aircrew coordination skills', *Proceedings of the 12th Annual Department of Defense Symposium*, 97–101.

George, A. (1980), *Presidential Decision Making in Foreign Policy: The effective use of information and advice*, Boulder, CO, Westview.

Goldstein, I. L. (1980), 'Events in the training world', *The Industrial/Organizational Psychologist*, **18**, 19.

Griffiths, T. J., Steel, D. H., Vaccaro, P. and Karpman, M. B. (1981), 'The effects of relaxation techniques on anxiety and underwater performance', *International Journal of Sports Psychology*, **12**, 176–182.

Hackman, J. R. and Morris, C. G. (1975), 'Group tasks, group interactions, and group performance effectiveness: A review and proposed integration', in L. Berkowitz (ed.), *Advances in Experimental Social Psychology: Vol. 8*, New York, Academic Press, pp. 45–99.

Hall, J. K., Driskell, J. E., Salas, E. and Cannon-Bowers, J. A. (1992), 'Development of instructional design guidelines for stress exposure training', *Proceedings of the 14th Inter-service/Industry Training Systems Conference*.

Hancock, P. A. (1986), 'The effect of skill on performance under an environmental stressor', *Aviation, Space, and Environmental Medicine*, **57**, 59–64.

Hayakawa, S. I. (1968), *Modern Guide to Synonyms and Related Words*, New York, Funk & Wagnalls.

Helmreich, R. L. (1984), 'Cockpit management attitudes', *Human Factors*, **26**, 583–589.

Helmreich, R. L. (1991), 'The long and short term impact of crew resource management and training', *Proceedings of the AIAA, NASA, FAA, and Human Factors Society Conference on Challenges in Aviation Human Factors: The National Plan*, 81–83.

Hussain, R. and Lawrence, P. (1978), 'The reduction of test, state, and trait anxiety by test-specific and generalized stress inoculation training', *Cognitive Therapy and Research*, **2**, 25–37.

Hytten, K., Jensen, A. and Skauli, G. (1990), 'Stress inoculation training for smoke divers and free fall lifeboat passengers', *Aviation, Space, and Environmental Medicine*, **61**, 983–988.

Janis, I. and Mann, L. (1977), *Decision Making*, New York, The Free Press.

Janis, I. (1982), 'Decision making under stress', in L. Goldberger and S. Breznitz (eds), *Handbook of Stress*, New York, The Free Press, 69–87.

Jensen, R. S. and Benel, R. A. (1977), *Judgement Evaluations and Instruction in Civil Pilot Training*, (DOT/FAA Report RD78–24), Washington, DC.

Johansson, N. (1991), 'Effectiveness of a stress management program in reducing anxiety and depression in nursing students', *Journal of the American College of Health*, **40**, 125–129.

Kantowitz, B. H. and Casper, P. A. (1988), 'Human workload in aviation', in E. L. Wiener and D. C. Nagel (eds), *Human Factors in Aviation*, New York, Academic Press.

Keinan, G. (1987), 'Decision making under stress: Scanning of alternatives under controllable and uncontrollable threats', *Journal of Personality and Social Psychology*, **52**, 639–644.

Krahenbuhl, G. S., Marett, J. R. and Reid, G. B. (1978), 'Task-specific simulator pretraining and in-flight stress of student pilots', *Aviation, Space, and Environmental Medicine*, **49**, 1107–1110.

Lazarus, R. S., Deese, J. and Osler, S. F. (1952), 'The effects of psychological stress upon performance', *Journal of Experimental Psychology*, **43**, 100–105.

Lippert, F. G., Shechter, J. and Burke, J. (1989), 'A study of the integration of aviation and operating room cognitive skills', in R. S. Jensen (ed.), *Proceedings of the Fifth International Symposium on Aviation Psychology*, Columbus, OH, 366–370.

Lofaro, R. J. (1992, May), 'Crew resource management: Past, Present, and Future', in R. J. Adams and C. A. Adams (eds), *Workshop on Aeronautical Decision Making (ADM), Volume II: Plenary Session with Presentations and Proposed Action Plan*, (Document No. DOT/FAA/RD-92/14, 11), Washington, DC, Federal Aviation Administration.

McGrath, J. E. (1964), *Social Psychology: A brief introduction*, New York, Holt, Rinehart, & Winston.

Meichenbaum, D. (1985), *Stress Inoculation Training*, New York, Pergamon Press.

Meichenbaum, D. and Jarembo, M. (1983), *Stress Reduction and Prevention*, New York, Plenum.

Moore, M. S. (1977), 'Complexities of human factors in aviation', *Aviation, Space, and Environmental Medicine*, **48**, 471–473.

Morgan, B. B., Jr. and Bowers, C. A. (in preparation). 'Teamwork stressors and their implications for team decisionmaking', to appear in R. A. Guzzo and E. Salas (eds), *Team Decision Making Effectiveness in Organizations*, New York, Josey-Bass.

Novaco, R. (1977), 'A stress inoculation approach to anger management in the training of law enforcement officers', *American Journal of Community Psychology*, **5**, 327–346.

Orasanu, J. (1992, May), 'Shared mental models in crew decision making', *The Federal Aviation Administration Aeronautical Decision Making Workshop*, Denver, COL.

Paul, S. E. (1992, August), 'Delta Airlines decision making training', in R. J. Adams and C. A. Adams (eds), *Workshop on Aeronautical Decision Making (ADM), Volume II: Plenary Session with Presentations and Proposed Action Plan*, (Document No. DOT/FAA/RD-92/14,11), Washington, DC, Federal Aviation Administration.

Prince, C. and Salas E. (1993), 'Training for teamwork in the military aircrew: Military

CRM programs and research', in E. Wiener, B. Kanki and R. Helmreich (eds), *Cockpit Resource Management*, New York, Academic Press.

Prince, C., Chidester, T. R., Bowers, C. and Cannon-Bowers, J. (1992), 'Aircrew coordination: Achieving teamwork in the cockpit', to appear in R. W. Swezey and E. Salas (eds), *Teams: Their training and performance*, New York, Ablex.

Rodale, J. I. (1978), *The Synonym Finder*, New York, Warner Books

Roget, P. M. (1988), *Roget's II The New Thesaurus*, New York, Berkeley Books.

Rosch, P. J. and Pelletier, K. R. (1989), 'Designing worksite stress management programs', in L. R. Murphy and T. F. Schoenborn (eds), *Stress Management in Work Settings*, New York, Praeger.

Ruffel Smith, H. P. (1979), *A simulator study of the interaction of pilot workload with errors, vigilance, and decisions*, (NASA Technical Memorandum No. 78482), Moffett Field, CA, NASA-Ames Research Center.

Sarason. I., Johnson, J., Berberich, J. and Siegal, J. (1979), 'Helping police officers to cope with stress: A cognitive-behavioral approach', *American Journal of Community Psychology*, **7**, 593–603.

Smith, K. (1992, May), 'Decision making in operational systems', *The Federal Aviation Administration Aeronautical Decision Making Workshop*, Denver, COL.

Smith, K. and Salas, E. (1991, March), 'Training assertiveness: Impact of active participation', *37th Annual Meeting of the Southeastern Psychological Association*, New Orleans, LA.

Smith, R. E. (1980), 'A cognitive-affective approach to stress management training for athletes', in C. H. Nadeau, W. Hollowell, K. M. Newell and C. G. Roberts (eds), *Skillfulness in Movement: Psychology of Motor Behavior and Sport*, Champaign, IL, Human Kinetics.

Stone, R. B. (1988), 'Airline pilots' perspective', in E. L. Weiner and D. C. Nagel (eds), *Human Factors in Aviation*, San Diego, CA, Academic Press.

Weitz, J. (1970), 'Psychological research needs on the problems of human stress', in J. E. McGrath (ed.), *Social and Psychological Factors in Stress*, New York, Holt, Rinehart, and Winston.

Wexley, K. N. and Latham, G. P. (1981), *Developing and Training Human Resources in Organizations*, Glenview, IL, Scott, Foresman and Company.

# Part 4

# The Delivery of Training

# 14 Crew Resource Management: Achieving enhanced flight operations

*William R. Taggart*

## Introduction

Crew Resource Management (CRM) continues to receive increasing attention throughout the aviation community. As with any new operational concept, there is the need for clear and precise definitions as to what CRM is, and what can be expected from implementing CRM concepts and techniques in a flight operations environment. CRM and its companion training programme, LOFT, differ from traditional training activities to which crews have been exposed over the years. Developing teamwork and interpersonal skills has not traditionally been a formal part of aviation training.

At the outset, we need to define Crew or Cockpit Resource Management. It is the effective use of all resources available to the flight crew, including equipment, technical/procedural skills, and the contributions of flight crew and others. The objective of CRM training is straightforward: to use all available resources to ensure safe and efficient flight operations, while at the same time providing the technical and pilot skills training that is needed to maintain proficiency in the most effective way possible.

In years past, superior technical proficiency was considered the prime factor in assessing a pilot's ability to perform safely. This became known as the 'right stuff'. Whoever possessed the 'right stuff' was regarded in the highest esteem. Yet analysis of air carrier accidents and incidents over the past 20 years has confirmed that at least 65% can be attributed to inadequacies in leadership qualities, communications skills, crew co-ordination and decision making. In these accidents and incidents a lack of the 'right stuff' was not a factor. Instead, these accidents and incidents involved inadequate management and use of the various resources available to the crew.

CRM is not new. Some form of CRM is manifested each time a pilot takes control of the cockpit. However, the formal recognition and study of its merits, inside the cockpit, is new. Pilots have all had formal training regarding the proper method to perform an ILS approach. Through practice and training, the necessary skills can be learned to safely perform an ILS approach to both company and regulatory standards. But combine that ILS approach with personal stress, low fuel, an inoperative auto pilot, a poor dispatch system, difficulties with ATC handling, and language difficulties, more than just technical skill will be required to complete the flight safely. Under these conditions, dedicated crew involvement, exemplary leadership, and interpersonal skills directed at problem solving are needed to maintain safety inside the cockpit.

## Early beginnings

In 1975, as part of the United States FAA's responsibility to facilitate the introduction of the human factors concepts into flight training, airlines began to deal with two issues in pilot training and checking. These issues were hardware requirements for total simulation and the redesign of training programmes to deal with complex human factors problems. It is the issue of human factors errors, as these relate to management and leadership in the cockpit, that is addressed in this chapter.

Why do we need CRM? The aviation operational environment is growing increasingly complex. Deregulation has spawned increases in traffic patterns and congestion. Having limited resources to expand the Air Traffic Control (ATC) system has led to greater demands of cockpit crews for coordination and safety awareness. Aeroplanes are growing more complex, and automation, while helping, creates its own set of demands on the crew. When we add to this the variety of aircraft types and instrument layouts, it becomes apparent that one person in the cockpit cannot handle all the activities alone.

As early as the Second World War, it was recognized that the limiting factor in the development and design of aeroplanes was the ability of the human to effectively operate and manage the resources provided. The human factors engineering lessons learned during this period were not lost and have been incorporated in research, development and design of military and transport category aircraft and space vehicles.

Because of improvements in materials, aircraft design and the reliability of jet engines over the past 20 years we have seen a dramatic improvement in the safety record of air carriers. However, human error remains the major cause of incidents and accidents in civil transport operations. The net result is 65–70% of all air carrier accidents had as one of the causal factors the lack of effective use of resources by the cockpit crew (Sears, 1986; Nagel, 1988). This statistic has been found to apply equally to both small and large operators, and also to the military community.

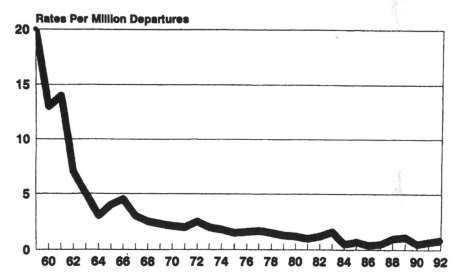

**Figure 14.1   Fatal accidents – worldwide commercial jet fleet (excludes sabotage and military action)**

Statistics are something that pilots rarely get excited about, but a review of the Boeing Commercial Airplane Company statistics indicates that almost three-quarters of all fatal commercial jet accidents have occurred in aeroplanes which were technically capable of a safe landing (Boeing, 1993).

The problem in these accidents sometimes began with a relatively minor mechanical malfunction, or with a less than ideal operating environment, which was then compounded into disaster. The significant contributing factor common to all these tragedies was the ineffective management of the resources available to the flight crew. These included both technical and human resources, most notably command and leadership skills.

The Boeing Accident Survey covers the years 1958–1992. Figure 14.1

311

shows accident rates per million departures. It is of interest that the rapidly descending line between 1959 and 1964 represents the introduction of jet aeroplanes into commercial aviation. It also reflects improvements gained through better equipment, advanced simulation devices and overall experience.

Although it is clear that equipment reliability has improved over the past 25 years, the number of accidents has stayed pretty much unchanged. On the other hand, if a trend line is plotted for the actual number of fatalities per year, that rate is actually increasing on a worldwide basis. Included among the reasons for this increase are the following: aeroplanes are larger and carry more passengers, the number of aircraft in use has grown dramatically, and these aircraft are being used for more hours per day.

A traditional method for attacking the accident rate problem is to improve the level of technology. Engineering and manufacturing methods are greatly improved. New composite materials are in use, and today's jet engines possess remarkable reliability. Aircraft systems have multiple levels of redundancy, and the quality of automated and safety systems continues to improve. However, in spite of these technological advances the accident rate has remained largely unchanged.

In addition to improved engineering and aircraft reliability, another method used to enhance safety is to look at procedures and standards. When accidents are analysed by phase of flight, over 70% of the accidents are found to occur during less than 5% of the flight time. The critical phases remain approach and landing, and take-off and initial climb.

Procedural responses have included increased levels of training and standardization through enhanced checking, improved flight manuals and required company procedures. And yet, although there have been significant improvements in these areas over the past 20 or so years, the accident picture remains much the same.

Traditional accident investigation methods have focused on finding and determining causal factors of accidents. Figure 14.2 depicts Boeing's assessment of these causal factors. These data imply that the crew is the primary cause in over 73% of the accidents Boeing has studied. The Boeing data are consistent with NTSB (National Transportation Safety Board) and NASA findings. In the May 1993 issue of the *ICAO Journal*, Maurino stresses that it is necessary to look at accident prevention from a systems perspective and develop suitable educational packages that include both technical training and human factors training. Research conducted by NASA, major airlines, and various research organizations has confirmed that a majority of air carrier accidents could be prevented by providing crew resource management training to the flight crew. Further evidence in support of this was recently published by the NTSB

Percentage of accidents

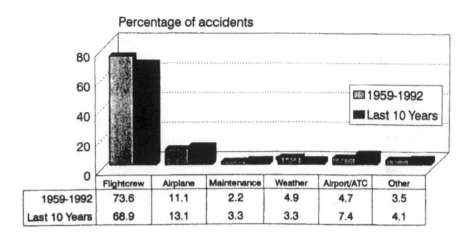

| | Flightcrew | Airplane | Maintenance | Weather | Airport/ATC | Other |
|---|---|---|---|---|---|---|
| 1959-1992 | 73.6 | 11.1 | 2.2 | 4.9 | 4.7 | 3.5 |
| Last 10 Years | 68.9 | 13.1 | 3.3 | 3.3 | 7.4 | 4.1 |

**Figure 14.2   Primary factors for hull-loss accidents – worldwide commercial jet fleet (excludes sabotage and military action)**
(Source: Boeing Aircraft Company)

following two commercial airline accidents. The accidents were similar in that they both involved windshear and crew coordination problems. NTSB findings indicated that cockpit resource management training attended by one of the crews played a positive role in preventing a much more serious accident. This crew's airline has been providing such training for several years.

A review of the NTSB aircraft accident report on the fatal crash of a Lockheed Electra in Reno, Nevada on 21 January 1985 clearly shows attitudes and cockpit resource management are directly related to the individual and crew performance.

The captain was characterized as 'always in command'. In addition a first officer described him as the type of captain who often checks first officers on their knowledge of equipment and procedures. The first officer and flight engineer differed considerably from the captain in two important dimensions that affect the nature of the interpersonal relationships of flight crew members in the cockpit: age and flight experience. This diversity can contribute, under routine circumstances, to deference by the junior crew members to the senior member. Under the type of critical conditions that Galaxy 203 experienced, the interaction could, and did become one-sided, as demonstrated by cockpit voice recorder conversations. The captain, who typically employed a commanding leadership style, took complete control of the actions of other flight crew members after the problem in the aeroplane developed.

Given the composition of the crew the NTSB felt that crew coordination or Cockpit Resource Management training may have enhanced the quality of the interaction of crew members as well as their ability to cope effectively with their increasing acute stress. The actions of the crew members suggest that a less dominating captain and a more assertive junior crew would have probably improved the flight crew's response both individually and overall to what quickly developed into an emergency.

NASA sponsored research projects suggest that the addition of CRM concepts to current training programmes in flight operations offers solutions to human factors related problems. There is the definite potential for improving the existing margin of aviation safety. And while improved technology and better operating procedures and policies offer partial solutions, the flight crew is still the focal point in any aircraft operation. Changes to existing training systems and the use of techniques such as LOFT offer one avenue for continuing improvement.

## FAA regulatory involvement

Federal Aviation Administration (FAA) involvement in CRM began in 1975 when a request was received seeking to implement a new type of training at Northwest Airlines. The FAA agreed to the proposed test programme, which was subsequently to be called Line Oriented Flight Training (LOFT).

In 1977 a regulatory change was proposed to permit any airline to utilize this LOFT training as part of their recurrent training programme. In 1978 a meeting of industry, FAA training personnel, instructors and FAA inspectors was held in an attempt to determine guidelines for LOFT. This resulted in US FAA Advisory Circular 120–35, published in May 1978. Regulations were amended to allow LOFT to be a part of any airline's training programme. This was a significant step in training and dealing with crew coordination problems – problems already noted to be prevalent in the accident statistics.

After the introduction of the advisory circular there was initially a slow response from industry in accepting the voluntary programme. In 1979 the Air Transport Division issued a letter to the FAA regions and the industry soliciting comments about how to advance training in a progressive way so that we could meet the challenge raised by previous accidents. The NASA/Industry Workshop, 'Guidelines for LOFT' at the NASA-Ames Research Center, Moffett Field, California was a critical first step in this programme.

The goal of the regulatory programme in the early 1980s was to work with the industry to develop a draft Notice of Proposed Rule-Making

(NPRM) that would address the issues of crew concept and CRM. On 13 January 1981 Charles Huettner (FAA) stated at the proceedings of the NASA/Industry Workshop, Moffett Field, California, 'we are embarking on an adventure into the flight training techniques of the future. In recent years a growing consensus has occurred in industry and government that training should emphasize crew coordination and the management of crew resources.'

Current regulations do not reflect recent advancements in aircraft technology or advancements in training methods and techniques. Certain training, checking and testing requirements in the regulations are out of date for airline operations of advanced technology aircraft. The FAA has been accommodating air carrier training needs by continuing to issue exemptions to current training requirements.

In June 1988, the National Transportation Safety Board recommended that all major air carriers:

> Review initial and recurrent flight crew training programmes to ensure that they include simulator or aircraft training exercises which involve cockpit resource management and active coordination of all crew member trainees and which will permit evaluation of crew performance and adherence to those crew coordination procedures.

The FAA approach for redesigning training programmes aims to maximize the benefits of advanced simulation and to deal with the increasing complexity of cockpit human factors. The FAA recognizes that training needs to address the failure of flight crew members to engage in proper crew coordination during normal and abnormal in-flight operations. The FAA contends that advances in aircraft and cockpit technology, along with increased complexity in the airspace system, will tend to increase demands on crew coordination and cockpit management.

From this perspective, it is evident that manual manipulation of the aircraft through the traditionally fixed package of manoeuvres trains and checks a pilot for less than his or her whole job. The FAA saw LOFT as a first step towards more complete crew training. In most cases, however, the focus of training was still on the individual crew member, and not on the crew itself as a unit of training. One of the early courses to involve the entire crew was developed at United Airlines, following a string of hull-loss accidents which were not due to any specific mechanical failure, or to problems which themselves prevented a safe conclusion of the flight. The United Airlines Command/Leadership/and Resource Management programme (C/L/R) was made available on a commercial basis to all operators in the early 1980s, but the industry was slow to respond. In fact, most of the early participants in these commercial seminars were international operators and corporate jet fleets. KLM was another early pioneer with their KHUFAC human factors course, while a few other carriers began to offer command courses for captains as part of their upgrade training.

In the second half of the decade, several large commercial airlines developed and implemented CRM seminars on their own. These programmes went beyond some of the earlier efforts and became more modular and operational in nature. Specific segments relating to topics such as situational awareness and stress management, for example, were now included in the training. At this point, CRM training began to move away from traditional management training seminars and became more operational and real-world centred. Part of the reason for this shift is that pilots tended to reject any form of training that could be labelled 'psychobabble.' Another reason is that as senior flight managers began to become more involved and committed to the training, they looked for additional ways to build the concepts into everyday operations.

In 1990, the FAA issued an advisory circular on CRM and comprehensive CRM training became a reality. In conjunction with the advisory circular on CRM training, a Special Federal Air Regulation (SFAR) was issued that would provide for updated training programmes modelled to the specific needs of an air carrier. This was the Advanced Qualification Program (AQP). AQP was developed to encourage new training concepts designed to address the challenge of integrated flight crew management of abnormal systems, adverse environment, and human factors problems.

The various components of AQP comprise:

1  A formal course in CRM.
2  A data collection and validation programme.
3  Use of crew concept for training and checking.
4  Integration of LOFT for all crew members.
5  Additional training emphasis on check-airmen.
6  Extended (or shortened) training cycles to meet the needs of specific crew members.

The AQP establishes an alternative method for meeting the training, checking, certification and qualification requirements for flight crew members, flight attendants, aircraft dispatchers, instructors and evaluators. The FAA developed AQP in response to recommendations made by a joint task force of industry and government representatives, formed (at the FAA's request) in 1987.

If these programmes for CRM and LOFT training can be continued and sustained throughout the industry, it will be possible to reduce the number of incidents and accidents due to flight crew factors. Integrated

Crew Resource Management and LOFT training implemented throughout the aviation industry offers the challenge and the opportunity to address these system problems.

## Assessing your own organization

Before initiating a CRM training programme, it is important to examine the existing status of the culture relative to CRM. Issues that need to be examined include readiness of senior level management to endorse and support CRM training, as well as the attitudes that currently exist among line crew members about CRM.

As an initial step, an organization considering CRM training can designate a 'steering' or planning committee charged with gathering initial information regarding CRM in the organization. This committee or task force should ideally be composed of individuals that represent a cross-section of flight operations. This cross-section can include flight operations management, ground school instructors, check-airmen, pilots union, captains and first officers. It is important that each person selected carry substantial credibility with the line population. Individuals are often selected on the basis of availability or on the basis of personal interest. It is critical that this initial work group carry the respect of their peers.

The committee can be charged with diagnosing the current status within the airline, and within flight operations in particular. Once the current level of CRM training is determined, and a set of training objectives for CRM have been specified, a plan can be developed to close the gap between actual status and planned objectives.

There are several advantages of using a work group approach for this first step. First, various segments of flight operations can be represented. This avoids the perception that CRM is a 'management project', or that it is the work of one or two individuals. Another benefit is that there are likely to be different perceptions about the current state of CRM effectiveness. Being able to debate and discuss different perceptions can result in a better diagnosis and planning effort.

The specific recommendations seen to be critical issues for CRM training include the following:

1  Demonstrating total commitment to the CRM programme.
2  Communicating the nature of the programme and what it does, and does not do before start-up.
3  Assessing the status of the organization before designing and initiating the programme.

4   Customizing the training to reflect the nature and needs of the organization.
5   Ensuring that the CRM course contains activities that actively engage the participants and provide feedback.
6   Recognizing that the initial CRM course is only the first step, and must be followed by additional training and reinforcement.
7   Integrating CRM training with Line Oriented Flight Training (LOFT).
8   Stressing the critical role of check-airmen and instructors.
9   Selecting check-airmen and instructors carefully and making sure that all are in agreement with CRM concepts.
10  Instituting quality control procedures to ensure that CRM and LOFT are accomplished effectively.
11  Considering extending the training to cabin crews and other elements in the organization.
12  Preparing to work with individuals who reject CRM concepts.
13  Evaluating and addressing multi-cultural issues within the organization.
14  Recognizing and addressing the existence of organizational barriers.

Another key part of assessing organization readiness is to identify limitations and restrictions that need to be considered in CRM planning. An organization might be inclined to postpone CRM training for its crew members due to the following:

1   Lack of simulator capability for LOFT training.
2   Not having anyone on-staff who is CRM-knowledgeable.
3   Not having the financial resources to develop a comprehensive training programme.
4   Not having enough time in the training schedule to add a CRM course.
5   Concern about the actual payoff for an investment in CRM training.

One way to deal with these negative perceptions and limitations is to recognize that CRM is an ongoing process and does not need a significant investment of funds. Effective courses have been developed in-house by a variety of operators and the ongoing research sponsored by NASA, the FAA and others promises to make much of the training information widely available.

Probably the best individuals who are CRM-knowledgeable are those crew members that command the respect of others on the line. Experience has proven that these individuals, with some structural assistance, can put together and deliver an effective CRM programme. In addition, there are frequently creative ways of working with training schedules to systematically integrate CRM into the training and development curriculum. From a financial perspective, there are usually operational improvements resulting from CRM training that go far beyond the initial monetary outlay involved.

## Implementing CRM crew skills

Once the assessment process has been completed, flight operations can determine how to best implement CRM training. There are various approaches to this, and there is no 'one best way' to do CRM training. One important principle to keep in mind is that CRM training will work best if it reflects the actual culture and needs of the organization.

In today's operating environment, the challenge that must be answered is how to 'get' crew coordination or teamwork to happen. To meet this challenge three distinct phases of CRM training are necessary. The first is the awareness phase, the second is a practice and feedback phase, and the third is the continual reinforcement phase. In addition to these phases of training, there must exist a training and checking operational environment which actively supports strong CRM concepts in everyday use. What this means in an operational context is that there is an emphasis in training programmes of all types towards using crews wherever possible, instead of individuals. There need to be quality control measures that go beyond traditional checking. And finally, there needs to be management support and commitment to the programme and its components. In sum, CRM needs to be integrated throughout all facets of training and checking in a logical and seamless manner. It was noted at the NASA/ MAC CRM conference in 1986 that we will know that CRM has been successful when the term disappears, but the concepts have been embedded throughout the training and checking programme.

The first phase, awareness, is usually accomplished through a seminar environment in a classroom setting. The length of these programmes usually ranges from 1–5 days. The baseline CRM seminar is often accomplished at a location which is independent and separate from the normal flight training environment. A local hotel might be used, or another facility that will result in a neutral setting apart from the distractions and the normal day-to-day interruptions that are a normal part of flight operations. One guideline to use is that even though the seminar may take place 'on the property' or nearby, the seminar will function as though it is 500 miles away. This way, distractions can be minimized for those attending the programme. It is usually better for the training goals if participants can devote their time and energies to the various issues in the seminar without being presented with distractions involving work, home, calls from the chief pilot, and so forth.

The final quality indicator is that the location must be appropriate for the training. Frequently the classroom is set up seminar style, with participants divided into teams or crews of 4–6 persons each. The participants represent a cross-section of fleets and crew positions. While some of the early programmes involved captains only, experience indicates that a higher quality programme occurs when there is a mixing of all positions and fleets.

An increasing number of organizations have chosen to involve others that interface with the flight crew during the course of normal flight operations. These individuals include dispatchers, flight attendants, mechanics, gate agents, ATC staff, and others. All parties gain from this interchange, though one must be careful not to dilute the impact of the learning on the flight crew. To accomplish these items, the CRM course may consist of a series of case studies, teamwork activities, and the building of crew concept and teamwork skills using various projects and team crew activities. The content of one such programme is outlined below.

One of the key considerations in planning for the CRM baseline seminar is deciding who will conduct the programme. Various approaches have been used. Some operators have elected to use outside consultants and educators to deliver the training. Others have used non-flying ground school personnel, and some have used retired crew members. One approach that is effective and which was adopted at United and Pan American is to use a combination of line airmen and training captains from the flight academy. In all cases, the training programme is better received if there are active, line-oriented crew members leading the seminar activities. Individuals selected should have credibility with line crew members as well as being committed to the principles that are being examined and tested in the seminar. Frequently outsiders will be viewed with scepticism, and this can form unnecessary barriers to the training process. Selecting seminar administrators that represent both management and line pilots is recommended.

Another benefit of this approach in selecting administrators for the training is that it ensures that the training is part of flight operations rather than being centred in the personnel department or part of other management development programmes. It is of vital importance that the CRM programme be delivered and administered by persons who have credibility with line crews, and who can address operational issues instead of just concentrating on academic and research information. Unless the course is seen as being operationally useful and relevant by line crews, there is a strong possibility that resistance to the training will be difficult to overcome.

The results at the end of the CRM seminar are a shift in both attitudes on the part of crew members to the behaviour of others, as well as their own behaviour. One important measure here is the degree to which self-deception can be removed, so that crew members can recognize that changes are needed on their part, not just on the part of others.

When crews can recognize that they are not infallible in the aeroplane and that external problems such as stress at home or fatigue do affect their own and others' behaviour and problem-solving ability, this can become an important key in improving aviation safety.

Different forms of learning that have been used in CRM training include instructor-centred, facilitator-centred, and self-directed. It has generally been found that experiential or crew-centred learning yields the best results since research into learning behaviour has shown that flight crews learn best when they are involved in the process in a direct, hands-on way. The seminar instructors act as administrators, as opposed to traditional instructors in a traditional teacher-tell setting. There is a quotation that is frequently heard to support involvement-centred learning: 'Tell me and I forget. Show me and I remember. Involve me and I understand.' Whatever form the initial baseline CRM training may take, the learning will have the most impact if the training meets the standards listed in Table 14.1

**Table 14.1   Initial CRM training**

| Quality Indicators |
| --- |
| • Experiential |
| • High personal involvement |
| • Skill-based |
| • Crew-centred |
| • Line relevant |
| • Emphasizes feedback and critique |
| • Learning is self-convincing, not imposed |

During the CRM seminar, it is important to point out that this is not training in amateur psychology, or in trying to change crew members' personalities. It is not a California-style high mountain desert experience in a hot tub. It also should not be the time to just watch videos and listen to someone at the front of the room. Instead, the focus is on how to equip oneself with a better set of skills to deal with the various types of crew members and situations that will be faced in the future.

At one airline, the participants themselves define their own personal objectives for the seminar, instead of using a canned and preset list presented by the seminar administrators. Typical CRM objectives of the participants include the following items:

- Improved teamwork
- Better communications
- Raise morale
- Study crew decision making
- Better collaboration
- Enhance captain's leadership
- Learning to manage better
- How to deal with the difficult crew member
- Learning critique
- Preparing for LOFT.

Given that there are different academic approaches to the study of CRM and that different training approaches to the subject have been used, it is useful to look at what CRM is and what it is not. CRM *is* the following:

- a comprehensive system for improving crew performance;
- crew concept training addressing the needs of the total crew population;
- a system which can be extended to all forms of crew member training;
- a focus on crew member attitudes and behaviours, and how safety is affected;
- an opportunity for individuals to examine their behaviour and make individual decisions on how to best improve cockpit teamwork;
- a training method using the crew as the unit of training, not the individual;
- active training where the participants experience and participate, instead of teacher-tell lecture style;
- an operational focus on safety improvement;
- self-convincing and hands-on.

CRM *is not*:

- a quick fix that can be implemented overnight;
- a training programme administered to only a few specialized 'fix-it' cases;
- a system that occurs independent of other training activities;
- a psychological assessment or assessment of personality profile;

- a system where crews are provided with a specific prescription on how to work with others on the flight deck;
- another form of individually-centred crew member training;
- a passive lecture-style classroom course;
- an attempt by company management to dictate and control cockpit behaviour.

A first step in preparing for CRM and other human factors training steps such as LOFT, is to 'set the stage' both at the management level, and with the line crew member population. At the outset, many crew members may have the attitude that 'yes, this training is useful, but we should give it to the 6 or 7% that really need it.' This same thinking applies to checking and training pilots. Somehow there seems to be in place the myth that 'because I am a check-airman, I am immune or simply don't need this kind of training.' Involvement of the pilot's association and other union groups is essential, and their inclusion in the planning process should be achieved as early as possible. The very best CRM programmes are those where unions are involved at an early stage, and where there is a good orientation programme before the training actually starts. Periodic newsletters and articles in flight operations bulletins are a good way to start. The importance of a good orientation and communications programme about the training cannot be over-emphasized.

CRM concepts include how flight crew members communicate with one another, how they make decisions as a crew, how leadership is exercised, how problems are assessed and dealt with, along with other crew-centred factors. Training in these and other areas is a new challenge to instructors and chief pilots who have been exclusively used to the technical side of flight training. The good news, however, is that when exposed to these concepts in an operational context, most immediately see the benefits and wish they had received this training earlier in their careers.

Many of the concepts contained in CRM are common sense. Recognizing this, some may ask why these concepts need to be part of a training curriculum. One reason is that the concepts need to find their way into everyday operations. Another reason is that most crew members tend to revert back to earlier patterns and responses when dealing with a tough situation. Good and effective training can provide a vehicle for changing these patterns. Another important focus of CRM is the concept that other crew members can contribute to routine cockpit problem solving and situational awareness. We all make mistakes, and an operational culture that encourages others to speak up and assert their views can help small errors from being magnified, and ultimately affecting the quality of the operation at hand.

## LOFT

The CRM seminar experience is similar to the foundation of a building; it is an essential structural part, but by itself the foundation has limited operational utility. If CRM training is to be operationally effective, it must be built into other training steps and activities in a systematic way. One recognized method to do this is through the use of LOFT as part of the crew members' periodic training.

An increasing number of simulators are now equipped with video cameras and recording equipment, so that a LOFT scenario can be videotaped for playback and critique during the post-flight (LOFT) debrief discussion. Several operators use portable video equipment to use in simulators that are leased, and not permanently video equipped. To quote one leading CRM expert:

> LOFT with video tape feedback is one of the most powerful tools we have for reinforcing desirable behaviour in cockpit resource management (Helmreich, 1986).

Studies suggest that the value of LOFT is enhanced if there is a good strong foundation in CRM concepts and awareness to build on. An organizational culture and pilot attitudes that support critique and feedback for training and improvement purposes are also important. LOFT training should be systematic and designed to simulate actual problem situations on the line that require good crew skills for effective resolution and decision making. It is best if LOFT scenarios are designed to be operationally relevant and believable, but also to be a good test of the cockpit crew's teamwork skills.

When crew members have learned and can appreciate the importance of open and direct critique for the purposes of operational review and analysis, a platform is in place for effective post-LOFT debriefing which can address more than traditional technical and stick-and-rudder skills.

One misconception that periodically occurs is that LOFT training means that the workload of the crew in the simulator is continuously increased until the crew is overloaded. This is not the purpose or intent of LOFT training. Indeed, this can actually help to defeat LOFT's training intent. LOFT training is systematic and is intended to simulate actual problem situations on the line that require good crew skills for effective resolution and decision making.

LOFT scenarios are best if they are straightforward and if the entry point is relatively simple. For example, choosing a departure airport that requires a good, effective pre-flight briefing even under the best of conditions, might be one way to begin. Providing an entry point that allows for different options for the crew to choose from is also useful, since one scenario can have a wide variety of outcomes and choices depending on the decisions and courses of action that the crew chooses. Again, the scenario should be realistic, and the situation should be one where the crew lives with whatever problems they have until the situation is either resolved, or the aeroplane (simulator) is back on the ground.

After the LOFT is completed, the manner in which the debriefing is handled by the LOFT instructor is of key importance if CRM skills are to be reinforced and improved. It is best if the instructor does not handle the debrief in a 'teacher-tell' manner, but instead operates as a resource to the crew members and highlights different portions of the LOFT that may be suitable for review, critique and discussion. It is best if the discussion is led by the crew themselves, using the instructor and the videotape as resources for use during their critique. If handled in this way, then perhaps crew-led debriefs will occur with increasing frequency on the line after a difficult segment, or one where crew critique and review is called for.

Organizations using LOFT are extremely enthusiastic about the potential of CRM training through LOFT as a means of effecting real change in the bottom line of safe and efficient operation. Of vital importance for the effectiveness of LOFT is the creation of a strong illusion of reality in the simulated trips. This requirement dictates that many individual details, such as pre-flight activities, trip paperwork, manuals, communications, etc. be carefully prepared. Previous experience with LOFT has shown that overlooking even the smallest detail can destroy this illusion. It is also important to distinguish between fidelity and reality. There have been examples of LOFT programmes where extremely high fidelity simulators have been used, but yet the exercise is unrealistic due to non-fidelity related issues such as paperwork and implausible scenarios.

The LOFT briefing should include mention of the role playing aspect of LOFT and its importance to overall LOFT effectiveness. The instructor's goal is to produce crew performance and behaviour that would be typical for an actual line flight in the same set of circumstances as those developed in the scenario. In keeping with this goal, it is essential that crews have access to all the resources they would have on an actual line trip.

Communications is another area vitally important to the assurance of realism in LOFT operations. All communications must be conducted in the manner normally found on a line flight (i.e. via radio from outside the 'aeroplane', via interphone or normal conversations between cockpit crew members, or, in the case of cockpit-cabin, via the usual aeroplane equipment for this purpose). All external communications (ATC, ground crew, etc.) must also be credible and realistic.

Where possible, the entire simulator phase of the flight should be recorded on videotape. During debriefing the videotape can be reviewed and discussed by the flight crew, with main emphasis being placed upon crew performance, including their use of the elements of CRM. Following review of the videotape, the tape is always erased and no record is retained.

The role of the instructor in LOFT should be viewed as that of communicator, observer and moderator in the debriefing process; he is not an instructor in the traditional sense during the simulator period. He is the 'coordinator' or manager of the flight, using appropriate radio calls or responses to direct the flight along the desired path; he must be prepared to accept and manage alternate courses of action that the crew may wish to follow. The instructor should remain as unobtrusive as possible within the physical limitations of the simulator. He should resist the temptation to 'instruct' and must not intrude in any way into the situation.

To a considerable extent, the conflict between 'training' and 'checking' in a LOFT programme can be offset by the manner in which the instructor sets the scene during the pre-flight briefing. The instructor should emphasize that:

1 LOFT is designed as a pure learning experience.
2 LOFT is a new training concept designed to access command responsibilities, crew coordination, communication, and Cockpit Resource Management.
3 Mistakes may well be made, just as they sometimes occur on the line, but the crew must carry on. To some extent, LOFT is an exercise in 'mistake management'.
4 There is frequently no book solution to a LOFT exercise – there may be no 'right' solution.
5 The instructor's role is to manage the training situation, not to 'teach' right solutions, or to 'test' the trainees.
6 There will be an opportunity for full self-analysis during the debriefing.
7 The instructor will take notes only to assist in the debriefing.
8 It is important that crews have a complete understanding of the 'rules' under which LOFT is conducted. These conditions should be presented in a thorough pre-flight briefing.

Experience has shown that crews frequently debrief themselves very effectively. Self-criticism and self-examination are almost always present in these situations and in many cases they are much more effective than instructor critique. Often, crews are more critical of themselves than the instructor would ever be. Thus, the instructor should do everything possible to foster this sort of self-analysis, while at the same time keeping it at a constructive level. In his role as moderator, the instructor can guide the discussion to areas that he has noted need attention. Questions about specific procedures, decisions, and mistakes should be asked. However, unless absolutely necessary, the instructor should avoid 'lectures' about what is right and what is wrong. Obviously, the instructor should avoid the embarrassment of crew members as much as possible. The instructor should assure crews participating in LOFT that their jobs are not in jeopardy every time they enter the simulator for a LOFT session. While 'satisfactory completion' is an inescapable aspect of LOFT, at the same time it is hard to imagine 'unsatisfactory training' if it is conducted appropriately.

Helmreich, Wilhelm and Gregorich (1988) recommended that it is important to separate formally the training and evaluation function of LOFT. They also felt that it is invaluable for crew members, especially captains, to receive the opportunity available during LOFT to gain more understanding of their behaviours and its consequences, and be able to explore new behavioural strategies in a 'no-fault' situation. Their recommendation was that the term LOFT be reserved for training periods where formal, mandated evaluation is explicitly omitted.

## The check-airmen

Another key extension of CRM and human factors training in general is into the line check environment, as well as new-hire orientation, position upgrade training, and flight attendant and dispatcher training. The check-airman (instructor or evaluation) group is particularly important, since they can benefit from additional training on how to deal with CRM issues encountered during checks, and with crew members going through their initial operating experience.

The importance of the proper training of check-airmen for the purpose of evaluating CRM crew behaviour was outlined in a study citing the central role of the check-airman/LOFT facilitator in implementing and reinforcing CRM concepts (Helmreich, 1987). The check-airmen/ LOFT facilitators are trained in the philosophy, principles and conduct of line checks, LOFT and CRM, to enable them to effectively observe and critique individual and crew performance during a line observation or LOFT scenario. Special training is received in CRM principles to observe and critique these areas in LOFT and line evaluations.

Techniques have been developed to provide special training in observation, evaluation and debriefing for these groups. The performance evaluation training focuses on the application of the Line/LOFT worksheet to both the line operation and the simulator environment. Staged videotaped LOFT simulations are used for practice in evaluating and rating 14 aspects of specific crew behaviour, in addition to overall behaviour. Additionally, instruction is given for using the worksheet as a tool for debriefing crews on crew coordination – Evaluating Cockpit Resource Management Training (Helmreich *et al.*, 1991).

Issues that might be addressed as part of the human factors component of a line check could include the following: What was the quality of the pre-flight briefing? Was it appropriate given the amount of time the crew has worked together and actual operating environment being faced by the crew? Was open communications established among all crew members? Issues during the in-flight stage might be: was the timing of communications proper, were communications relevant, concise, targeted, and complete, including verification? Was assertion and advocacy present? Where disagreements happened, did crew members propose a course of action or solution to the problem at hand, or did they just express their discomfort and concern?

For interpersonal styles and actions there might be additional areas for consideration. We know that crew members have their own personal style, and that this determines how they operate on the ground and in the air. Research into personal characteristics and styles has found them to be strongly related to the ways crews function on the flight deck. An important implication of this research is that awareness of our own personal styles, and how they are perceived, may help us work more effectively with others.

In workload and planning, discussion topics might include: How is workload distributed and communicated? What happened during overload situations? Were the crew members actually ahead of the aeroplane, or did they find themselves behind and in a 'catch-up situation'?

In decision making, issues are: Was there the appropriate participation in decision making by crew members? How were the alternatives discussed and weighed before a final decision was reached? How were conflicts among crew members resolved? Finally, in terms of crew atmosphere and coordination, the crew should be able to look at the overall workload as well as the current climate on the flight deck itself.

The Line/LOFT worksheet serves additional, operational functions as a measure of fleet standardization and 'quality control' for check-airmen and instructors. The analysis of the ratings given by a check-airman, when compared with a standard, provides information on the evaluative skills of the rater (Helmreich *et al.*, 1991).

## Research efforts and results

In the research field of human factors and performance in aviation and aerospace, NASA had taken the lead in the early 1970s at its Ames research facility in California. Their current 'human factors in aviation safety' programme began in 1973 when a series of interviews were conducted with airline crew members. A typical comment was 'My company trains pilots well but not captains.' In a classic NASA study Ruffel Smith (1979) looked at the way a crew functions when faced with both normal and difficult situations. During this full-mission simulation study, it became apparent that the way in which leadership was exercised in using available resources greatly contributed to the success or failure of the crew.

Since 1985 an attitude and demographic survey, the Cockpit Management Attitudes Questionnaire (CMAQ), has been distributed to cockpit crew members at several airlines and military operations (Helmreich, 1984). The purpose of this survey was to gain baseline data on attitudes concerning CRM prior to and after completing formal CRM training. This initial research on attitudes of crew members showed there were significant differences in attitudes as a function of crew position, aircraft fleet and the merged pilot groups of some airlines (Helmreich and Wilhelm, 1986). In 1993, this survey was updated and expanded to include issues involving multicultural dynamics (Merritt, 1993). Preliminary findings indicate that there are strong and significant differences across cultural boundaries in terms of how crews respond to CRM training. For example, a very collectivist culture will respond well to various types of group exercises that are rejected by members of a very individualistic culture.

It was found that about two-thirds of crew members expected the training to be moderately or extremely useful. By position, captains thought training would be least useful, followed by first officers, with the flight engineers the most positive. On some questions there were minorities of individuals with highly divergent opinions. On some of these questions, 5% of captains strongly disagreed that crew members should be sensitive to the personal problems of other crew members, and 8% agreed strongly that first officers should not question actions or decisions of captains, except when they directly threaten safety of flight. Of all crew members, 8% believed that successful crew management was primarily a function of the captain's technical proficiency and 7% strongly disagreed that the captain's responsibilities include coordination of cabin crew activities. These data led to the conclusion that CRM training would be of great value for these individuals.

In one airline which had not hired new pilots for 17 years, almost 70%

disagreed with the statement that training is an important responsibility of the captain. In survey questions on briefing and debriefing, 7% did not agree that pre-flight briefing was important and 70% did not agree with the importance of the post-flight debriefing.

Although only 8% of those surveyed in 1986 had received CRM training, the trends showed favourable attitudes on the part of those who perceived having changed their behaviours. Those who said they had changed are more sensitive to others, accept more responsibility for training and coordination of cabin activities, and recognize the value of crew debriefing. In the overall analysis a clear pattern emerged: those who thought the training would be a waste of time or of minimal use held the least favourable attitudes.

The results of the data collected and evaluated by Helmreich indicate there is a positive shift in attitudes for 90% of the crew members. Figure 14.3 illustrates how crews from five different airlines rated their CRM training. This combined with measured attitude shifts, speaks well for improving flight safety.

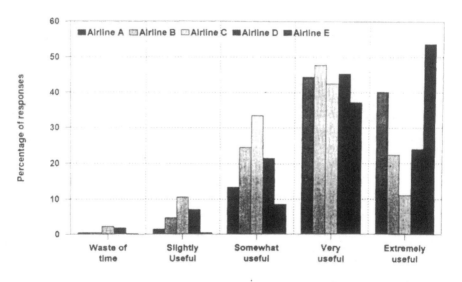

**Figure 14.3  Rated usefulness of CRM training: data from five airlines**
(Source: NASA/UT Crew Research Project)

One of the key supporting parts of many CRM programmes is the Line/LOFT checklist (Helmreich *et al.*, 1991) which is used as a data collection tool for the NASA/UT database. This worksheet has gone through several improvement stages since 1986, and is used to track crew performance over time. It has also been used by several operators as a template for designing advanced crew performance and evaluation systems.

The Line/LOFT checklist also calls for data elements to be scaled instead of simply being graded on a pass/fail, or satisfactory/unsatisfactory basis. The issue of scaling has received mixed reviews at several air carriers, depending on their own culture and their historical methods of assessing crew performance. The advantage of scaling is that performance on a range of manœuvres, procedures, flying skill, systems knowledge, etc. can be assessed from a viewpoint of strengths and weaknesses. Opportunities for improvement can be identified, and trends can be analysed over time. However, experience has shown that once scaling has been initiated and used by instructors and evaluators, the reaction is positive.

To provide an overview of the types of items being looked at, the following is a list of the current worksheet categories:

1   Briefing
2   Communications
3   Inquiry/questioning
4   Assertion/advocacy
5   Decisions communicated and acknowledged
6   Crew self-critique
7   Concern for accomplishment of tasks
8   Interpersonal relationship/group climate
9   Overall vigilance
10  Preparation and planning
11  Distractions avoided/prioritized
12  Workload distributed and communicated
13  Overall crew effectiveness
14  Overall technical proficiency
15  Overall workload
16  Management of abnormals/emergencies
17  Conflict resolution.

## Applying research results to real-world training

Crew Resource Management research has resulted in numerous payoffs in flight training and standardization for air carrier flight operations. It is clear that the focus of CRM is extending steadily beyond the confines of the flight deck and includes others that need to interface with the flight crew (Helmreich, in press).

Nonetheless, data collected from structured line observations indicated that there was one specific area of human factors and teamwork that was going untouched. First officers upgrading to the left seat received the traditional ground school training followed by simulator work and certification, but even the qualification LOFT simulator session tended to concentrate on individual skills and linear decision making. It was felt that there were additional training opportunities that could help equip new captains with specific skills appropriate to solving challenges and making decisions as a first time aircraft commander.

It was decided to use the NASA/UT/FAA database of air crew performance as a guide to select those topics that would be most germane for captain upgrade CRM training. Another objective was to conduct the training in a manner that would create the highest level of fidelity without actually using a simulator. The training needed to be operational, hands-on, and include other departments that routinely interface with captains for purposes of problem solving during routine flight operations.

Law (in press) reports on an analysis of over 2100 actual crew member observations. This work follows earlier studies which first identified crew position-specific performance factors. This 1992 NASA/UT/FAA analysis of actual crew observations looked at significant crew member behaviour using the behavioural markers and the LLC checklist items. In terms of crew position and critical behaviours, Table 14.2 summarizes the results.

#### Table 14.2 The key items

| Captains | First officers |
| --- | --- |
| Briefings | Inquiry and assertion |
| Leadership | Preparation, planning, vigilance |
| Interpersonal skill | Technical proficiency |
| Communications | Leadership |

An attitude data set based on the Cockpit Management Attitudes Questionnaire (CMAQ) (Helmreich *et al.*, 1991) indicates changes in crew attitudes across several scales, communications and coordination, command responsibility, and recognition of stress effects. This instrument has been undergoing refinement over the last ten years, and the most recent version has been renamed the Flight Management Attitudes Questionnaire (FMAQ 93). Among the scales on the CMAQ, the one that has shown the least substantive change is that for command authority. Changes in communications and coordination are usually positive as a result of a good baseline CRM course. It appears that if further gains are to be achieved in command and the use of captain's authority, then perhaps a focused training programme designed specifically for new captains might be in order.

A survey for measuring attitudinal and training needs items was developed for use at several carriers for administration to check-airmen and flight instructors. Among the items on this survey were several pertaining to the quality of training and preparation for new captains. One of the recurring indications from these data at several carriers was the response – new captains receive good technical training but are not well trained to solve actual operational problems on the line. It became clear from responses to several of the items on the survey that additional interpersonal skill development and problem-solving training was a need for new captains. It should be stressed that these findings are common to air carriers that participate in the NASA/UT/FAA research project.

It was decided to use the key elements of briefings, leadership, interpersonal skill and communications as central themes for the upgrade team skills workshop for captains. The methods used include simulations (role plays), short lectures to introduce selected research data, and video vignettes to model sound examples of critical skills and captain behaviours.

It also became apparent that for the programme to be effective, it would need the involvement of others in the airline that interface with captains on a regular basis. Figure 14.4 shows the various elements of the airline

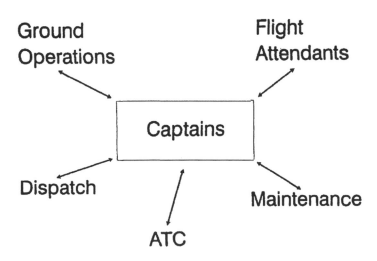

**Figure 14.4 Elements of the airline that participate in the programme**

that participate in the training programme. A typical programme consists of the eight or ten pilots going through the upgrade programme. At the conclusion of ground school and before simulator training, a one-day team skills workshop is held where flight attendants, ground operations agents, mechanics, maintenance supervisors, dispatchers and FAA representatives, are brought in to work with the new captains in a variety of problem-solving situations.

A series of case studies are used for the purpose of analysing actual problem-solving situations and how resources outside of the cockpit were effectively or ineffectively used. Several of the case studies include Avianca Flight 052 at Cove Neck, New York in 1990, Air Ontario Flight 1363 in Dryden, Ontario in 1989, United Airlines Flight 232 at Sioux City, Iowa, in 1989, and American Airlines Flight 132 at Nashville, Tennessee in 1988. As a brief example of how these case studies are written for analysis, the following is a segment of one example that is used in the seminar.

---

### American Airlines Flight 132

Nashville, Tennessee
3 February 1988

On 3 February 1988, AA flight 132, an MD-83, departed DFW at 1445 destined for Nashville's Metropolitan Airport. Operating as a regularly scheduled passenger flight, there were 120 passengers, four flight attendants and two pilots on board.

In addition to passenger luggage, the aeroplane was loaded with 6365 pounds of air freight, including two packages of hazardous materials: a 20 pound cylinder of oxygen, and a 104 pound fibre drum of textile treatment chemicals. The contents of the drum included 5 gallons of hydrogen peroxide solution, an oxidizer, and 25 pounds of a sodium orthosilicate-based mixture, a granular corrosive solid material.

No labels or markings had been affixed to the drum to indicate that it contained hazardous materials, and the shipping paper did not describe the drum as hazardous. Both the oxygen cylinder and the fibre drum were loaded into the middle cargo compartment of the aeroplane, and the fibre drum, which had no markings or arrows for orientation, was laid on its side.

The captain was notified in writing that the 20 pound oxygen cylinder had been loaded and he knew its location. The captain was not aware of any other hazardous cargo on the aeroplane.

After departing DFW, the flight was uneventful until it was on final ILS approach to Nashville. At 11 minutes 6 seconds before

---

landing, the seatbelt sign was switched on and shortly thereafter, the third flight attendant performed passenger safety seatbelt checks. She noticed 'no odd smells or smoke as reported later'. She returned to the front of the aeroplane and entered the lavatory. When she heard several chimes, she exited the lavatory and saw the first flight attendant checking seat rows. She observed that 'the cabin was slowly filling with light smoke'.

A few minutes before landing, a passenger in seat 17-E saw smoke rising from the floor 'through what appeared to be a seam where the carpeting meets the sidewall'. He summoned the fourth flight attendant, alerted her to the situation, and pointed to a location two rows ahead. At this time, he also noticed a burning electrical smell, but did not notice if the floor was hot. The fourth flight attendant noticed a 'haze' and detected an 'irritating odor of something burning'. At this time the aeroplane was descending from 6000 feet to 4000 feet.

The fourth flight attendant quickly went to the rear of the aeroplane to call the cockpit crew on the interphone. She told the second flight attendant to get a fire extinguisher and determine the source of the smell. At 5 minutes before touchdown, the fourth flight attendant called the first officer on the interphone and told him, 'We've got smoke in the cabin' and '. . . we don't know where it's coming from.'

Participants in the workshop have the opportunity to analyse the quality of the problem solving that occurred from a variety of perspectives from within the airline. Comparisons are made to actual examples from flight operations that illustrate where problem solving has been effective or has broken down on occasion.

Following several of these, a shift is made to using actual situations that require use of the Minimum Equipment List to determine the best course of action. Each team analyses the situation and decides on a course of action. In many cases two teams will come to radically different decisions and solutions for the problem presented.

Features of this training that seem to contribute to its initial success are as follows:

1   The training content is based on research identifying critical and significant new captain behaviours. The fact that the airline was a participant in the NASA/UT/FAA research project helps to lend credibility to the data presented.

2   The participants in the training include operational people from a

wide cross-section of airline flight operations and support departments. In many cases, this was the first training of substance that participants have experienced with other elements of the airline.

3   All problem-solving experiments and simulations are taken from actual incidents, irregularities and other occurrences in real line operations. The importance of reality-based training cannot be overstated.

This effort at using actual research data in the design of a training programme for new captains has resulted in a programme that has a high level of crew member buy-in, and has resulted in a decision to expand this type of inter-departmental training and problem solving to other areas of the airline. It is a recognition of the importance of teamwork and crew coordination skills that go beyond the traditional boundaries of the cockpit door.

A strong and telling testimonial to the value of this training is that the airline discussed here invited the line pilot instructors administering the programme to present it to the top 70 or so managers and officers of the airline. The result was a one-day programme that did much to build an increased appreciation of the coordination demands that face captains on the line.

Future directions and initiatives include the challenge of building effective training programmes for dispatch and for operations control that embody these same types of features and components for enhanced teamwork. Reality-based training that is operational, interactive, and challenging may be one important ingredient to enhance and maintain safety and effectiveness in an increasingly complex aviation system.

## Programme validation

Currently, NASA is engaged in collecting programme validation information from the Line/LOFT worksheets. Information is collected on a de-identified, anonymous basis. The database that is being built will help to provide important information and trends on the effectiveness of LOFT and CRM training. This will provide assistance and direction on how to modify and improve training and checking activities.

Robert Helmreich and John Wilhelm at the University of Texas at Austin have been collecting this information for NASA. Preliminary results from several CRM programmes are encouraging. Results from changes in attitudes on the part of crew members show that attitudes can indeed change, resulting in improved behaviour and teamwork on the flight deck. A first look at the data indicates that in terms of overall usefulness, about 80% have found the training to be very useful or extremely useful. Over 85% of the participants believe that this training will improve flight safety. These initial indicators are encouraging and some of the positive changes in attitudes can be summarized as follows:

1   Training is seen as an important responsibility of the captain.
2   The captain's responsibilities include the coordination of cabin crew activities.
3   There is recognition of the importance of the pre-flight briefing.
4   Each pilot and crew member should monitor other crew members for stress and fatigue.
5   The pilot flying the aeroplane should verbalize plans and intentions.
6   It is recognized that critique and debriefing are important.

There are also compelling data that CRM and LOFT do actually influence day-to-day line operations. Figure 14.5 illustrates the actual changes that have occurred at one specific airline over a multi-year period.

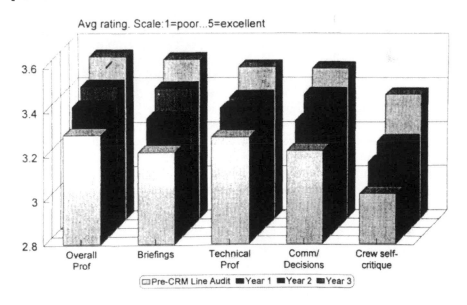

**Figure 14.5   Observed effects of CRM training – baseline and three subsequent years**
(Source: NASA/UT Crew Research Project)

Note that overall proficiency as well as technical proficiency increased over time along with specific CRM-related behaviours. The value of this information is that trends can be detected, along with areas that deserve an additional training focus. For this particular carrier, the decision was made to concentrate recurrent training in the following year on the subject of workload management and distribution. This decision was based on the research data that the organization was receiving. This is CRM in one of its best forms – an evolving and mature system that uses data to help determine further improvement possibilities.

## Summary

The evolution of CRM is a continuing process that continues to grow in importance and contribution to the aviation industry. Current efforts are noticeably different from the early fledgling programmes. The focus of CRM and LOFT has shifted from being a narrow in-the-cockpit orientation, to one that emphasizes the extended crew beyond the confines of the cockpit door. The use of continuing research into CRM issues has enabled the industry to refine existing programmes and to develop new ones.

Flight crew members have always valued good, relevant training that contributes to improved competency and performance on the flight deck. The continuing challenge for CRM is to demonstrate and prove that good teamwork and interface skills also enhance individual performance. High quality research combined with concrete training that mirrors the real world will provide flight crews with the tools needed to function safely in an increasingly complex operating environment. Technology alone is not the answer, and better checklists and procedures will not solve the dilemma of how to improve safety. One answer lies in solid technical and procedural training, integrated with human factors training such as CRM and LOFT. These efforts will provide real-world tools that pilots will value and use.

## References

*Boeing Airliner*, The Boeing Company, Seattle, April–June, 1993.

Cooper, G.E., White, M.D. and Lauber, J.K. (eds) (1980), *Resource Management on the Flightdeck: Proceedings of a NASA/Industry Workshop*, Moffett Field, CA, NASA-Ames Research Center, NASA Report No. CP-2120.

Helmreich, R.L. (1984), 'Cockpit management attitudes', *Human Factors*, **26**, 583–589.

Helmreich, R.L. and Wilhelm, J.A. (1986), *Preliminary Findings: Survey of Pan American Flight Crewmembers*, NASA/UT Report 86–6.

Helmreich, R.L., Wilhelm, J.A. and Gregorich, S.E. (1988), *Notes on the Concept of LOFT: An agenda for research*, NASA/UT Technical Manual 88–1.

Helmreich, R.L., Chidester, T.R., Foushee, H.C., Gregorich, S.E. and Wilhelm, J.A.

(1989), *How Effective is Cockpit Resource Management Training? Issues in Evaluating the Impact of Programs to Enhance Crew Coordination*, NASA/UT Technical Report 89–2.

Helmreich, R.L., Chidester, T.R., Foushee, H.C., Gregorich, S.E. and Wilhelm, J.A. (1989), *Critical Issues in Implementing and Reinforcing Cockpit Resource Management Training*, NASA/UT Technical Report 89–5.

Helmreich, R.L., Wilhelm, J.A., Gregorich, S.E. and Chidester, T.R. (1990), 'Preliminary Results from the Evaluation of Cockpit Resource Management Training: Performance Ratings of Flightcrews', *Aviation Space and Environmental Medicine*, **61**(6), 576–579.

Helmreich, R.L., Wilhelm, J.A., Kello, J.E., Taggart, W.R. and Butler, R.E. (1991), *Reinforcing and Evaluating Crew Resource Management: Evaluator/LOS Instructor Reference Manual*, NASA/University of Texas Technical Manual 90–2.

Helmreich, R.L. and Foushee, H.C. (1993), 'Why crew resource management: The history and status of human factors training programs in aviation', in E. Wiener, B. Kanki and R. Helmreich (eds), *Cockpit Resource Management*, New York Academic Press, pp. 3–45.

*ICAO Journal* (1993), **48**(4), May.

Law, J.R. (in press), 'Position specific behaviors and their impact on crew performance: Implications for training', *Proceedings of the Seventh International Symposium on Aviation Psychology*, Columbus, OH, Ohio State University.

Merritt, A. (1993), 'The influence of national and organizational culture on human performance', presented to the Australian Aviation Psychology Association Seminar, Sydney, October.

National Transportation Safety Board, Aircraft Accident Report (NTSB-AAR-79–7), Washington, NTSB Bureau of Accident Investigation (United Airlines – Portland).

Orlady, H.W. and Foushee, H.C. (eds), *Cockpit Resource Management Training: Proceedings of the NASA/MAC Workshop*, Moffett Field, CA, NASA-Ames Research Center, NASA Conference Publication 2455.

Sears, R.L. (1986), 'A new look at accident contributors and the implications of operational and training procedures, (Boeing), IATA presentation, Geneva, Switzerland.

Ruffel Smith, H.P. (1979), *A simulator study of the interaction of pilot workload with errors*, Moffett Field, CA, NASA-Ames Research Center, TM-78482

Taggart, W.R. (1987), 'CRM – a different approach to human factors training', *ICAO Bulletin*, **42**(5), May.

# 15  Improving aviation instruction

*Ross Telfer*

'You want to observe me, right, then see if the observations correlate with how well my students do, compared to other people's? And then you want to generate some instruments that will help me get better at teaching, right?'

I answered (eagerly as I recall) that she was right, and that she had grasped the import of the study. Then she said something like this.

'You know, when I teach, it's like an act of love. Between me and my students, there's a relationship that creates a current that explains whether they learn something or not. You won't be measuring that current because you or your damn instruments can't see it and probably don't even think it's there. Besides, if you're going to try to observe an act of love, you're little better than a voyeur. And if you're going to use the observations to make people get better at lovemaking, you're a pervert.'

I recall having made some lame retort such as 'if you can make love with 120 students at a time in a lecture hall, you're worth observing.' But she was already half way out of the door she was slamming (Huberman, 1987, p. 11).

## Instructional theory

The generation of valid and reliable instructional theory is problematic in educational organizations in general, and even more so in the aviation environment. Theory has to provide a basis for predicting what will occur, and to be capable of evaluation by that stringent criterion. In the sciences they may have erred in the instances of early astronomy or alchemy: but the instructional researcher can only envy the relative simplicity of their task. In the social sciences, there are myriad variables to consider, such as task, personality, environment, ability, experience, attitude and motivation. Additionally, the physical scientist usually works with a unitary theory (such as gravity) in contrast to the overlapping, complementary or conflicting theories about human behaviour.

When theory is applied to practice in areas such as engineering or navigation, usually there has been a proving period in small-scale trials. These are followed by general acceptance, with further research and development providing a valued guide. Actual application is usually determined by the rigour of the research. By being sufficiently representative in their sampling and methodology, studies can provide results capable of universal application. This scenario, however, is sounding less and less like the situation in training and instruction with which we are familiar.

This chapter argues that instructional and training research in aviation has a very narrow and inconsistent scientific base. Trial and error experience provides guides from which industry is reluctant to stray without obvious cost benefits. Uncritical application of the results of studies in schools and other educational settings has not helped the cause of instructional research. Aviation is a specialist application, requiring an eclectic selection of general instructional and training theory.

For a start, aviation instruction can be distinguished by its very high stakes. Because of this social responsibility, aviation instruction is a professional activity as instructors use technical or specialized knowledge and skill in service of the public welfare. In our society, aviation instruction is one of the few educational activities having equivalent accountability to that of the unfortunates involved in bridge design in the Roman Empire. The architect was required to stand directly beneath the span as the first Roman legions clattered over with all their accoutrements. Today's flight instructor has similar, if less direct, professional accountability to society. This accountability places particular constraints upon the nature and types of instruction which can be utilized, and the standard of attainment which is required.

There is, however, some common ground between aviation instruction and education or training in general. To the extent that it seeks to convey knowledge, skills, attitudes or values, aviation instruction can draw from educational research which has been subject to critical scrutiny to determine application. For example, topical overviews which synthesize the results of instructional studies are available from Brandt (1992) and Reynolds (1992). They provide classifications and groupings which simplify the task of selecting research from an appropriate sample and with comparable content or process. Contemporary methodologies for conducting meta-analyses indicate that such overviews will become increasingly available.

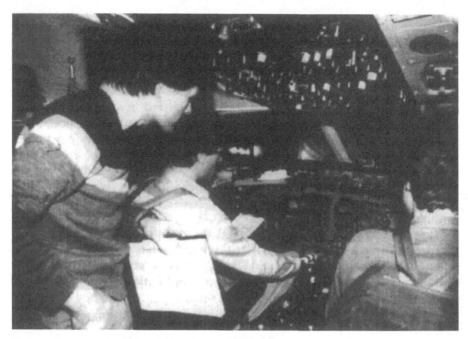

**Figure 15.1   Simulator training in Swissair**
(Source: Swissair)

Unfortunately, most of the results of such syntheses cannot be applied directly to aviation instruction. What was found in a group of eight-year-olds studying art in Chicago, or in an analysis of PhD students involved in medical research in London, may have little to do with the problem Bloggs is experiencing in judging the flare and roundout on landing, or in overcoming an awareness deficiency which is threatening his command training.

What is different about aviation instruction? How can we isolate the research that applies?

## Aviation instruction and training as a special field

First, aviation differs from other forms of, and especially school, education in its structure, mission, focus, budget and staffing. Because of their special task, aviation instructors have less autonomy than school teachers, having a clearly defined body of knowledge, skills and values to transmit with a focus on immediate transfer. Their operating budget forces upon them an emphasis on efficiency (value for dollar) rather than effectiveness (take your time and do it well), the rigour translating into attrition rates rather than remedial instruction as the consequence of trainee failure. Unlike teachers, flight instructors, in particular, are rarely permanently occupying a career position. This tour-of-duty approach militates against a public perception of professionalism in flight instruction, especially when the multi-barred epaulettes of pilots active in flight operations carry greater prestige and remuneration.

Apart from these differences, there are a number of characteristics of aviation instruction which distinguish it as a form of teaching. First, it almost invariably involves interaction with adult learners. This makes *andragogy* (adult learning) rather than *pedagogy* (child learning) the appropriate underlying learning theory. Because adult learners are characterized by different depths of experience, motivation, commitment and abilities, appropriate methods of instruction are different from those used with children. When we consider cognitive and behavioural theory as those which best contribute to an understanding of effective flight instruction, we need to remember that we are applying them to adult learners. This view is supported by research showing that experienced pilots have developed different ways of learning from those they employed as school students (Telfer, 1991).

**Figure 15.2   The instructor–trainee empathy which characterizes both the art and craft of flight instruction**
(Source: Swissair)

Second, there can be a strong element of personal charisma and art in flight instruction (Telfer and Bent, 1992). From the trainee's viewpoint, instructors have enviable confidence, knowledge and ability. Experienced pilots have usually developed special insights and colourful anecdotes which accompany their structured teaching. Add a dash of empathy, patience and good humour and the quintessential teacher emerges, but usually better at some aspects of instruction than others. The ability to tap the reservoir of precedent is a valuable instructional method. Analysis of experience and a considered response to practical circumstances are the attributes of a quality instructor. Indeed, flying schools and airlines could benefit from an in-house, loose-leaf manual in which experienced instructors are encouraged to record 'what works' for the benefit of beginning instructors and for use in instructor refresher courses. Such entries would include common learning problems and training solutions which can be applied. Many of these entries would be applicable only within that organization because of the special circumstances encountered there owing to the nature of the environment, the staff, the syllabus, aircraft or students. Memory aids such as imagery, mnemonics or special learning codes would also be entered as an aid to instructors and their students facing similar situations. There would, however, be common entries across the industry, and these would form a start towards recording the art of the profession, for linking to the craft accumulating from scientific enquiry.

How can we accelerate progress towards the sagacity of experience in our young and inexperienced instructors? Can we leave it to chance that the variables will be shuffled so that they fall into place to form the attributes learners appreciate? The craft of instruction offers a solution as underlying theory to guide practice. If based on a solid foundation of relevant and valid research, the craft of instruction is a means of ensuring that all instructors reach an acceptable level of professional competence.

A means of clarifying the application of such theory is to consider instruction in aviation as a three stage model. First is *presage* in which task analysis, lesson planning and preparation of the student precede the second stage, the actual *process* of instruction. Finally, there is the *product* which results, subject to assessment, diagnosis, remediation or further process. Hence the term *3P model* to describe this means of perceiving aviation instruction (Biggs and Telfer, 1987).

## Presage

*Behavioural* and *cognitive* theory have the most fruitful application to the 3P model. The first, behavioural theory, links *presage* and *product* through instructional objectives and evaluation. When behavioural objectives are used the outcome of training is unambiguously identified for both instructor and trainee as these objectives are derived from careful task analysis. We have to be confident that each of the exercises, topics, assignments or sorties combine to form the commercial or airline transport pilot we are aiming to produce. For example, the *Spectrum* flight training programme at the University of North Dakota; the British Aerospace Flying College Airline Pilot Course; and the Lufthansa *Futura* scheme are based upon a detailed and systematic analysis of the competencies required of a pilot. At the University of Newcastle we commenced the degree programme (including integrated flight training) with a statement of the competencies required in the graduating student. These competencies were then translated into objectives with criterion performance levels. The performance objective describes the desired trainee performance, the conditions under which it is to occur, and criteria which will be used in testing.

As an example of how these objectives can affect the nature of instruction and evaluation, consider the difference between the following four objectives for an *ab initio* flight training session:

- *Objective 1:* To teach an understanding of aircraft controls.
- *Objective 2:* To teach the ways in which an aircraft is maintained in straight and level flight.
- *Objective 3:* To regain straight and level flight from a turn, climb or descent.
- *Objective 4:* After an introduction to the effects of controls, the student will be able to return the aircraft to straight and level from slight turns, dives and climbs.

The first objective communicates the instructor's intention, but in very vague terms. When does someone 'understand' something? Do we ever fully 'understand' aircraft controls? Which controls are we talking about?

With this objective, a number of different lessons could be taught, to various levels of pilot competence. Worse, in each case the instructors involved would believe, and argue, that they had done the job well.

The second objective again provides information about what the instructor plans to do, but it also is unclear in important areas. Which ways will be taught? In what circumstances? We do not really know the standards which will be attained by trainees each time this lesson is taught.

The third objective introduces a subtle but vital twist. We are now considering what the *trainee*, rather than the instructor, will do. This change in perspective has a major advantage: it clarifies how the lesson can be evaluated by examining the performance of the trainee. The problem, however, is that there is quite a range of potential circumstances involved. How steep and sustained a turn, climb or descent? With what preparation or assistance?

The fourth objective removes these problems. It maintains the use of the trainee as the subject, but indicates the conditions under which the objective is to be met (after an introduction to controls) and the extent of the recovery (from 'slight' dives, climbs and turns). It could be argued that this statement still retains some imprecision: but it is difficult to argue that it is inferior to the three objectives which precede it. The use of the trainee as a focus is a technique which bridges behavioural and cognitive theory. The instructor considers what the learner has to do, know or value (rather than what the instructor plans to do); plans collaboratively with the trainee to achieve that goal; and continues to involve the student in a cooperative approach to evaluation and its implication for further learning.

Behavioural objectives aid communication and standards in instruction. They are especially useful in assessing progress when performance criteria are included. These criteria include reference to the *time* allocated (such as so many times per lesson; within two minutes; per circuit, etc.); the *standard* required (100% accuracy; within five degrees; within one hundred feet, and so on); the *aids* permitted (without reference to instruments; without a calculator; given a model aircraft, for example); and any other *restrictions* (such as from memory or without reference to a manual or textbook).

As the term implies, the key to behavioural objectives is in the choice of a verb to describe a student's behaviour as a result of the instruction. Verbs such as 'appreciate', 'understand' or 'know' are replaced by words such as 'control', 'explain', 'correct' or 'tell'. A list of such verbs is available (Telfer and Biggs, 1988, p. 126).

One criticism of such an approach is that it trivializes learning and fails to account for high-order objectives such as problem solving, airmanship and judgement. That indictment can only be valid if the range of actual target behaviours is so narrow as to justify it. Think how a pilot demonstrates airmanship. What are the constituents of that behaviour? Those components help to form the detailed training objectives. Trivialization will only occur when insufficient analysis is made of the training task, and the omissions carry through to the performance objectives and, ultimately, to the product of the programme.

Analysing the teaching task has another advantage. By listing the detailed competencies, the instructor enables the trainee to be given an opportunity to take advantage of any prior experience or training. There is little point in forcing every trainee to do the same course and spend the same number of days in training if they have different backgrounds, abilities and experience. Behavioural objectives are a simple way of individualizing instruction while maintaining systematic and demonstrable standardization.

Specification of competencies, as we will see later in this chapter, also links directly with ease and reliability of evaluation. There will be an objective test available to ascertain that each of the trainees meets the same standard and can be accredited with achieving that part of the course.

**Figure 15.3 Individualized Instruction by means of computer-based learning**
(Source: Swissair)

*Process*

Comprehensive reviews of educational research provide the sources for a selection of conclusions with implications for the process of aviation instruction (Reynolds, 1992; Waxman and Walberg, 1991). Apart from the need for a *trainee-centred approach*, discussed above, they support the method of involving the student in planning instruction. *Advance organizers* can be used to keep the student interested and informed about the progress of instruction, especially the goals and experiences to follow in the next lessons. Rather than being of the form 'Next week we'll be doing . . .', they can take the form of, 'What do you think would happen if . . .? Next lesson, we're going to find out.'

Natural *enthusiasm* can be fostered and then utilized to motivate the student to prepare for a lesson. Fortunately for the instructor, enthusiasm is extremely infectious. The requirement is only that the instructor sincerely enjoys the challenges and opportunity to teach. A benefit of enthusiasm is the link with *modelling*: in which trainees consciously and unconsciously style their dress, speech and behaviour on the instructor. In terms of occupational socialization, the instructor becomes one of the most significant influences on the *ab initio* trainee. This powerful influence can be constructively channelled by an instructor who has the *empathy* to be able to interpret the trainee's verbal and non-verbal communication. An apt description of this attribute is 'with-it-ness': being able to pick up the nuances of group interaction, an individual's silence, an unusual distraction, and so on.

Empathy pays off for the instructor in other ways, too. The with-it instructor will be aware of the trainee's *self-concept* and can help to provide the vital ingredients to restore balance. Airlines increasingly recruit and employ on the basis that they are gaining a career employee who will repay expensive training by demonstrating commitment and the quali-fications for upward mobility. Where the potential for command is demonstrated in the young pilot, self-concept provides an instructional vehicle for ensuring the potential is met. Positive feedback, careful selection of realistic goals, recognition of learning plateaux, awareness of individual differences, and especially, the discerning debrief: these are some of the instructional skills for enhancing the trainee's self-concept. When praise is used for reinforcement, it can become more meaningful when a '. . . because . . .' extension is added to explain just why the exercise was completed with merit, or why the manoeuvre was smoothly accomplished. The link between self-concept and a student's approach to learning is further discussed later in this chapter.

*Indirectness* refers to the successful method of guided self-discovery in which the instructor prompts the student towards the solution, but does not supply it. The time constraints of aviation's emphasis on efficiency rarely offers this luxury, but occasionally it can be used as an assignment between lessons. The questions could be such as 'For next week, see if you can draw some diagrams demonstrating what the secondary effects of the controls would be'; or, 'Given what we know about hypoxia, at what flight level would it become a hazard without oxygen masks'; or 'Why would a designer want to provide an aircraft with in-built instability?'

To maintain interest and to cater for different learning preferences, effective instructors *vary* their approach. Especially in mass briefs, students should not be able to predict the approach that will be taken by the instructor. To use an overhead projector in every lesson is just as defective as using a lectern or chalkboard continuously. There are three means of varying instruction: the instructor's style; the media and materials of instruction; and the interaction between students and instructor. An insightful experience is for instructors to have themselves videotaped while teaching. Gestures, words, intonations may be repeated unconsciously. Vary the tone, volume, speed and body-language to suit the lesson and the audience at the time. Exploit the range of visual, aural and tactile aids available. Aviation is a rich source of models, aircraft components, photographs, videotapes, cockpit voice recordings, actual ATIS messages, meteorology radar pictures, accident reports, flight plans, visiting experts, charts and more. There is no excuse for boring repetition of chalk and instructor talk. A harsh but informative criterion is for an instructor to record then analyse a lesson to determine, then justify, who did the most talking.

The final source of variation is the interaction between trainee and instructor. The most common form is one instructor talking to either a single trainee or a class of trainees. Other variants are a trainee doing the talking to either the class or to the instructor, groups of trainees working together then reporting back to the class; individuals working alone, and combinations of these. These variations offer a means of integrating the development of leadership, communication, judgement and other social skills in trainees.

Periodic *summarizing* enables trainees to consolidate their learning by review and clarification. In support of variability, the instructor does not have to do all the summarizing, either. There is no reason why the trainee should not provide a review, rather than the instructor. This will provide a form of testing which will help the instructor's planning of the next instructional phase, and adjust the pacing of instruction.

Because of individual differences, students approach learning in various ways. While there is evidence that this is less true of experienced pilots (Telfer, 1991), it is known that there are three basic approaches drawn from pilots' motivations and the strategies they employ. Research has provided a means of identifying these approaches, the Pilot Learning Process Questionnaire (PLPQ), which enables instructors and trainees to make adjustments for optimal learning (Moore and Telfer, 1992). The first of the approaches has been termed *surface learning* because it is usually motivated by anxiety to do the minimum possible to get through the check or test. The method is based on reproducing course notes or manual content, usually by rote learning methods. After the test, there is usually little recall of the knowledge.

*Deep learning* is intrinsically motivated, so trainees personalize their learning. They consider what has to be learned, reflecting on the possible applications, implications and problems which are inherent. Deep learners usually discuss these matters with others, read widely, and relate the learning to what they already know and general practice in aviation. They aim for understanding and professional competence.

*Achievement learning* is based upon ego-enhancement and competition for success. These learners aim for top marks and high grades, setting themselves high standards and thriving on competition. They organize their learning by gathering useful resources (such as past examination papers), and will often consult with those who have already undertaken the test or check to identify likely questions or problems.

The instructional implications of these approaches should be apparent, especially in the link with self-concept. Surface learners need motivation to become deeper (unless the topic is simply the acquisition of meaning-

**Table 15.1  Instructing trainees for long-term learning**

| | |
|---|---|
| **Presage:** | Use performance objectives linked with on-the-job behaviour. |
| | As a criterion, use higher standards from industry rather than the syllabus. |
| | Commence by a pre-test to ascertain what is already known. |
| | Consult training records to identify any problems or remediation needed. |
| | Use advance organizers and collaborative planning to establish shared responsibility for learning. |
| | Ensure expectations are clear. |
| | Prepare structured self-guided learning materials. |
| **Process:** | Encourage trainee participation and initiative. |
| | Model desired behaviours and attitudes: especially enthusiasm. |
| | Link new knowledge or skill back to what is already mastered, and forward to ultimate on-the-job behaviour. |
| | Vary media of instruction and type of interaction. |
| | Utilize full range of questioning skills. |
| | If time permits, be indirect and use guided discovery. |
| | Provide any rules, principles, concepts or memory codes to help the student to store the new knowledge. |
| | Provide problems, simulations or decisions which require the student to apply and use the new knowledge or skill. |
| | Require the trainee to demonstrate the knowledge or task on several, spaced occasions. |
| | Gradually introduce distractions (such as radio calls, simultaneous tasks, or demanding operation conditions). Consolidate the new learning's part in the big picture. |
| | Reinforce trainee's self-concept by emphasizing achievement. |
| | Preview next lesson and decide upon preparatory activities. |
| | Note on training record any reminder for future instruction. |
| **Product:** | Conduct the evaluation stage as part of a wider assessment of skill or knowledge, linking with prior learning. |
| | Use the full range of testing methods, according to purpose. |
| | Make forms of self-assessment available for the trainee. |
| | Relate the evaluation to the objective, and to the big picture. |
| | Use the performance objective as the evaluation criterion. |
| | Lead into the next topic by posing questions and arousing curiosity. |

less data for a one-time circumstance), and achievement needs to be linked with an intrinsic motivation. The intrinsically-motivated achiever is a formidable learning combination.

The skills taught in aviation instruction may be termed 'motor control' or the 'psycho-motor domain' of learning. Regardless, they are a major part of flight instruction. For the instructor, two informative theories of skill-learning are the *closed-loop* explanation (Adams, 1971) and *schema theory* (Schmidt, 1988). The former argues that students develop a blueprint or 'perceptual trace' of a skill as the instructor provides helpful feedback enabling trainees to differentiate good from bad executions of the skills. This is where the 'because' extensions on praise become so important.

The latter argues that memory has two states which are vital in learning a skill: *recall* enables the skill to be produced when required, and *recognition* tests the way the skill is executed to ensure it is appropriate. The two explanations are not mutually exclusive, having in common the importance of instructor feedback to enable the skill to be refined then stored so that it can be recalled when needed.

There is support from several sources to support the view that skills are learned in stages (Anderson, 1982). There are differences in the labels given, but the process is clear. In the first *cognitive* level the student discovers what is required and what the skill involves. Then comes a demonstration, practice, feedback and more practice until the desired standard is achieved in this *fixative* level. Finally, with more practice the skill becomes automated so that it is *autonomous*, able to run off on its own when required. As an example, consider the acquisition of cross-wind landing skills by a recreational pilot.

## Product

The final stage of instruction involves discovering the extent to which trainees have attained the performance objectives. This will identify the areas in which the training has been successful and unsuccessful. This information has two applications: one to the trainee's learning, and the other to the course of instruction. The former (termed 'summative evaluation') enables the instructor and the trainee to consider the next objective (taking into account the test results), and the latter ('formative evaluation') enables any modification of syllabus or teaching method before the next trainee undertakes the exercise, block or topic.

For adult learners, especially when the assistance of computer technology is available, self-assessment is an important part of the learning cycle. Finding out what you do not know is a vital step, and the preparation of a self-test quiz could be seen as part of lesson preparation by the instructor.

Most air forces favour the so-called Systems Approach to Training (SAT), premised upon competency-based training (Given, 1993). As a variant of performance-based instruction, the ultimate criterion in the evaluation of the product of SAT is on-the-job performance. The weakness is our ability to assess competencies. In part, this weakness currently comes from the tendency to use tests that are economic rather than effective. This places an unfortunate emphasis on so-called objective tests (multiple-choice or true/false) at the expense of other means of demonstrating not only if candidates can differentiate between correct, not so correct and incorrect answers. Essay, short answer, practical and oral tests also have disadvantages as well as advantages (Telfer and Biggs, 1988, p. 112), but need to be part of the instructor's repertoire of evaluation options. Other, less conventional but equally applicable alternatives include simulator scenarios, decision-making role plays and critical incidents. These also have the advantage of high student interest and close resemblance to the actual task performance.

**Figure 15.4    Performance-based training for air traffic controllers**
(Source: University of Tasmania)

The other unfortunate consequence to conventional evaluation in aviation is that the overwhelming majority of objective tests are norm-

**Table 15.2   Some differences between norm-referenced and criterion-referenced tests**

| Characteristic | Norm-referenced | Criterion-referenced |
|---|---|---|
| Results | Based on group average | Based on performance standard |
| Learning measured | In varying degrees | Achieved or not achieved |
| Application | Broad, general | Specialized, narrow |
| Planning | How much learned | What learned |
| Uses | Versatile | Limited |
| Range of results | Wide | Narrow |
| Standard | Average | Performance level |
| Score | Per cent, rank, score | Pass or fail |
| Measure | Relative | Absolute |
| Intent | Rank students | Determine competence |
| Domain | Cognitive | Cognitive/psychomotor |

referenced, rather than related to performance criteria. Some of the major differences between the two forms of testing are shown in Table 15.2.

Norm-referenced tests are those in which the results of an individual are compared to the performance of others. 'Bloggs came last again', 'Julie was top again', 'Smith's mark was 60', are all examples of norm-referenced results. The criticism of this practice is that in aviation we seek competence: an ability to meet performance criteria not magic numbers. That is why we should do most of our evaluation by means of criterion-referenced tests. Individuals are evaluated in terms of whether they have met the prescribed standard of performance. But we do this already, you protest (probably with some vehemence). Yes, that is the way flight tests are conducted.

What, however, occurs in theory subjects for pilot licensing purposes? How will new areas of study such as human factors or crew coordination be evaluated? An area of ongoing international concern is the matter of licensing and professional assessment in aviation. The problems are not restricted to aviation nor are they new: but there is little evidence that regulatory agencies are prepared to take the expensive steps necessary to solve them. It would have to be assumed that cost, rather than ignorance, has led to the disrepute of pilot and instructor licensing tests. Among the major concerns are the narrowness of sampling of knowledge, skills and attitudes; testing only a narrow range of practice situations. Competence is translated as assessment of acquired knowledge rather than more extensive assessment of practical skills. Knowledge of the true/false answer is equated with appropriate qualification.

Personal or professional qualities such as honesty, judgement, airmanship, stability and ability to cope with change or emergency may be subjugated to more easily-measured characteristics. What really determines fitness for professional practice in aviation?

## Change in aviation organizations

Another perspective upon instruction (especially in train-the-trainer courses) is that, essentially, it is about change. As a result of a structured educational experience we expect students or pilots to behave differently from their natural tendencies when faced with a certain circumstance. For example, participants in an airline CRM course responded positively to an evaluation survey, agreeing that the course was professionally planned and presented. They agreed, almost unanimously, that the concepts were valuable, and that they had enjoyed active participation in the course. When asked, however, whether the CRM would change their operational behaviour, the answer was negative. There is a gap between intent, design and implementation (Moore, Telfer and Wilkinson, in progress).

There is a difference between staff development and enduring change. Providing training programmes for pilots and instructors can initially affect instructional practice in the briefing room or cockpit. It may thus directly relate to a change in trainee learning outcomes. The ultimate goal, however, is a change in the values of the instructors or pilots. The end of the process is the change in their beliefs and attitudes towards instruction, learners or cockpit interaction.

Change is thus not a matter of introducing a workshop, consultant or new training programme. Because it is long term, problematic and gradual, there are implications for the ways in which aviation training should be managed. These implications include the following:

- the increase in the workload of both participants and presenters;
- some anxiety about coping with new methods, especially the risk of failure;
- a consequent reluctance to abandon traditional approaches, especially if the change is of a larger magnitude; and,
- uneven adaptation because of personality and other individual differences among participants.

Methods which can be used to minimize these hazards are presented in Table 15.3. Clearly, the provision of a training workshop is simply the beginning of the change process. For the instruction or training to be effective, the full cycle of organization needs to follow. This involves consideration of the individual needs of participants, formal and informal support, follow-up refreshers and logistical resources. It is possible to

**Table 15.3   Enhancing the change sought by training programmes**
(After Guskey, 1986)

- Provide the programme's intentions in simple and concrete terms.
- Indicate clear advantages over current methods.
- Seek and consider pilot or instructor concerns about how the new practices will affect them.
- The leading facilitator or change agent must have high credibility in the organization.
- Management provides various indicators of support (apart from resources), such as attendance at workshops, an office or 'home room' from which the new programme is coordinated, company communications, and informal interest.
- Regular feedback is provided on evidence of the positive effects of the new methods as a result of the training programme.
- Provide sustained support in the introductory period of trial and experimentation, an opportunity for instructors to discuss their problems and achievements.
- Organize a refresher course after initial training.

identify three major participants in the process of change: the individual, the peer group and the organization. Each is a focus for considering how the process of change can be enhanced.

Process and content are complementary and merit equal prominence. An elaborate decision-making simulation has little value if the conclusion for the participants is not one of the superiority of a group decision over that of any individual. Change is a systematic and sustained process, not an episodic intervention. Supportive leadership, management and collegial relations are important variables. There is evidence of the importance of the structure of the organization for individuals' commitment to change (Richardson, 1990). Unfortunately, there appears to be little evidence that flying schools or airlines approach their training task with change in mind. For example, in their report on the outcomes of Crew Resource Management (CRM) training, Helmreich and Wilhelm (1991) describe a minority of participants as 'boomerangs' because they show a negative change in attitudes as a result of attending a CRM course. Apart from personality, the authors see group dynamics in a form of 'dyssynergy' as 'critical determinants of reaction to training'. Learning outside school is usually socially shared. CRM takes place in a social system and 'each person's ability to function depends on what others do and how several individuals' mental and physical performances mesh' (Resnick, 1987, p. 13).

Resnick uses the analogy of US Navy ships being piloted into San Diego harbour. Six people involved in three different jobs are involved. On deck, two take visual sightings and call them out to two who relay them by telephone to a specialist on the bridge. He records and repeats the bearings to another who plots the positions and projects the course to the next fix and beyond to the next landmarks for sightings. The cycle is repeated every three minutes. No individual can pilot the vessel alone. The vital knowledge is distributed among the group. Common to CRM and similar aviation workshops are the socially-shared intellectual work and joint accomplishment of tasks. Elements of the skill, be it decision making, leadership or communication, accrue meaning only in the total context. Active participation is vital for the aggregate result.

If instructors can be involved as researchers or collaborators in studies there are many advantages, especially as a means of filling the gaps in instructional knowledge. They can also test their assumptions; become professionally active; become more critical and authoritative in their judgement of others' work; and will probably interact more with their students and colleagues.

To-date, aviation instruction, at both *ab initio* and airline levels, has tended to follow historical precedent. This is not surprising, for it has predictive validity. Simply, it works. Why it works is a question which merits scrutiny. The motivation and expertise of those involved in the instructional process are key variables, and where one or both of these is lacking, the limits of traditional practice are exposed.

For aviation instruction and practice to sustain improvement it needs to be professionalized. There needs to be a conscious, critical development of a theoretical base for practice, and there needs to be parity of esteem for those who seek to make a career in aviation instruction in the classroom or aircraft. Neither will come easily, for they involve many levels of the aviation industry, consultants and academia working together. The need for.effectiveness, as well as efficiency, may prove to be the catalyst.

# References

Adams, J.A. (1971), 'A closed-loop theory of motor learning', *Journal of Motor Behaviour*, **3**, 11–149.

Anderson, J.R. (1982), 'Acquisition of a cognitive skill', *Psychological Review*, **89**(4), 369–406.

Biggs, J.B. and Telfer, R.A. (1987), *The Process of Learning*, 2nd edn, Sydney, Prentice Hall.

Brandt, R. (1992), 'On research on teaching', *Educational Leadership*, **49**(7), April, 14–19.

Given, K.C. (1993), 'The evaluation of training', in R. A. Telfer (ed.), *Aviation Instruction and Training*, Aldershot, Ashgate, 188–202.

Guskey, T.R. (1986), 'Staff development and the process of teacher change', *Educational Researcher*, **15**(5), 5–20.

Helmreich, R.L. and Wilhelm, J.A. (1991), 'Outcomes of crew resource management training', *The International Journal of Aviation Psychology*, **1**(4), 287–300.

Huberman, M.N. (1987), 'How well does educational research really travel?' *Educational Researcher*, **16**(1), 5–21.

McLaughlin, M.W. (1990), 'The Rand change agent study revisited: Macro perspectives and micro realities', *Educational Researcher*, **19**(9), 11–16.

Moore, P.J. and Telfer, R.A. (1992), 'Approaches to learning in a vocational context,' *Conference of the Australian and New Zealand Association for Research in Education*, Geelong, Deakin University.

Moore, P.J., Telfer, R.A. and Wilkinson, R.L. (in preparation), 'An evaluation of cockpit resource management training in Qantas'.

Resnick, L.B. (1987), 'Learning in school and out', *Educational Researcher*, **16**(9), 13–20.

Reynolds, A. (1992), 'What is competent beginning teaching? A review of the literature', *Review of Educational Research*, **62**(1), Spring, 1–35.

Richardson, V. (1990), 'Significant and worthwhile changes in teaching practice', *Educational Researcher*, **19**(7), 10–18.

Schmidt, R.A. (1988), *Motor Control and Learning: A Behavioural Emphasis*, 2nd edn, Champaign, Human Kinetics Publishers.

Telfer, R.A. (1991), 'How airline pilots learn', *Aviation Psychology Symposium of the New Zealand Psychological Society*, Palmerston North, Massey University.

Telfer, R.A. and Bent, J. (1992), 'Producing a workshop for training airline instructors', *The Journal of Aviation/Aerospace Education and Research*, **2**(2), Spring, 31–38.

Telfer, R.A. and Biggs, J.B. (1988), *Psychology and Flight Training*, Ames, IA, Iowa State University Press.

Waxman, H.C. and Walberg, H.J. (1991), *Effective Teaching: Current Research*, Berkeley, CA, McCutcheon.

357

# Index

Printed and bound by CPI Group (UK) Ltd, Croydon, CR0 4YY

23/10/2024

01777675-0012